AF888260

Methods in Molecular Biology

Series Editor
John M. Walker
School of Life and Medical Sciences
University of Hertfordshire
Hatfield, Hertfordshire, UK

For further volumes:
http://www.springer.com/series/7651

For over 35 years, biological scientists have come to rely on the research protocols and methodologies in the critically acclaimed *Methods in Molecular Biology* series. The series was the first to introduce the step-by-step protocols approach that has become the standard in all biomedical protocol publishing. Each protocol is provided in readily-reproducible step-by step fashion, opening with an introductory overview, a list of the materials and reagents needed to complete the experiment, and followed by a detailed procedure that is supported with a helpful notes section offering tips and tricks of the trade as well as troubleshooting advice. These hallmark features were introduced by series editor Dr. John Walker and constitute the key ingredient in each and every volume of the *Methods in Molecular Biology* series. Tested and trusted, comprehensive and reliable, all protocols from the series are indexed in PubMed.

Schmidtea Mediterranea

Methods and Protocols

Edited by

Luca Gentile

Department of Animal Physiology, University of Osnabrück, Osnabrück, Germany

Humana Press

Editor
Luca Gentile
Department of Animal Physiology
University of Osnabrück
Osnabrück, Germany

ISSN 1064-3745 ISSN 1940-6029 (electronic)
Methods in Molecular Biology
ISBN 978-1-0716-3274-1 ISBN 978-1-0716-3275-8 (eBook)
https://doi.org/10.1007/978-1-0716-3275-8

© The Editor(s) (if applicable) and The Author(s), under exclusive license to Springer Science+Business Media, LLC, part of Springer Nature 2023
This work is subject to copyright. All rights are solely and exclusively licensed by the Publisher, whether the whole or part of the material is concerned, specifically the rights of translation, reprinting, reuse of illustrations, recitation, broadcasting, reproduction on microfilms or in any other physical way, and transmission or information storage and retrieval, electronic adaptation, computer software, or by similar or dissimilar methodology now known or hereafter developed.
The use of general descriptive names, registered names, trademarks, service marks, etc. in this publication does not imply, even in the absence of a specific statement, that such names are exempt from the relevant protective laws and regulations and therefore free for general use.
The publisher, the authors, and the editors are safe to assume that the advice and information in this book are believed to be true and accurate at the date of publication. Neither the publisher nor the authors or the editors give a warranty, expressed or implied, with respect to the material contained herein or for any errors or omissions that may have been made. The publisher remains neutral with regard to jurisdictional claims in published maps and institutional affiliations.

This Humana imprint is published by the registered company Springer Science+Business Media, LLC, part of Springer Nature.
The registered company address is: 1 New York Plaza, New York, NY 10004, U.S.A.

Preface

When I first approached the world of planarians, back in the early 2000s, there were a few laboratories worldwide actively using this incredible animal model and no conferences or meetings dedicated exclusively to it. Thanks to the induced pluripotent stem cell revolution, a revived awareness in the intimate biology of the stem cells and in their contribution to regeneration sparked new interest in the planarian model, which is one of the very few bilateral animals known to possess pluripotent stem cells in the adult. With new lymph flowing and a much larger planarian community, it is imperative not only to acknowledge the biodiversity of planarians, but to provide the scientists with recipes, methods, and protocols that are optimized on their planarian species of choice. Therefore, in this book, we offer a collection of methods that are fine-tuned on *Schmidtea mediterranea*. The chapters encompass a variety of experimental protocols aimed to refine established methods, such as in situ hybridization, immunohistochemistry, cell dissociation, and flow cytometry, to detail pipelines for the analysis of large datasets, as in genomics and transcriptomics, and to offer novel protocols in their most exhaustive dress possible, so that both experts in the field and newcomers are given the best possible toolbox for their everyday lab work: the *Schmidtea mediterranea* toolbox. In hoping to see very soon similar books offering specific tools for the other planarian species of interest, I wish to thank all the people who contributed to this book, the authors of course, John Walker for his precious advice, and my wife Linda Barlassina, who helped me a lot in the editing work.

Osnabrück, Germany ***Luca Gentile***

Contents

Contributors

TERESA ADELL • *Department of Genetics, Microbiology and Statistics and Institute of Biomedicine, Universitat de Barcelona, Barcelona, Catalunya, Spain; Institut de Biomedicina de la Universitat de Barcelona (IBUB), Universitat de Barcelona, Barcelona, Catalunya, Spain*

TOM ARTOIS • *Centre for Environmental Sciences, Hasselt University, Diepenbeek, Belgium*

KERSTIN BARTSCHERER • *Department of Biology and Center for Cellular Nanoanalytics (CellNanOs), Osnabrück University, Osnabrück, Germany*

LISANDRA BENÍTEZ-ÁLVAREZ • *Departament de Genètica, Microbiologia i Estadística, and Institut de Recerca de la Biodiversitat (IRBio), Universitat de Barcelona, Barcelona, Catalonia, Spain*

KAROLIEN BIJNENS • *Centre for Environmental Sciences, Hasselt University, Diepenbeek, Belgium*

FRANCESC CEBRIÀ • *Department of Genetics, Microbiology and Statistics, Faculty of Biology, University of Barcelona, Catalunya, Spain; Institute of Biomedicine of the University of Barcelona (IBUB), Catalunya, Spain*

MEENALOSINI VIMAL CRUZ • *Department of Computer Science, Keene State College, Keene, NH, USA; Department of Information Technology, Georgia Southern University, Statesboro, GA, USA*

TIM DEMTRÖDER • *RNA Biochemistry, University of Bayreuth, Bayreuth, Germany*

DANIEL DOLS-SERRATE • *Departament de Genètica, Microbiologia i Estadística, and Institut de Recerca de la Biodiversitat (IRBio), Universitat de Barcelona, Barcelona, Catalonia, Spain*

LEONARD DREES • *Department of Tissue Dynamics and Regeneration, Max Planck Institute for Multidisciplinary Sciences, Göttingen, Germany*

ELENA EMILI • *Department of Biological and Medical Sciences, Oxford Brookes University, Oxford, UK*

DANIEL FONT-MARTÍN • *Department of Genetics, Microbiology and Statistics, Universitat de Barcelona, Barcelona, Catalunya, Spain; Institut de Biomedicina de la Universitat de Barcelona (IBUB), Universitat de Barcelona, Barcelona, Catalunya, Spain*

SUSANNA FRAGUAS • *Department of Genetics, Microbiology and Statistics, Faculty of Biology, University of Barcelona, Catalunya, Spain; Institute of Biomedicine of the University of Barcelona (IBUB), Catalunya, Spain*

OLGA FRANK • *Max Planck Institute for Molecular Cell Biology and Genetics, Dresden, Germany*

HELENA GARCÍA-CASTRO • *Department of Biological and Medical Sciences, Oxford Brookes University, Oxford, UK*

LUCA GENTILE • *Planarian Stem Cell Laboratory, Max Planck Institute for Molecular Biomedicine, Münster, Germany; Hasselt University, Campus Diepenbeek, Agoralaan, Diepenbeek, Belgium; Pluripotency and Regeneration Laboratory, Department Animal Physiology, University of Osnabrück, Osnabrück, Germany*

MARKUS A. GROHME • *Max Planck Institute for Molecular Cell Biology and Genetics, Dresden, Germany*

RICHARD NELSON HALL • *Department of Bioengineering, Stanford University, Stanford, CA, USA*
MARTIJN HELEVEN • *Centre for Environmental Sciences, Hasselt University, Diepenbeek, Belgium*
VINCENT JAENEN • *Centre for Environmental Sciences, Hasselt University, Diepenbeek, Belgium*
IANA V. KIM • *RNA Biochemistry, University of Bayreuth, Bayreuth, Germany; Centre for Genomic Regulation (CRG), Barcelona Institute of Science and Technology (BIST), Barcelona, Spain*
CLAUS-D. KUHN • *RNA Biochemistry, University of Bayreuth, Bayreuth, Germany*
LAIA LERIA • *Departament de Genètica, Microbiologia i Estadística, and Institut de Recerca de la Biodiversitat (IRBio), Universitat de Barcelona, Barcelona, Catalonia, Spain*
NICOLE LINDSAY-MOSHER • *Hospital for Sick Children, Program in Developmental and Stem Cell Biology, Toronto, ON, Canada*
ELI ISAEL MACIEL • *Department of Molecular & Cell Biology, University of California, Merced, CA, USA; Quantitative and Systems Biology Graduate Program, University of California, Merced, CA, USA*
Mª. DOLORES MOLINA • *Department of Genetics, Microbiology and Statistics, Faculty of Biology, University of Barcelona, Catalunya, Spain; Institute of Biomedicine of the University of Barcelona (IBUB), Catalunya, Spain*
ALYSSA M. MOLINARO • *Pape Research Institute, Oregon Health & Science University, Portland, OR, USA*
CLARISSA J. NOBILE • *Department of Molecular & Cell Biology, University of California, Merced, CA, USA; Health Sciences Research Institute, University of California, Merced, CA, USA*
CLAUDIA ORTMEIER • *Planarian Stem Cell Laboratory, Max Planck Institute for Molecular Biomedicine, Münster, Germany*
NÉSTOR J. OVIEDO • *Department of Molecular & Cell Biology, University of California, Merced, CA, USA; Health Sciences Research Institute, University of California, Merced, CA, USA*
EUDALD PASCUAL-CARRERAS • *Department of Genetics, Microbiology and Statistics, Universitat de Barcelona, Barcelona, Catalunya, Spain; Institut de Biomedicina de la Universitat de Barcelona (IBUB), Universitat de Barcelona, Barcelona, Catalunya, Spain*
BRET J. PEARSON • *Hospital for Sick Children, Program in Developmental and Stem Cell Biology, Toronto, ON, Canada; Pape Research Institute, Oregon Health & Science University, Portland, OR, USA*
JASON PELLETTIERI • *Department of Biology, Keene State College, Keene, NH, USA*
MATTHEW PITTENDREIGH • *Department of Computer Science, Keene State College, Keene, NH, USA*
ANDREAS PITTROFF • *RNA Biochemistry, University of Bayreuth, Bayreuth, Germany*
KALEIGH POWERS • *Department of Biology, Keene State College, Keene, NH, USA*
JOCHEN C. RINK • *Department of Tissue Dynamics and Regeneration, Max Planck Institute for Multidisciplinary Sciences, Göttingen, Germany; Max Planck Institute for Multidisciplinary Sciences, Göttingen, Germany*

MARTA RIUTORT • *Departament de Genètica, Microbiologia i Estadística, and Institut de Recerca de la Biodiversitat (IRBio), Universitat de Barcelona, Barcelona, Catalonia, Spain*
ELKE F. ROOVERS • *Center for Molecular Medicine, University Medical Center Utrecht, Utrecht University, Utrecht, The Netherlands*
MARIA ROSSELLO • *Department of Genetics, Microbiology and Statistics and Institute of Biomedicine, Universitat de Barcelona, Barcelona, Catalunya, Spain; Institut de Recerca de la Biodiversitat (IRBio), Universitat de Barcelona, Barcelona, Catalunya, Spain; Department of Biological and Medical Sciences, Oxford Brookes University, Oxford, United Kingdom*
EMILI SALÓ • *Department of Genetics, Microbiology and Statistics, Universitat de Barcelona, Barcelona, Catalunya, Spain; Institut de Biomedicina de la Universitat de Barcelona (IBUB), Universitat de Barcelona, Barcelona, Catalunya, Spain*
KAREN SMEETS • *Centre for Environmental Sciences, Hasselt University, Diepenbeek, Belgium*
JORDI SOLANA • *Department of Biological and Medical Sciences, Oxford Brookes University, Oxford, UK*
ASHLEY VALLE AREVALO • *Department of Molecular & Cell Biology, University of California, Merced, CA, USA; Quantitative and Systems Biology Graduate Program, University of California, Merced, CA, USA*
MIQUEL VILA-FARRÉ • *Department of Tissue Dynamics and Regeneration, Max Planck Institute for Multidisciplinary Sciences, Göttingen, Germany*
HANH THI-KIM VU • *European Molecular Biology Laboratory, Heidelberg, Germany*
BO WANG • *Department of Bioengineering, Stanford University, Stanford, CA, USA*
URI WEILL • *Department of Tissue Dynamics and Regeneration, Max Planck Institute for Multidisciplinary Sciences, Göttingen, Germany*

Chapter 1

Building Phylogenies from Transcriptomic Data

Lisandra Benítez-Álvarez, Laia Leria, Daniel Dols-Serrate, and Marta Riutort

Abstract

Transcriptomic data (obtained from RNA sequencing) has become a very powerful source of information to reconstruct the evolutionary relationships among organisms. Although phylogenetic inference using transcriptomes retains the same core steps as when working with few molecular markers (viz., nucleic acid extraction and sequencing, sequence treatment, and tree inference), all of them show significant differences. First, the needed quantity and quality of the extracted RNA has to be very high. Although this may not represent a challenge when working with certain organisms, it may well be a headache with others, especially for those with small body sizes. Second, the tremendous increase in the quantity of sequences obtained requires a high computational power for both treating the sequences and inferring the subsequent phylogenies. This means that transcriptomic data can no longer be analyzed using personal computers nor local programs with a graphical interface. This, in turn, implies the requirement of an increased set of bioinformatic skills from the researchers. Finally, the genomic peculiarities of each group of organisms, such as the level of heterozygosity or the percentage of base composition, also need to be considered when inferring phylogenies using transcriptomic data.

Key words Freshwater planarians, Sampling, Transcriptome, Phylotranscriptomics, Molecular phylogeny

1 Introduction

The present procedure is designed to infer phylogenies from transcriptomic data in planarians (Fig. 1). We present the basic methodologies and some advice regarding sampling, storage, RNA extraction, and sequencing. Additionally, we describe with full detail the bioinformatic workflow, including in-house developed scripts. Although we explain how to collect field samples and fix them to keep the RNA in a good state until its posterior extraction, the transcriptome sequencing and all the bioinformatic procedures can equally be applied to any transcriptome obtained from laboratory strains or simply downloaded. We give some details on the best libraries to be used to sequence the transcriptomes, recommending

Luca Gentile (ed.), *Schmidtea Mediterranea: Methods and Protocols*, Methods in Molecular Biology, vol. 2680,
https://doi.org/10.1007/978-1-0716-3275-8_1,
© The Author(s), under exclusive license to Springer Science+Business Media, LLC, part of Springer Nature 2023

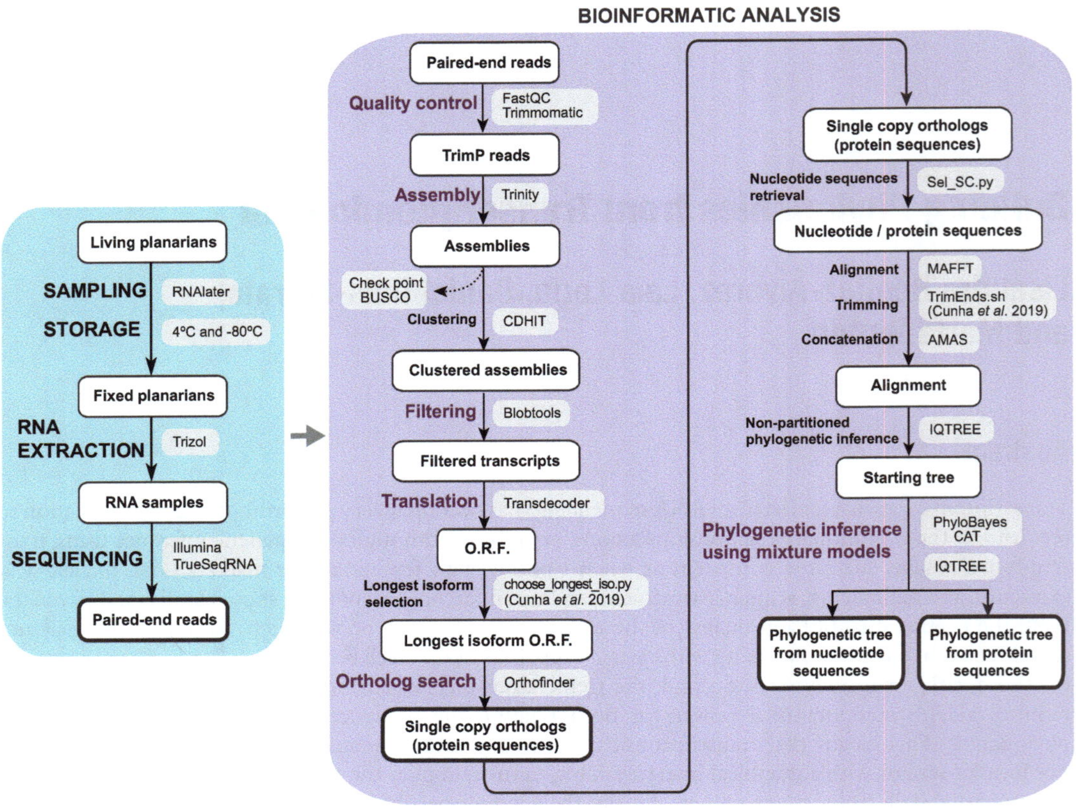

Fig. 1 Workflow of the main procedures: sampling, storage, RNA extraction, sequencing (green panel), and bioinformatic analysis (lilac panel). In gray bubbles are highlighted the main reagent, condition, technology, or software necessary to perform each step indicated by the arrows. In the lilac panel are highlighted the main steps of the bioinformatic workflow: (1) quality control, (2) assembly, (3) filtering, (4) translation, (5) orthologue search, and (6) phylogenetic inference

paired-end reads with a library that guarantees the elimination of ribosomal RNAs. In our case the libraries were obtained and sequenced in a specialized company, but they can be built and sequenced in-house if the adequate laboratory and skills are available. To run the bioinformatic protocol it is necessary to have access to powerful computational resources as well as basic skills to work with the terminal in Unix systems and shell script programming. This protocol has six main steps (Fig. 1): (1) screening the quality of the reads and filtering using FastQC [1] and Trimmomatic [2], (2) de novo assembly of transcriptomes with Trinity [3, 4], (3) filtering of contaminants based on BlobTools results [5, 6], (4) translation to protein using TransDecoder [7], (5) orthologue search with OrthoFinder [8], and (6) phylogenetic inference with maximum likelihood and Bayesian inference methodologies using the mixture models implemented in IQ-TREE [9] and PhyloBayes [10], respectively.

This step-by-step procedure enables the obtention of phylogenetic trees from transcriptomic data, without requiring expertise in bioinformatics, only basic skills are necessary.

Phylogenetic tree inference methodologies constitute a growing and complex field, in the present chapter we only provide a "black box" strategy for obtaining correctly inferred trees. Nonetheless, we refer the readers to specific literature [11–16] to understand the procedures and correctly interpret the results. We only use concatenated dataset analyses. Other types of analyses, such as coalescence based on each gene, are possible and may be more adequate in certain circumstances that we do not discuss here.

2 Materials

2.1 Sampling

1. Plastic tray.
2. Watercolor brushes.
3. Ice blocks and portable freezer.
4. Plastic screw cap tubes (2 mL, 15 mL, and 50 mL).
5. Permanent markers.
6. Petri dishes.
7. Tweezers.
8. Lamp.
9. Tissues.
10. Gloves.
11. Pasteur pipettes.
12. Automatic pipette.
13. Pipette tips (1 mL).
14. Sterile razor or scalpel blades.
15. Absolute ethanol (molecular biology grade).
16. RNase away™ surface decontaminant.
17. RNAlater stabilization solution.

2.2 Storage

Refrigerator (4 °C), freezers (−20 °C and −80 °C).

2.3 RNA Extraction

1. RNase away™ surface decontaminant.
2. Trizol (or any guanidine thiocyanate-based extraction method).
3. Chloroform (molecular biology grade).
4. Isopropanol (molecular biology grade).
5. Ethanol 70% (molecular biology grade).
6. Diethyl pyrocarbonate (DEPC)-treated water.

7. Tissue homogenizer.
8. Refrigerated centrifuge.
9. Pipette tips with and without filter.
10. 1.5 mL tubes with safe lock.
11. Rack (1.5 mL tubes).
12. Box with ice.

2.4 Library Construction and Sequencing

Library preparation can be performed in an external company or in the lab using commercial kits. Illumina sequencing services are offered in multiple companies or may be available at research center services.

2.5 Bioinformatic Analysis

Basic skills to work in a terminal in Unix systems and shell script programming are required to run the bioinformatic protocol.

Hardware

Access to a powerful computational cluster running under Linux systems.

≥24 cores

>126 GB RAM memory

>1 TB available disk space

It is recommended to use a queuing system for submitting jobs.

Software

1. Conda [17] available at https://anaconda.org/, https://docs.conda.io/projects/conda/en/latest/user-guide/install/, https://conda.io/projects/conda/en/latest/user-guide/tasks/manage-environments.html#activating-an-environment.
2. FastQC [1] available at https://github.com/s-andrews/FastQC, https://www.bioinformatics.babraham.ac.uk/projects/fastqc/Help/.
3. MultiQC [18] available at https://multiqc.info/docs/#running-multiqc.
4. Trimmomatic [2] available at https://github.com/timflutre/trimmomatic.
5. Trinity [3, 4] available at https://github.com/trinityrnaseq/trinityrnaseq/releases, https://github.com/trinityrnaseq/trinityrnaseq/wiki.
6. assembly-stats [19] available at https://github.com/rjchallis/assembly-stats.
7. CDHIT [20, 21] available at https://github.com/weizhongli/cdhit, http://weizhong-lab.ucsd.edu/cd-hit/.

8. bwa [22] available at https://github.com/lh3/bwa.
9. Samtools [23, 24] available at https://www.htslib.org, https://github.com/samtools/samtools.
10. BLAST+ [25] available at https://blast.ncbi.nlm.nih.gov/Blast.cgi.
11. BlobTools [5, 6] available at https://github.com/blobtoolkit/blobtools2/tree/2.6.1, https://github.com/DRL/blobtools.
12. Seqtk [26] available at https://github.com/lh3/seqtk.
13. TransDecoder [7] available at https://github.com/TransDecoder.
14. choose_longest_iso.py [27, 28] available at https://github.com/tauanajc/Cunha_Giribet_2019_ProcRSocB/blob/master/choose_longest_iso.py.
15. OrthoFinder [8] available at https://github.com/davidemms/OrthoFinder.
16. capt_SCseq_by_sp.py [29] available at https://github.com/lisy87/dugesia-transcriptome/tree/main/scripts.
17. interleave_to_oneline.pl [29] available at https://github.com/lisy87/dugesia-transcriptome/tree/main/scripts.
18. sel_sc.py [29] available at https://github.com/lisy87/dugesia-transcriptome/tree/main/scripts.
19. MAFFT [30] available at https://mafft.cbrc.jp/alignment/software/linux.html.
20. trimEnds.sh [28] available at https://github.com/tauanajc/Cunha_Giribet_2019_ProcRSocB/blob/master/trimEnds.sh.
21. trimAL [31] available at http://trimal.cgenomics.org/, https://github.com/scapella/trimal, https://vicfero.github.io/trimal/.
22. AMAS.py [32] available at https://github.com/marekborowiec/AMAS.
23. IQ-TREE [9] available at http://www.iqtree.org/.
24. PhyloBayes [10] available at http://www.atgc-montpellier.fr/phylobayes/.
25. newick_utils [33] available at https://github.com/tjunier/newick_utils/.
26. FigTree available at https://github.com/rambaut/figtree/releases.
27. Optional: Transcriptome phylogeny workflow available at https://github.com/lisy87/dugesia-transcriptome.

3 Methods

3.1 Sampling (See Note 1)

To manipulate animals in the field:

1. Use RNase away to clean the space and set a portable laboratory in the plastic tray.
2. Prepare a small Petri dish to do the absolute ethanol fixation and another to do the RNAlater fixation.
3. Have a dedicated brush for each step (manipulating the animals in their water, ethanol fixation, and fixing in RNAlater, have them clearly identified) to avoid killing animals accidentally.
4. Select the animals to be fixed in the Petri dishes with river water.
5. Cut the selected animal's tail.
6. Transfer the small tail piece to the Petri with ethanol and the rest to the Petri containing RNAlater.
7. Leave the two fragments there for a few seconds.
8. Transfer each piece to the tubes containing the corresponding fixation liquid. If the portion of the animal to be fixed in RNAlater is very big, cut it into small pieces. Make sure you correctly identify the tubes with the name of the individual and the fixation liquid.
9. Place the tube in the portable refrigerator with ice or ice packs and try to keep them cold during all the transport (either in car, plane, etc.).

In case you decide to take the living animals to your laboratory:

1. Fill 50 mL conical bottom tubes with around 45 mL river water.
2. Do not overload the tube (around 10 worms/tube).
3. Place the tubes in the portable refrigerator with ice or ice packs and try to keep them cold during all the transport.
4. Open the tubes to aerate the animals at least twice a day.

3.2 Storage

The samples in RNAlater can be stored at room temperature (15–25 °C) for 1 week, 4 weeks at 2–8 °C, and indefinitely at −20 °C or −80 °C. For long-term storage it is recommendable at −80 °C.

Caution: do not freeze samples in RNAlater solution immediately. Store the samples at 4 °C or room temperature at least one night; the solution must penetrate the tissue before freezing.

Extract with pipette all the solution from the tube before freezing at −80 °C, at this temperature the solution crystallizes, and the crystals may break the tissue.

3.3 RNA Extraction (See Note 2)

Protocol of RNA extraction with Trizol following the manufacturer's instructions:

1. Place the fixed planaria in a 1.5 mL tube.
2. Add 500 μL of Trizol.
3. Use a tissue homogenizer to break the tissues.
4. Centrifuge 10 min at 12,000 × *g*. Use a refrigerated centrifuge at 4 °C.
5. Transfer the upper phase to a new 1.5 mL tube.
6. Add 100 μL of chloroform.
7. Invert the tube 10–15 times.
8. Incubate 3 min on ice.
9. Centrifuge 15 min at 12,000 × *g*. Use a refrigerated centrifuge at 4 °C.
10. Transfer the upper phase to a new tube.
11. Add 250 μL of isopropanol.
12. Invert the tube 10–15 times.
13. Incubate 10 min on ice.
14. Centrifuge 10 min at maximum speed. Use a refrigerated centrifuge at 4 °C.
15. Remove supernatant.
16. Add 500 μL of ethanol at 75%.
17. Centrifuge 5 min at 7,500 × *g*. Use a refrigerated centrifuge at 4 °C.
18. Remove supernatant.
19. Dry pellet at room temperature (not above 25 °C).
20. Add 20 μL of DEPC water preheated at 65 °C.
21. Dilute the pellet without pipetting; just tap gently with your fingers on the tube.
22. For long-term storage keep at −80 °C. But try to avoid freezing and thawing multiple times for that will easily degrade the RNA.

3.4 Library Construction and Sequencing

The bioinformatic workflow presented here is designed to analyze Illumina paired-end reads obtained from a Truseq, stranded, and ribo-zero library. We recommend using this sequencing strategy for the transcriptome obtention (*see* **Note 3**). Nonetheless, library preparation methodologies and kits evolve rapidly and researchers may consider at each experiment what are their best options when preparing their RNA to obtain a transcriptome.

Follow the indications of the sequencing service to process the samples for library preparation and sequencing.

3.5 Bioinformatic Analysis

The following steps are sufficient to execute the workflow. However, if you wish to expand your understanding of the workflow you can find a set of complementary scripts to run available at https://github.com/lisy87/dugesia-transcriptome/tree/main/scripts.

In addition, we provide the transcriptomic raw data of six *Dugesia* species that have been used to test and ensure the proper functioning of the bioinformatic workflow. They are available at NCBI-SRA's repository with the following accession codes: *D. subtentaculata* (SRR17642690), *D. vilafarrei* (SRR17642676), *D. aurea* (SRR17642743), *D. corbata* (SRR17642750), *D. benazzii* (SRR17642753), and *D. hepta* (SRR17642736). At different steps we show the results obtained for this dataset. You can download it and follow the bioinformatic workflow to check everything is working fine for you.

3.5.1 Install Conda

Install conda in the cluster. Consult conda's documentation (*see* **Note 4**).

3.5.2 Download the Data

Download the transcriptomic data.

The raw data format is *fastq.gz. If you have paired-end data, you will have two files per sample: e.g., samplename_1.fastq.gz and samplename_2.fastq.gz (*see* **Note 5**).

Save all the *fastq.gz files in the same folder.

Use the *wget* command to download it directly from a server:

```
wget http://path_to_raw_data
```

Or, if you have the raw data in a local machine, you can use the *scp* command to copy it to the cluster:

```
scp -rp PATH_from_your_machine/ user@ip.cluster:~/PATH_to_the_cluster
```

3.5.3 Quality Assessment of the Raw Data

1. Install FastQC and MultiQC with conda.
2. Check the quality of the data with FastQC using the default parameters (*see* **Note 6**):

```
fastqc samplename_1.fastq.gz
```

3. Use MultiQC to generate a single report for all samples (Fig. 2) (*see* **Note 7**). In a folder with all outputs of FastQC run:

```
multiqc .
```

4. Check the results for all samples. It is important that all samples show homogeneous and acceptable values for the following parameters:

Fig. 2 Sequence quality histograms (quality score by position) from FastQC report of 6 *Dugesia* species summarized with MultiQC

- Sequence quality >15.
- GC content homogeneous among samples and similar to the reported for the studied taxon.
- N content per base <5.
- Correct sequence length (it must be mostly the length defined by the library).
- Sequence duplication levels <20%.
- Overrepresented sequences <0.1%.
- Adapter content (it is not a problem; you will remove them in the next step).

Consult the FastQC help page for information about the different modules.

3.5.4 Adapter Removing and Filtering (See ***Note 8****)*

1. Install Trimmomatic with conda.
2. Remove adapters and filter the reads using Trimmomatic. For paired-end Illumina reads, use the following options:

```
trimmomatic PE -threads 8 -phred33 -trimlog samplename_trimmo.log samplename_1.fastq.gz
samplename_2.fastq.gz  samplename_1.trimP.fq.gz samplename_1.trimUP.fq.gz
samplename_2.trimP.fq.gz samplename_2.trimUP.fq.gz ILLUMINACLIP:TruSeq3-PE-
2.fa:2:30:10:1:keepBothReads LEADING:3 TRAILING:3 SLIDINGWINDOW:4:15 MINLEN:36
```

If new adapter sequences or Unique Dual indexes are provided by the sequencing service, add them to the TruSeq3-PE-2.fa file.

3. Retain the *trimP.fq.gz files for downstream analysis.

Optional: Run FastQC again to corroborate the improvement in quality of your data.

*3.5.5 Assembly with Trinity (See **Note 9**)*

Follow the steps carefully to obtain the filtered transcripts:

1. Install Trinity with conda in a new environment with python ≥3.6.
2. Use Trinity to assemble the transcriptome of each sample independently. Use default options.

```
Trinity --seqType fq --max_memory ">=100GB" --left samplename_1.trimP.fq.gz --right
samplename_2.trimP.fq.gz --CPU ">=24" --output samplename.trinity_out
```

Only change these options if you are sure about its functionality and how it can affect your analysis. If that is the case, consult Trinity's documentation for a detailed explanation on the parameters.

3. Retain the Trinity.fasta and Trinity.gene_trans_map files for all samples from the output folders (e.g. samplename.trinity_out).
4. Change the names in a way that you can easily identify all samples, for example, samplename.trinity.fasta and samplename.gene_trans_map. You have to correctly identify every assembly and its corresponding gene_trans_map.
5. Install assembly-stats with conda.
6. Look at the stats of every assembly with assembly-stats or use the script TrinityStats.pl from the Trinity package to obtain the stats by gene and isoforms.

```
assembly-stats -t -u samplename.trinity.fasta > assemblies_stats_samplename.tsv
PATH_to/trinityrnaseq-v2.9.1/util/TrinityStats.pl samplename.trinity.fasta >>
assemblies_stats_samplename.tsv
```

7. Check the completeness of the assemblies with BUSCO. Use the Metazoa database (ODB10) to predict the orthologous genes in planarian transcriptomes. Install the latest version of BUSCO with conda. Be sure to install all dependencies, and run the following command:

```
busco -i samplename.trinity.fasta -l metazoa_odb10 -o busco.out -m transcriptome -c ">=24"
```

3.5.6 Transcript Clustering

Use CDHIT-EST to cluster each transcriptome using a sequence identity threshold of 0.99.

1. Install CDHIT-EST with conda and run the following command:

```
cd-hit-est -i samplename.trinity.fasta -o samplename.trinity.cdhit.fasta -c 0.99 -G 1 -M
">=120000" -T ">=24" -d 0 -g 1
```

3.5.7 Map Reads Against Assembly

Mapping the reads against the assembly is a way to explore the quality of the assembly and one of the steps to perform the Blob-Tools analysis.

1. Install bwa and samtools with conda in a new environment with python ≥3.6.
2. Map the forward and reverse reads of every sample against its corresponding assembly. Use the following commands:

```
bwa index samplename.trinity.cdhit.fasta
bwa mem -t ">=24" -o samplename_Trinity.cdhit.sam samplename.trinity.cdhit.fasta
samplename_1.trimP.fq.gz samplename_2.trimP.fq.gz
samtools view -S samplename_Trinity.cdhit.sam -bo samplename_Trinity.cdhit.bam
samtools sort -o samplename_Trinity.cdhit.sort.bam samplename_Trinity.cdhit.bam
samtools index samplename_Trinity.cdhit.sort.bam
```

The intermediate files samplename_Trinity.cdhit.sam and samplename_Trinity.cdhit.bam can be deleted if necessary, to save space.

3.5.8 Map Transcripts Against Nucleotide Database

1. Download the update_blastdb.pl script from ncbi page.
2. Use it to download the nucleotide database:

```
wget
https://www.ncbi.nlm.nih.gov/IEB/ToolBox/CPP_DOC/lxr/source/src/app/blast/update_blastd
b.pl
chmod u+x update_blastdb.pl
update_blastdb.pl --showall
update_blastdb.pl --decompress nt
```

3. Install BLAST+ with conda and run the following command:

```
blastn -db PATH_to_nucleotide_database/nt -query samplename.trinity.cdhit.fasta -
num_threads ">=24" -max_target_seqs 10 -max_hsps 1 -outfmt '6 qseqid staxids bitscore
std' -evalue 1e-25 > samplename.blastn.outfmt6
```

3.5.9 Filter Transcripts with BlobTools

1. Install BlobTools. Follow the steps described in the GitHub page of the program.
2. Put the following files in a folder:

```
The assembly: samplename.trinity.cdhit.fasta
The bam file of mapping (reads VS assembly): samplename_Tri-
nity.cdhit.sort.bam
The blastn file (assembly VS nt): samplename.blastn.outfmt6
```

3. Run BlobTools using the following commands:

```
#Creating blobDB
PATH_to_BlobTools_directory/blobtools create -i samplename.trinity.cdhit.fasta -b
samplename_Trinity.cdhit.sort.bam -x bestsumorder -t samplename.blastn.outfmt6 -o
./samplename .blobt.out
#Creating view
PATH_to_BlobTools_directory/blobtools view -i *.blobDB.json  -x bestsumorder
#Creating plot
PATH_to_BlobTools_directory/blobtools plot -i *.blobDB.json -x bestsumorder
```

4. Explore the graphics in the output of BlobTools and check the taxonomic affiliation of the sequences and their coverage (Fig. 3a) as well as the total percentage of mapped reads and the percentage assigned to different taxonomic groups (Fig. 3b). You can summarize these values in a table for all samples. The percentage of no-hit can be high because there is not much information for Platyhelminthes in the NCBI.

3.5.10 Capture Filtered Transcripts

1. Select the names of the transcripts with hit against Platyhelminthes and with no-hit from the samplename.blobt.out.blobDB.table.txt output file.

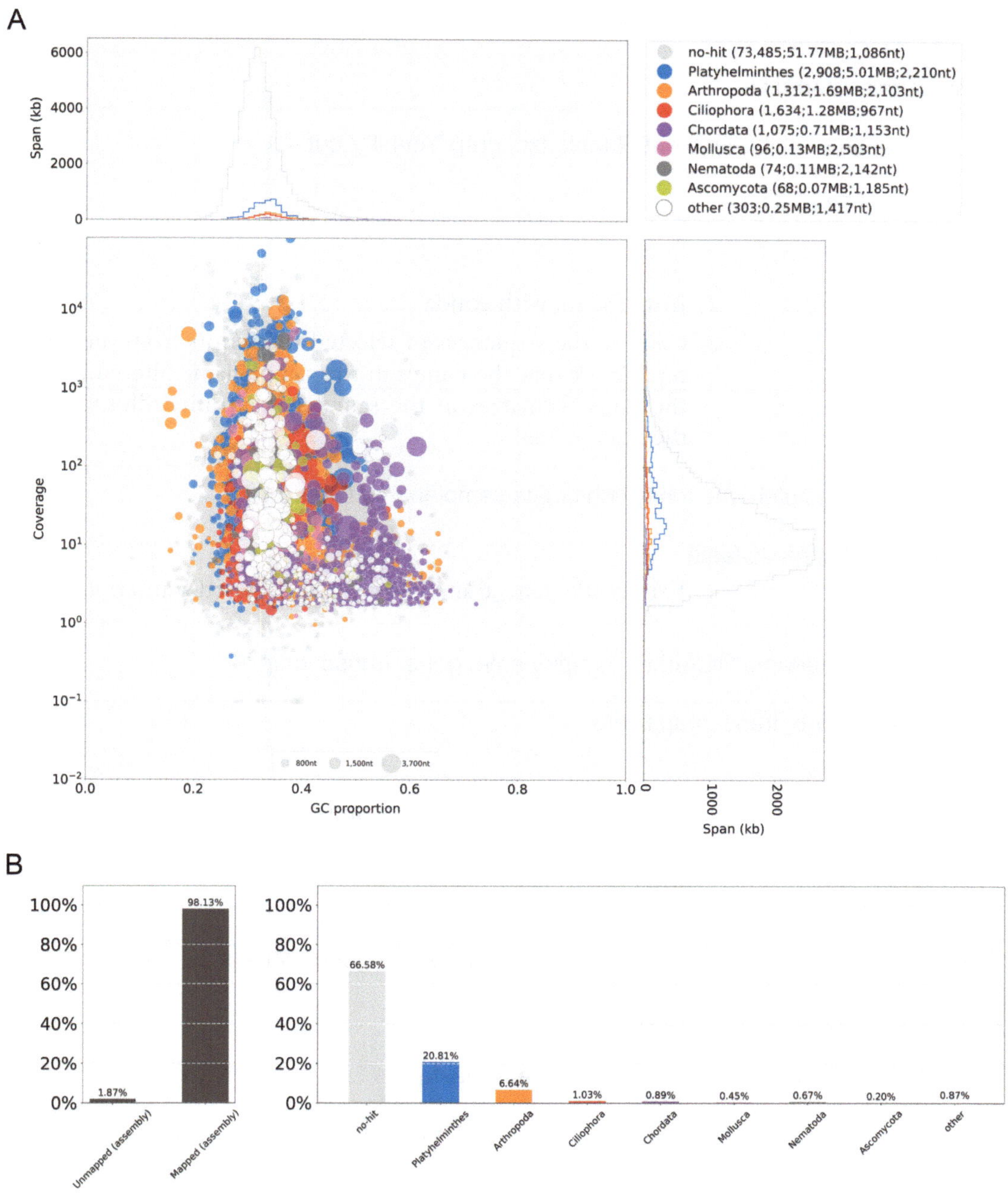

Fig. 3 BlobTools plots for *Dugesia aurea* sample. (**a**) BlobPlot, dimensional scatterplots showing the GC content and coverage. The sequences are represented by circles with diameters proportional to sequence length. The assignment to taxonomic groups is identified with colors. (**b**) ReadCovPlot histogram showing the proportion of unmapped/mapped reads against the assembly and percentage of mapped reads by taxonomic group

```
cat samplename .blobt.out.blobDB.table.txt | grep "Platyhelminthes" | cut -f 1 >
samplename.filtered.txt
cat samplename .blobt.out.blobDB.table.txt | grep "no-hit" | cut -f 1 >>
samplename.filtered.txt
```

2. Install seqtk with conda.
3. Capture the sequences of selected transcripts from the assembly. Check that the names in the samplename.filtered.txt have the same format as in the samplename.trinity.cdhit.fasta and they can match.

```
seqtk subseq samplename.trinity.cdhit.fasta samplename.filtered.txt >
samplename.filt.transc.fasta
```

4. Extract the gene_transc_map for the filtered transcripts.

```
grep -wFf samplename.filtered.txt samplename. gene_transc_map >
samplename.gene_trans_map.filtered
```

3.5.11 Translate Transcripts to Protein

1. Install TransDecoder with conda in a new environment with python ≥3.6.
2. Run:

```
TransDecoder.LongOrfs -t samplename.filt.transc.fasta --gene_trans_map
samplename.gene_trans_map.filtered
TransDecoder.Predict -t samplename.filt.transc.fasta
```

3. Retain the *.pep and *.cds files for downstream analysis.

3.5.12 Select the Longest Isoform

1. Download the choose_longest_iso.py script:

```
wget
https://github.com/tauanajc/Cunha_Giribet_2019_ProcRSocB/blob/master/choose_longest_i
so.py
```

2. Use python2 to run the script as follows:

```
python PATH_to_script/choose_longest_iso.py -l -i= samplename.filt.transc.fasta.pep -o=
sample_A.filt.transc.pep.longest.fasta
```

3.5.13 Orthologue Search

1. Install OrthoFinder following the indications in the web page (*see* **Note 10**).
2. Put all the fasta files with the longest isoforms selected (one by sample) into a folder.
3. Run OrthoFinder just as follows:

```
PATH_to/OrthoFinder/orthofinder -f ./ -t ">=24" -M msa -S diamond_ultra_sens
```

For more information about OrthoFinder's parameters, please consult its web page.

3.5.14 Extract the Protein Sequences of Single-Copy Genes (SC)

Follow the path to Orthofinder/Results_"date"/Single_Copy_Orthologue_Sequences/ and copy the protein sequences of the SC genes directly from this folder.

3.5.15 Extract the Nucleotide Sequences of Single-Copy Genes (SC)

Pay attention to all the steps in this section. Especially to those regarding the format in the names of the files.

1. Capture the name of the sequences that form each SC:

```
grep -Fwf Orthogroups/Orthogroups_SingleCopyOrthologues.txt
Orthogroups/Orthogroups.tsv > SingCopy_ID.txt
```

2. Generate the species/sample list file.

```
ls -1 PATH_to_sequences/ | grep ".fasta" > species_list.txt
sed -i 's/all_that_is_not_the_name_of_the_sample/_seq.txt/' species_list.txt
```

3. Capture the names of single-copy sequences by sample.

   Download the script capt_SCseq_by_sp.py available at https://github.com/lisy87/dugesia-transcriptome/tree/main/scripts

   Run the following script:

```
python3 capt_SCseq_by_sp.py-f=species_list.txt -i= SingCopy_ID.txt
```

4. Copy the *.cds files obtained from TransDecoder's run (**step 2** in Subheading 3.5.11) into a folder.

5. Convert the *.cds files from interleave format to sequential. You can use any tool of your liking for that, or you can use the script interleave_to_oneline.pl available at https://github.com/lisy87/dugesia-transcriptome/tree/main/scripts.
6. Put the *_seq.txt files obtained in **step 3** of this section in the same folder with the *.cds files and the Orthogroups_SingleCopyOrthologues.txt file, obtained from OrthoFinder run (**step 7** in Subheading 3.5.13).
7. Download the sel_sc.py script available at https://github.com/lisy87/dugesia-transcriptome/tree/main/scripts.
8. Run the sel_sc.py script in the folder that contains all these files: *.cds, *_seq.txt, and Orthogroups_SingleCopyOrthologues.txt.

```
python3 sel_sc.py
```

   Refer to the help of the script for more information about it:

```
python3 sel_sc.py --help
```

9. Retain the OG*.fasta files for downstream analyses. Every file contains the nucleotide sequences of all the samples for the single-copy gene in fasta format (*see* **Note 11**).

3.5.16 Alignment (See ***Note 12****)*

1. Install MAFFT with conda.
2. Align the single-copy genes independently using MAFFT:

```
mafft --auto --maxiterate 1000 --thread ">=24" OG*.fasta > OG*_alig.fasta
```

3. Convert the OG*_alig.fasta files from interleaved to sequential format using the interleave_to_oneline.pl script.

3.5.17 Trim the Ends

1. Download the script trimEnds.sh.

```
wget https://github.com/tauanajc/Cunha_Giribet_2019_ProcRSocB/blob/master/trimEnds.sh
```

2. Use the script directly to trim the protein sequences:

```
trimEnds.sh OG*_alig.oneline.fasta
```

3. Edit the script to trim the nucleotide sequences; for that change the aa letters by a, c, g, t (lines 31 and 44) in the original script.
4. Run:

```
trimEnds_edit.sh OG*_alig.oneline.fasta
```

3.5.18 Delete Poorly Aligned Regions (See ***Note 13****)*

1. Install trimAL with conda.
2. Run trimAL using the automated1 option:

```
trimal -in OG*_alig.oneline.trimmed.fasta -out OG*_alig.oneline.trimmed.trimal.fasta -
automated1
```

3.5.19 Concatenation

1. Install AMAS with conda.
2. Put all the OG*_alig.oneline.trimmed.trimal.fasta files in a folder.
3. Be sure that the names of the sequences are the same in all files. They have to match perfectly. If that is not the case, use this bash command:

```
cat OG*_alig.oneline.trimmed.trimal.fasta | sed 's/_DN/@/' | cut -d "@" -f 1 >
OG*_alig.oneline.trimmed.trimal.samename.fasta
```

4. Run AMAS in the folder with all the fasta files:

```
#For protein sequences

AMAS.py concat -i *.fasta -f fasta -d aa -t concat_prot -p concat_prot.part -u fasta -y raxml -c
8

#For nucleotide sequences

AMAS.py concat -i *.fasta -f fasta -d dna -t concat_nuc -p concat_nuc.part -u fasta -y raxml -c
8

# Obtaining the phylip format

AMAS.py convert -i concat_prot.fasta -f fasta -d aa -u phylip -c 8

AMAS.py convert -i concat_nuc.fasta -f fasta -d dna -u phylip -c 8
```

From those command lines a partition file and two concatenate files (fasta and phylip format) are obtained for nucleotide and protein data.

3.5.20 Phylogenetic Inference

1. Exploratory analysis without any partition scheme.

   Install IQ-TREE with conda in a new environment and run:

   ```
   iqtree -s concat_nuc.fasta -m MFP -bb 10000 -nt ">=24"

   iqtree -s concat_prot.fasta -m MFP -bb 10000 -nt ">=24"
   ```

   Use newick_utils to visualize and handle the trees from the terminal (Fig. 4):

   ```
   nw_display *.treefile
   ```

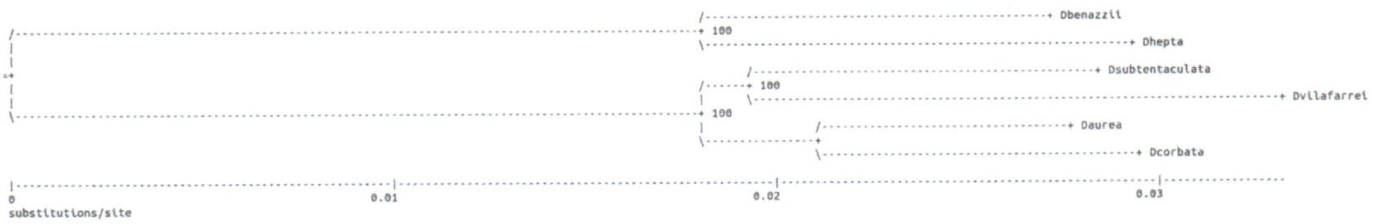

Fig. 4 Screenshot of a tree obtained with the maximum likelihood method, visualized with newick_utils in the terminal. The bootstrap values are shown on the nodes. Scale bar represents the substitutions per site

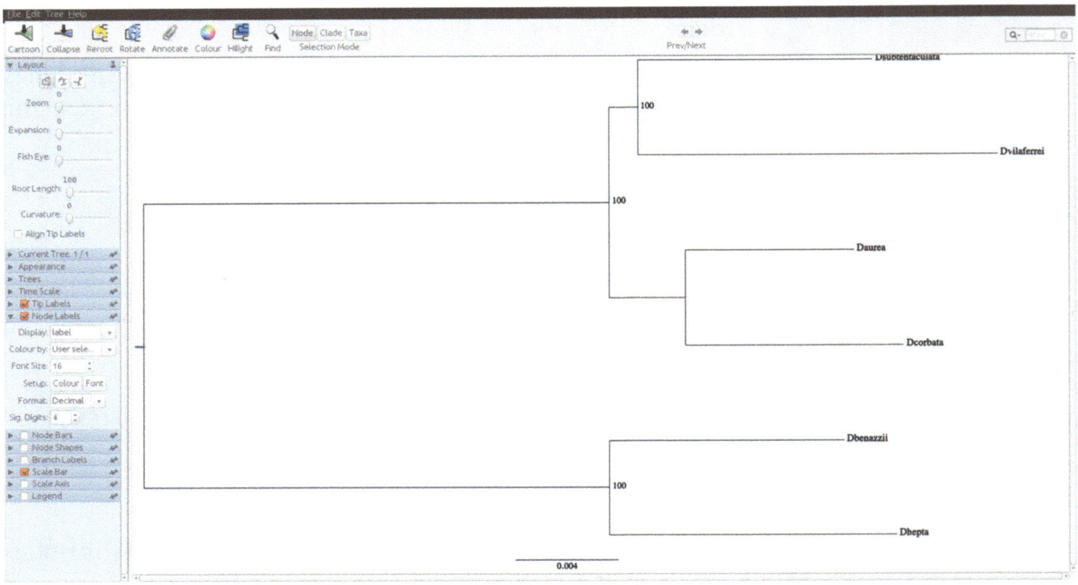

Fig. 5 Screenshot of a tree obtained with the maximum likelihood method, visualized with FigTree. The bootstrap values are shown on the nodes. Scale bar represents the substitutions per site. The options in the upper and left panels enable the tree's customization, with several exporting formats fit for publication

Reroot the tree with the outgroup using newick_utils:

```
nw_reroot *.treefile "outgroup" > *.treefile.rooted
```

If the outgroup is formed by more than one terminal, use their names separated by spaces.

```
nw_display *.treefile.rooted
```

Or FigTree in a local machine to work with a graphic interface (Fig. 5).

Retain the *.treefile files as starting trees for downstream analyses.

2. Phylogenetic Inference using mixture models (*see* **Note 14**).
   Using mixture models with nucleotide data in IQ-TREE:

```
iqtree -s concat_nuc.phylip -m "MIX{JC,HKY,GTR}+G4" -bb 1000000 -nt ">=24"
```

   Using site-specific frequency models with protein data in IQ-TREE:

```
iqtree -s concat_prot.phylip -m LG+C20+F+G -ft concat_prot.fasta.treefile -bb 1000000 -nt
">=24"
```

   Using Bayesian CAT models implemented in PhyloBayes for protein data.
   Install phylobayes-mpi with conda and run:

```
mpirun -n ">=24" pb_mpi -d concat_prot.phylip -catfix C20 -lg -t concat_prot.fasta.treefile
chain_0
mpirun -n ">=24" pb_mpi -d concat_prot.phylip -catfix C20 -lg -t concat_prot.fasta.treefile
chain_1
```

   Using Bayesian CAT models implemented in PhyloBayes for nucleotide data:

```
mpirun -n ">=24" pb_mpi -d concat_nuc.phylip -cat gtr -t concat_nuc.fasta.treefile chain_0
mpirun -n ">=24" pb_mpi -d concat_nuc.phylip -cat gtr -t concat_nuc.fasta.treefile chain_1
```

   These last two analyses can take a long time. Check the chain periodically:

```
tail chain_0.trace
tail chain_1.trace
```

   Stop the chains softly if you estimate that the chains have reached convergence:

```
echo 0 > chain_0.run
echo 0 > chain_1.run
```

   Check the convergence of parameters:

```
tracecomp -x 1000 chain_0.trace chain_1.trace
```

   Use the following criteria to evaluate them:

maxdiff < 0.1 and minimum effective size>300: good run

maxdiff < 0.3 and minimum effective size > 50: acceptable run

```
bpcomp -x 1000 chain_0 chain_1
```

maxdiff < 0.1: good run

maxdiff < 0.3: acceptable: gives a good qualitative picture of the posterior consensus

0.3 < maxdiff < 1: the sample is not yet sufficiently large, and the chains have not converged yet, but this is on the right track.

maxdiff = 1 even after 10,000 points; this indicates that at least one of the runs is stuck in a local maximum.

If the values of maxdiff parameter are not good, restart the runs using the same commands as in the previous steps with the same name for the chains and the analysis will continue retaining the previous results.

Once the parameters indicate the runs have converged, visualize the trees (bpcomp.con.tre file) in newick_utils or FigTree.

At the end of the process, four trees based on mixture models are obtained. An ML tree and a Bayesian tree for each dataset, nucleotide and protein. Just as with Sanger data-based trees, it is possible to take one of these trees and represent the bootstrap and probability values obtained in the four analyses. (Fig. 6).

4 Notes

1. Sampling

The best option to have a good-quality nucleic acid extraction is always to work with living planarians. However, when these have to be collected and transported for a long time over long distances to the laboratory, it is better to fix them in situ lest we want to risk losing them in the process.

To obtain a good yield of high-quality RNA, the best fixation is accomplished using RNAlater. Another important point to take into account when sampling in the field is the fact that some localities may be occupied by more than one species, some being externally extremely similar to our target species. If we are interested in one specific species, then it is highly recommended to cut a small piece of the animal (commonly the tail) in situ and fix it in a microtube with absolute ethanol, while fixing the rest in another microtube containing RNAlater. Both tubes must be clearly identified to know which are the two pieces of each individual. It is a good practice to use tubes with screw caps of different colors for each

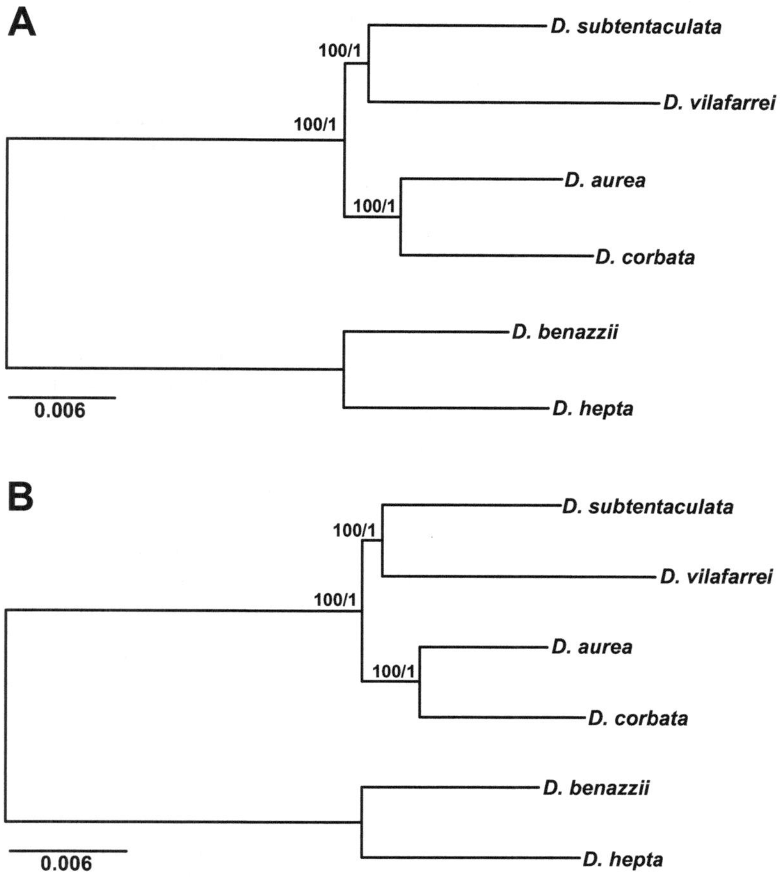

Fig. 6 Maximum likelihood trees based on the example dataset of the *Dugesia* genus. Inferred using mixture models and 1,000,000 replicates of ultrafast bootstrap approximation implemented in IQ-TREE. (**a**) Tree obtained from protein data (LG + C20 + F + G); (**b**) tree obtained from nucleotides (MIX{JC,HKY, GTR} + G4). Numbers over nodes indicate bootstrap/posterior probability support values. The probability values were obtained from analyses in PhyloBayes running two chains with (**a**) protein data (-catfix C20 -lg) and (**b**) nucleotide data (-cat -gtr)

fixation procedure (we regularly use red caps for alcohol and yellow ones for RNAlater).

Once in the laboratory we can check the identity of the species by obtaining DNA from the ethanol preserved fragment and sequencing its cytochrome oxidase I gene (COI) to give an example.

2. Extraction of nucleic acids in planarians

The experience of many years in our laboratory demonstrates that the use of commercial kits based on column technology is not a good strategy for nucleic acid extraction from planarian tissue. It is our hypothesis that the pigments form a molecular structure with the nucleic acids that is very difficult to separate. The impossibility to separate pigments from RNA has been reported previously [34]. Actually, when we extract DNA or RNA from pigmented planarians, it is very frequent to obtain a dark solution from the final step of the protocol, probably due to the formation of this complex between pigments and nucleic acids. For *Dugesia ryukyuensis*, an absorption spectrum of body pigment characterized by three absorption peaks of 256, 367, and 463 nm has been reported [35], the first peak being in the same range of nucleic acid absorption (260 nm) which may interfere the quantification of DNA and RNA in spectrophotometer-based apparatuses. We have witnessed this phenomenon when quantifying dark samples using a Nanodrop machine, showing high absorbance and overestimating concentration values when compared to quantification by Qubit. We think that the macromolecular complex formed by the nucleic acids and pigments cannot be separated by the current column methods and, in the final elution step, this structure does not properly elute. As a consequence, the yield of the extraction based on column methods is very low in planarians. We recommend using traditional methods of nucleic acid extraction not based on columns, such as phenol/chloroform for DNA [12] and Trizol for RNA.

3. Library construction and sequencing

Transcriptome sequencing can be achieved in different companies nowadays. Most RNAseq is still performed by Illumina sequencing (short reads) although some long read technology (as Nanopore, Oxford) options are beginning to be offered. The latter ensures full-length transcript sequences without the need of assembling the short reads, allowing an accurate identification of isoforms and alleles. Nonetheless, they are way more expensive options.

Library construction may preferably avoid including the highly repetitive and abundant ribosomal RNAs. This can be achieved using a polyA-based capture kit to include only protein coding RNAs or, presently more common, a Ribo-Zero library kit preparation. The latter uses probes that selectively bind rRNAs to eliminate them before sequencing, but keeps noncoding RNAs (ncRNAs) that may also be of interest. Also using stranded information identifies from which of the two DNA strands a given RNA transcript was derived. This information provides increased confidence in transcript annotation.

4. Software installation

A high number of informatic programs are necessary to execute the bioinformatic workflow. We describe briefly how to install and use the software implicated in every step. Please, consult the documentation of all of them and study the options carefully.

Use conda to access and install packages. We recommend creating new environments to install the software. In some cases, it is mandatory because the software has specific requirements. Please, consult the conda's manual as the first step.

Not all the software or the latest versions are available in conda. In those cases, carefully consult the installation guide of the software documentation. Take into account the characteristics of the cluster you are using.

5. Raw data from Illumina technology

For paired-end Illumina raw data, you obtain two files by sample: samplename_1.fastq.gz (forward/right reads) and samplename_2.fastq.gz (reverse/left reads). Please see [36] for information about fastqc format and [37] for the gzip data compression utility.

From this point we use "samplename" to denote a hypothetical sample.

6. Working with loops to automate steps

Here, we describe the steps to build a phylogeny with transcriptomic data. If you are working with a high number of samples, you can use loops to automate steps to avoid working with the samples one by one. If you have skills with bash, python, or perl, you can do it easily. However, if you are a beginner with these languages and the command line, you can use the scripts available at https://github.com/lisy87/dugesia-transcriptome/tree/main/scripts.

7. Quality control of the raw data

The MultiQC software allows us to summarize the results of FastQC of all samples in a nice html interface. Use this tool to compare the quality scores of your samples.

Take into account that the GC percentage in planarians is lower when compared to most organisms [38], for instance, GC content in the transcriptome of *Schmidtea mediterranea* is around 35% [39].

8. Remove adapters

Use Trimmomatic program to remove universal Illumina adapters and low-quality bases from the ends (LEADING:3 TRAILING:3). The sliding window method is used to trim bases with a mean quality under 15 (SLIDINGWINDOW:4:15). In addition, reads with lengths shorter than 36 bp are dropped (MINLEN:36). From paired-end reads, Trimmomatic outputs the paired (*trimP.fq.gz) and the unpaired (*trimUP.fq.gz) files for both forward and reverse reads. Retain the paired forward and reverse reads (*trimP.fq.gz) for downstream analyses.

After trimming, it is advisable to run FastQC again on the paired reads to corroborate the improvement in quality after Trimmomatic filtering.

9. Assembly with Trinity

Trinity is one of the most popular softwares to assemble a de novo transcriptome without a reference genome. Trinity has shown better performance against other transcriptome assemblers such as Bridger or SOAPdenovo-Trans [40]. Here, we use Trinity with default options, but it is recommended to consult the broad extension of information available about Trinity and the detailed pipelines described in its GitHub page.

10. Installing OrthoFinder

We recommend creating an environment with python ≥3.6 and install NumPy and SciPy. Then, follow the directions on the web page of OrthoFinder to download and install it.

Please pay attention to all specifications of OrthoFinder and the characteristics of your system. You may have to use the installer designed for glibc-2.15 libraries.

In later versions of OrthoFinder, the software first checks if the number of files that the user can open at the same time is enough to open all the required files to run. If not, OrthoFinder fails immediately and the user has to adjust the limits (see https://github.com/davidemms/OrthoFinder/issues/384). However, if the cluster works under a queuing system, it is possible that this failure happens even for a user with high limits of opened files. In this case, the job has to be submitted directly to the cluster without using the queuing system.

11. Select nucleotide sequences of single-copy orthogroups

After running the script sel_sc.py, you get a fasta file for each of the species with all the OGs (samplename_sc.fasta) and a fasta file for each of the OGs with all the species (OG*.fasta), ready to be aligned.

You get as many "samplename_sc.fasta" files, as samples you have, and as many "OG*.fasta" files as single-copy orthogroups you have. We recommend to move the OG*.fasta files to another folder to work in.

12. Alignment

All through this section be sure to keep the original input files in all intermediate steps and save the files with an informative name related to the step, e.g., OG000625.fasta –> OG000625_alig.fasta –> OG000625_alig_oneline.fasta –> OG000625_alig_oneline_trim.fasta. Follow the same steps for the protein and nucleotide datasets.

13. Delete poorly aligned regions

Please, consult the manual of trimAl and the usage options for the command line. We recommend using the -automated1 option to choose the best options.

14. Phylogenetic inference using mixture models

When using probabilistic-based phylogenetic inference methods, it is compulsory to implement an evolutionary model [13]. Initially, the same model was applied to all the datasets; later on some strategies to partition the datasets and apply different models depending on the fragments of analyzed sequences were developed. Nowadays, and especially ever since the upraise of genomic datasets, the use of mixture models has picked up in the field of phylogenomic analyses, replacing the traditional partition scheme by gene or codon. In this new strategy the phylogenetic inference algorithm evaluates for each site the more adequate model to be applied and groups similar sites in categories [14] making it unnecessary to give beforehand any partition scheme. Overall, the good performance of this method when applied to phylogenetic inference has been demonstrated [15, 16] and its implementation in a variety of software makes it easy to use.

To apply these models to the phylogenetic inference with protein and nucleotide data, we will use IQ-TREE and PhyloBayes. The IQ-TREE software has an excellent implementation of mixture models as well as a posterior mean site frequency (PMSF) model that could be likened to PhyloBayes' CAT model. However, PMSF is less consuming than the latter both timewise and memory-wise [12]. Furthermore, IQ-TREE benefits from the ultrafast bootstrap approximation [41] which in turn helps make this analysis relatively easy without an excessive consumption of either time or computer resources. The implementation of the CAT model in PhyloBayes [14] allows it to work with both protein and nucleotide data implementing a Bayesian approach. Alas, PhyloBayes is much more needy than IQ-TREE both computationally and timewise and the analyses may take much longer.

References

1. Andrews S (2010) FastQC: a quality control tool for high throughput sequence data
2. Bolger AM, Lohse M, Usadel B (2014) Trimmomatic: a flexible trimmer for illumina sequence data. Bioinformatics 2114–2120: 2114. https://doi.org/10.1093/bioinformatics/btu170
3. Haas BJ, Papanicolaou A, Yassour M et al (2013) De novo transcript sequence reconstruction from RNA-seq using the trinity platform for reference generation and analysis. Nat Protoc 8:1494–1512. https://doi.org/10.1038/nprot.2013.084
4. Grabherr MG, Haas BJ, Yassour M et al (2011) Trinity: reconstructing a full-length transcriptome without a genome from RNA-Seq data. Nat Biotechnol 29:644. https://doi.org/10.1038/NBT.1883
5. Laetsch DR, Blaxter ML (2017) BlobTools: Interrogation of genome assemblies [version 1; peer review: 2 approved with reservations]. F1000Research 6. https://doi.org/10.12688/f1000research.12232.1

6. Challis R, Paulini M (2021) blobtoolkit/blobtools2: v2.6.1
7. Haas B, Papanicolaou A (2019) TransDecoder 5.5.0
8. Emms DM, Kelly S (2019) OrthoFinder: phylogenetic orthology inference for comparative genomics. Genome Biol 20:238. https://doi.org/10.1186/s13059-019-1832-y
9. Minh BQ, Schmidt HA, Chernomor O et al (2020) IQ-TREE 2: new models and efficient methods for phylogenetic inference in the genomic era. Mol Biol Evol 37:1530–1534. https://doi.org/10.1093/molbev/msaa015
10. Lartillot N, Philippe H (2004) A Bayesian mixture model for across-site heterogeneities in the amino-acid replacement process. Mol Biol Evol 21:1095–1109. https://doi.org/10.1093/molbev/msh112
11. Lemey P, Salemi M, Vandamme A-M (2018) The phylogenetic handbook: a practical approach to phylogenetic analysis and hypothesis testing, 2nd edn. Cambridge University Press
12. Wang H-C, Minh BQ, Susko E, Roger AJ (2018) Modeling site heterogeneity with posterior mean site frequency profiles accelerates accurate phylogenomic estimation. Syst Biol 67:216–235. https://doi.org/10.1093/sysbio/syx068
13. Holder M, Lewis PO (2003) Phylogeny estimation: traditional and Bayesian approaches. Nat Rev Genet 4:275–284. https://doi.org/10.1038/nrg1044
14. Quang LS, Gascuel O, Lartillot N (2008) Empirical profile mixture models for phylogenetic reconstruction. Bioinformatics 24:2317–2323. https://doi.org/10.1093/bioinformatics/btn445
15. Venditti C, Meade A, Pagel M (2008) Phylogenetic mixture models can reduce node-density artifacts. Syst Biol 57:286–293. https://doi.org/10.1080/10635150802044045
16. Schrempf D, Lartillot N, Szöllősi G (2020) Scalable empirical mixture models that account for across-site compositional heterogeneity. Mol Biol Evol 37:3616–3631. https://doi.org/10.1093/molbev/msaa145
17. Anaconda Software Distribution (2020) Anaconda
18. Ewels P, Magnusson M, Lundin S, Käller M (2016) MultiQC: summarize analysis results for multiple tools and samples in a single report. Bioinformatics:3047–3048. https://doi.org/10.1093/bioinformatics/btw354
19. Challis R (2017) rjchallis/assembly-stats. Zenodo
20. Fu L, Niu B, Zhu Z et al (2012) CD-HIT: accelerated for clustering the next-generation sequencing data. Bioinformatics 28:3150–3152. https://doi.org/10.1093/bioinformatics/bts565
21. Li W, Godzik A (2006) Cd-hit: a fast program for clustering and comparing large sets of protein or nucleotide sequences. Bioinformatics 22:1658–1659. https://doi.org/10.1093/bioinformatics/btl158
22. Li H, Durbin R (2009) Fast and accurate short read alignment with Burrows-Wheeler Transform. Bioinformatics 25:1754–1760. https://doi.org/10.1093/bioinformatics/btp324
23. Danecek P, Bonfield JK, Liddle J et al (2021) Twelve years of SAMtools and BCFtools. Gigascience 10. https://doi.org/10.1093/gigascience/giab008
24. SAMtools (2020) SAMtools, Version 1.11
25. Camacho C, Coulouris G, Avagyan V et al (2009) BLAST+: architecture and applications. BMC Bioinform 10. https://doi.org/10.1186/1471-2105-10-421
26. Li H (2012) Seqtk. https://github.com/lh3/seqtk
27. Fernández R, Laumer CE, Vahtera V et al (2014) Evaluating topological conflict in centipede phylogeny using transcriptomic data sets. Mol Biol Evol 31:1500–1513. https://doi.org/10.1093/MOLBEV/MSU108
28. Cunha TJ, Giribet G (2019) A congruent topology for deep gastropod relationships. Proc R Soc B Biol Sci 286:20182776. https://doi.org/10.1098/rspb.2018.2776
29. Benítez-Álvarez L, Leria L, Fernández R et al (2023) Phylotranscriptomics interrogation uncovers a complex evolutionary history for the planarian genus Dugesia (Platyhelminthes, Tricladida) in the Western Mediterranean. Mol Phylogenet Evol 178:107649. https://doi.org/10.1016/j.ympev.2022.107649
30. Katoh K, Standley DM (2013) MAFFT multiple sequence alignment software version 7: improvements in performance and usability. Mol Biol Evol 30:772–780. https://doi.org/10.1093/molbev/mst010
31. Capella-Gutiérrez S, Silla-Martínez JM, Gabaldón T (2009) trimAl: a tool for automated alignment trimming in large-scalephylogenetic analyses. Bioinformatics 25:1972–1973. https://doi.org/10.1093/bioinformatics/btp348
32. Borowiec ML (2016) AMAS: a fast tool for alignment manipulation and computing of summary statistics. PeerJ 4:e1660. https://doi.org/10.7717/peerj.1660

33. Junier T, Zdobnov EM (2010) The Newick utilities: high-throughput phylogenetic tree processing in the Unix shell. Bioinformatics 26:1669–1670. https://doi.org/10.1093/bioinformatics/btq243
34. Stubenhaus BM, Dustin JP, Neverett ER et al (2016) Light-induced depigmentation in planarians models the pathophysiology of acute porphyrias. eLife e14175:10.7554/eLife.14175
35. Hase S, Wakamatsu K, Fujimoto K et al (2006) Characterization of the pigment produced by the planarian. Pigment Cell Res, Dugesia ryukyuensis. https://doi.org/10.1111/j.1600-0749.2006.00306.x
36. Support Illumina. https://emea.support.illumina.com/bulletins/2016/04/fastq-files-explained.html. Accessed 26 Jul 2021
37. Gzip. https://www.gzip.org/. Accessed 26 Jul 2021
38. Martín-Durán JM, Ryan JF, Vellutini BC et al (2017) Increased taxon sampling reveals thousands of hidden orthologs in flatworms. Genome Res 27:1263–1272. https://doi.org/10.1101/gr.216226.116
39. Abril JF, Cebrià F, Rodríguez-Esteban G et al (2010) Smed454 dataset: unravelling the transcriptome of *Schmidtea mediterranea*. BMC Genomics 11. https://doi.org/10.1186/1471-2164-11-731
40. Hölzer M, Marz M (2019) De novo transcriptome assembly: a comprehensive cross-species comparison of short-read RNA-Seq assemblers. Gigascience 8. https://doi.org/10.1093/gigascience/giz039
41. Hoang DT, Chernomor O, von Haeseler A et al (2018) UFBoot2: improving the ultrafast bootstrap approximation. Mol Biol Evol 35: 518–522. https://doi.org/10.1093/molbev/msx281

Chapter 2

Isolation and Library Preparation of Planarian piRNAs

Iana V. Kim, Tim Demtröder, and Claus-D. Kuhn

Abstract

In planarian flatworms, piRNAs and SMEDWI (*Schmidtea mediterranea* PIWI) proteins are both essential for the animals' impressive regenerative ability and for their survival. A knockdown of SMEDWI proteins disrupts the specification of the planarian germline and impairs stem cell differentiation, resulting in lethal phenotypes. As the molecular targets of PIWI proteins and thus their biological function are determined by PIWI-bound small RNAs, termed piRNAs (for PIWI-interacting RNAs), it is imperative to study the wealth of PIWI-bound piRNAs using next-generation sequencing-based techniques. Prior to sequencing, piRNAs bound to individual SMEDWI proteins must be isolated. To that end, we established an immunoprecipitation protocol that can be applied to all planarian SMEDWI proteins. Co-immunoprecipitated piRNAs are visualized by using qualitative radioactive 5′-end labeling, which detects even trace amounts of small RNAs. Next, isolated piRNAs are subjected to a library preparation protocol that has been optimized for the efficient capture of piRNAs, whose 3′-ends carry a 2′-O-methyl modification. Successfully prepared piRNA libraries are subjected to Illumina-based next-generation sequencing. Obtained data are analyzed as presented in the accompanying manuscript.

Key words piRNAs, PIWI proteins, Small RNAs, Immunoprecipitation, Small RNA library preparation, SMEDWI proteins, Planarians, RNA radiolabeling, Urea PAGE

1 Introduction

PIWI-interacting RNAs (piRNAs) are a class of animal-specific, small noncoding RNAs bound by the PIWI clade of the Argonaute protein family [1–3]. The primary role of piRNA-bound PIWI proteins is to protect the integrity of the animal germline genome by repressing transposable elements either through nucleolytic cleavage or through epigenetic changes [4, 5] . However, over the past decade it became clear that the function of the piRNA pathway extends beyond silencing of transposable elements. It was also found to regulate the stability and expression of coding genes [6] and to maintain transcriptome profiles by mRNA surveillance [7–9]. Moreover, piRNAs were found to constitute an antiviral defense system [10]. In the planarian flatworm *Schmidtea*

Luca Gentile (ed.), *Schmidtea Mediterranea: Methods and Protocols*, Methods in Molecular Biology, vol. 2680,
https://doi.org/10.1007/978-1-0716-3275-8_2,
© The Author(s), under exclusive license to Springer Science+Business Media, LLC, part of Springer Nature 2023

mediterranea, piRNAs and PIWI proteins (SMEDWIs) are enriched in adult pluripotent stem cells, termed neoblasts [11–14] . Disruptions of the piRNA pathway (by knocking down SMEDWI-2 or SMEDWI-3) are detrimental to the ability of neoblasts to self-renew and to orchestrate regeneration, which results in a lethal phenotype [11, 12, 14] . Moreover, SMEDWI-1 is required for proper germline specification in the sexually reproducing planarian species *Dugesia ryukyuensis* [15]. Strikingly, the molecular mechanisms underlying the importance of PIWI proteins for planarian regeneration and tissue homeostasis remain unknown. In recent years, however, considerable progress has been made towards a systematic characterization of the piRNA pathway in planarian flatworms, which included the development of experimental tools to study planarian piRNAs [14, 16–21] .

Here, we describe an immunoprecipitation protocol that allows for the isolation of individual SMEDWI proteins and their associated piRNAs (Fig. 1). In addition, we outline a protocol that does not require immunoprecipitation techniques, yet still allows for the enrichment of piRNAs. Isolated piRNAs are then used to construct complex piRNA libraries for subsequent next-generation sequencing applications (Figs. 2 and 3). Coupled with a bioinformatics analysis pipeline that we describe in an accompanying manuscript, this protocol facilitates the study of the biological role of the piRNA pathway in planarian flatworms. In addition, a protocol for radioactive 5′-end labeling of piRNAs and for their visualization is included.

2 Materials

2.1 Preparation of Whole-Cell Lysate from Planarian Flatworms

1. Lysis buffer: 30 mM HEPES (pH 7.7) (Sigma, H989), 150 mM NaCl, 10 mM KCl, 4 mM $MgCl_2$, 1 mM DTT, 0.5% Triton X-100, cOmplete EDTA-free protease inhibitor (Roche, 4693159001). Lysis buffer should be prepared freshly for each experiment by adding DTT, Triton-X-100, and protease inhibitor right before use.
2. Potter-Elvehjem Tissue Grinder with PTFE pestle, 5 mL (Corning).
3. 0.20 μm cellulose acetate syringe filter (LLG labware).
4. Bradford protein assay reagent (Bio-Rad).
5. UV spectrophotometer.

2.1.1 Immunoprecipitation of SMEDWI Proteins and Co-Bound piRNAs

1. 1.5 mL DNA low-binding tubes (Eppendorf).
2. Antibodies:

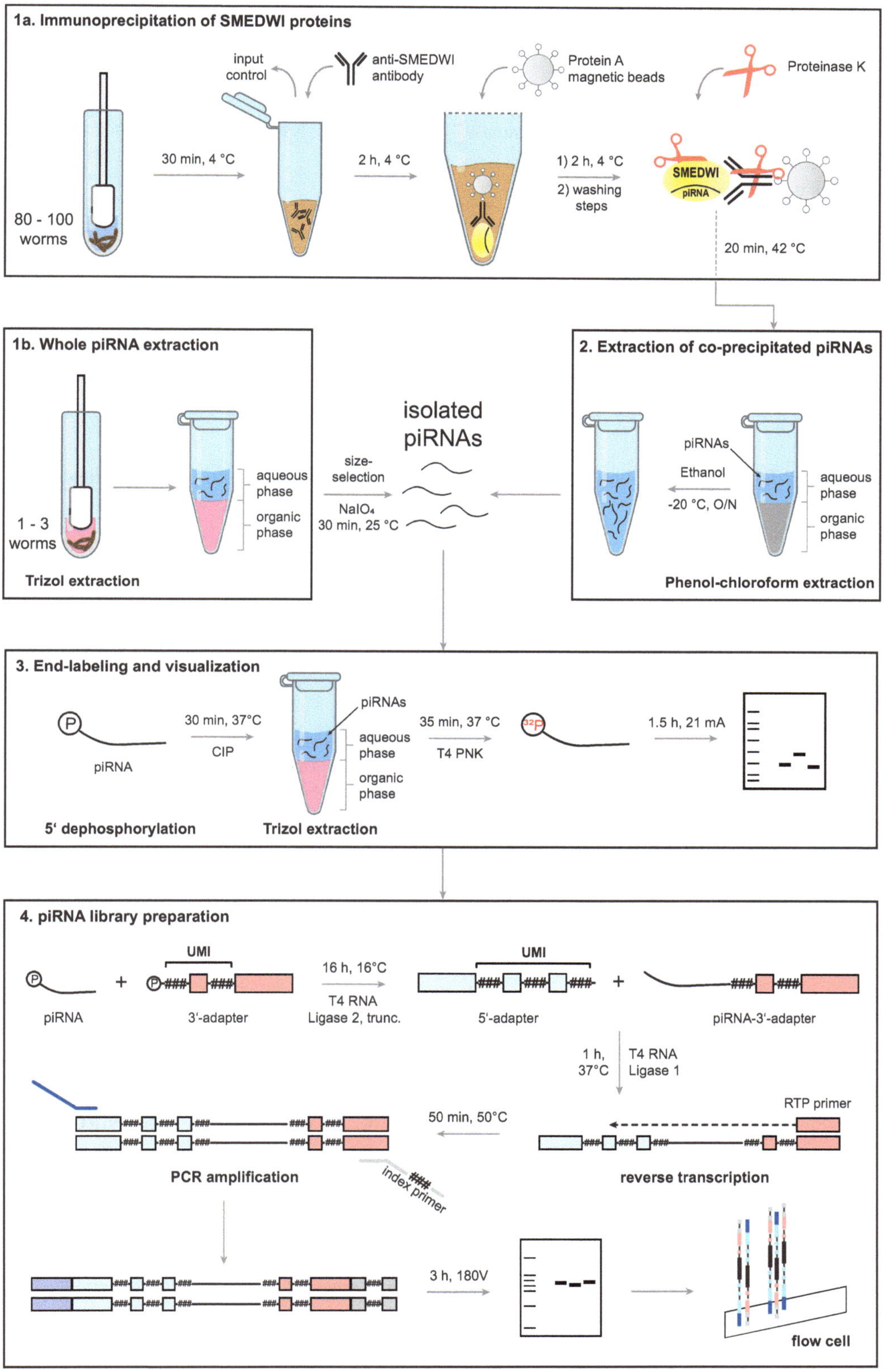

Fig. 1 Schematic overview of the methods described in this chapter: immunoprecipitation of PIWI proteins and co-bound piRNAs, alternative piRNA extraction and enrichment, and piRNA library preparation

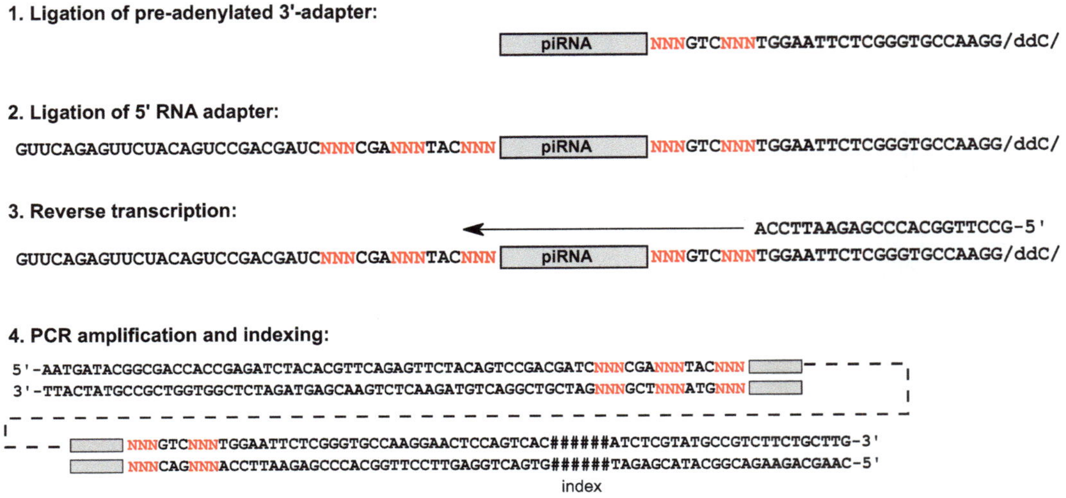

Fig. 2 The piRNA library preparation workflow including the utilized adapters and primer sequences

Antibody	Target peptide	Supplier
Anti-SMEDWI-1 antibody	NEPEGPTETDQSLS [21]	BioGenes, Berlin, Germany
Anti-SMEDWI-2 antiserum	KKDEEGVEKEK	BioGenes, Berlin, Germany
Anti-SMEDWI-3 antibody	N-terminal 200 amino acids of SMEDWI-3 [14]	BioGenes, Berlin, Germany
Rabbit pre-immune serum	–	BioGenes, Berlin, Germany

3. Protein A magnetic beads, Dynabeads (Invitrogen).
4. Magnetic stand, DynaMag (Invitrogen).
5. Low-salt wash buffer: 30 mM HEPES (pH 7.7) (Sigma, H989), 150 mM NaCl, 10 mM KCl, 4 mM $MgCl_2$, 5 mM EDTA, 1 mM DTT, 0.1% Triton X-100.
6. High-salt wash buffer: 30 mM HEPES (pH 7.7) (Sigma, H989), 300 mM NaCl, 10 mM KCl, 4 mM $MgCl_2$, 5 mM EDTA, 1 mM DTT, 0.1% Triton X-100.
7. Refrigerated tabletop centrifuge.

2.1.2 Extraction and Enrichment of piRNA from Planarian Lysate

1. Dounce homogenizer for 1.5 mL reaction tubes.
2. RNA Clean & Concentrator-5 kit (Zymo Research).
3. TRIzol reagent (also known as TRI Reagent or TriFast) (Invitrogen, Peqlab).
4. Recombinant, RNase-free DNase I (Roche 04716728001).
5. 10× DNase I buffer (Roche).

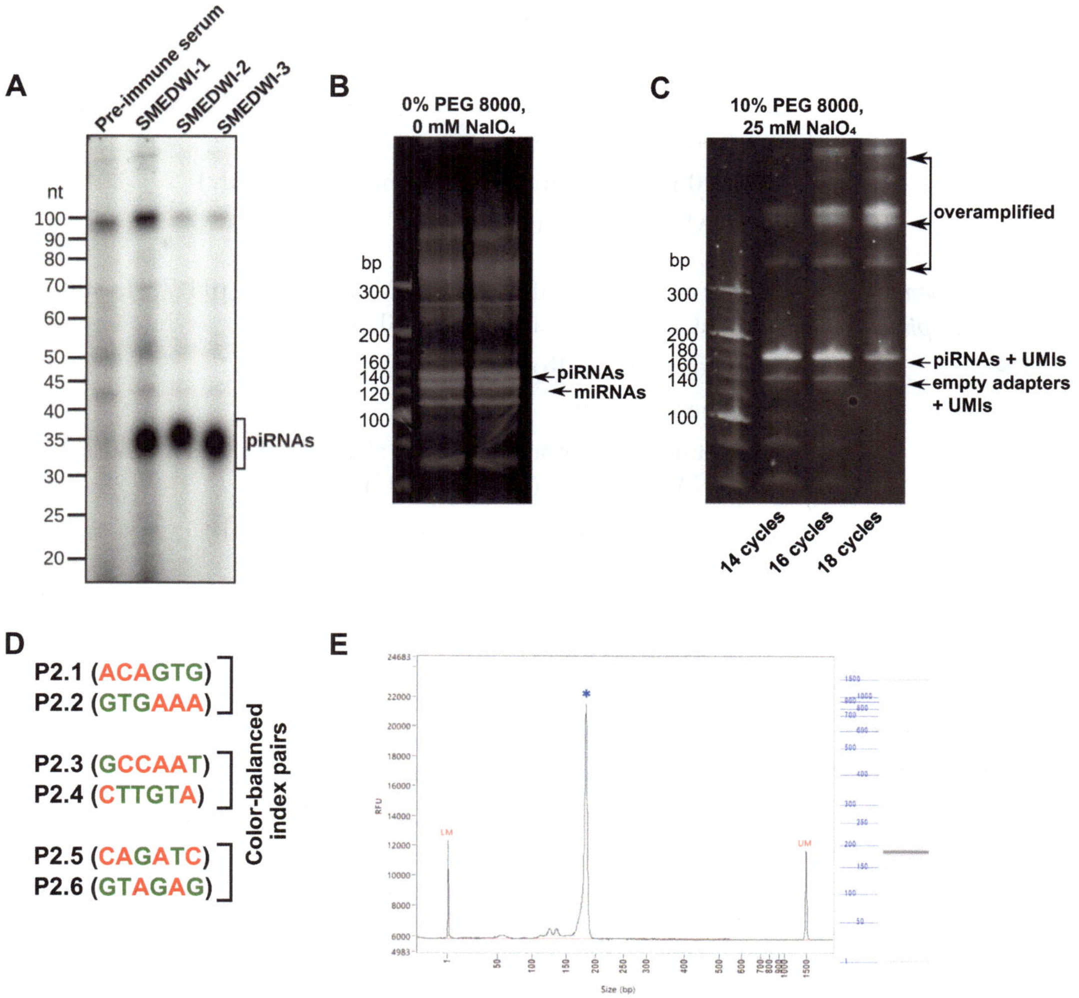

Fig. 3 (**a**) piRNAs were immunoprecipitated from total worm lysate using anti-SMEDWI-1, anti-SMEDWI-2, and anti-SMEDWI-3 antibodies. Immunoprecipitated RNAs were labeled with [γ-^{32}P]-ATP and separated on a 10% urea PAGE. Pre-immune serum was used as negative control [14]. Image used with permission by Cold Spring Harbor Press. (**b**) Amplified piRNA libraries resolved on a 6% urea PAGE. By using "0% PEG 8000 and 0 mM $NaIO_4$," piRNAs seem equally abundant as microRNAs, which is not the case in planarians [14]. To increase the efficiency of piRNA library preparation PEG 8000 needs to be added during adaptor ligation (*see* Fig. 3c). Please note that the displayed gel shows results of library preparation without UMIs. (**c**) Final amplification of the piRNA libraries using 14 cycles of PCR. The appearance of high-molecular-weight products above 300 bp (marked "overamplified") is a sign of overamplified libraries. This effect is predominantly detected after 16 and 18 cycles of PCR. (**d**) Color-balanced 6-base index pairs. (**e**) Fragment analyzer electropherogram of piRNA libraries prepared from whole worms. The peak marked by an asterisk denotes the cloned piRNA library. It has a size between 172 and 177 bp, consisting of 30–35 bp piRNAs and 142-bp adapter and primer sequences

6. 100% ethanol, ice-cold. Store at −20 °C,
7. 75% ethanol, ice-cold. Store at −20 °C.
8. RNase-free water.

9. 20 mg/mL Glycogene (Roche).
10. Chloroform (Carl Roth).
11. 5× borax/boric acid buffer (pH 8.6): 150 mM borax, 150 mM boric acid.
12. 200 mM sodium periodate (prepared freshly).
13. 3 M sodium acetate (pH 5.2).

2.2 Extraction of Co-Immunoprecipitated piRNAs

1. Proteinase K buffer: 200 mM Tris–HCl (pH 7.5), 300 mM NaCl, 25 mM EDTA, 1.5% SDS.
2. Proteinase K (Roche).
3. Thermomixer.
4. Phenol-chloroform-isoamyl alcohol (P/C/I) solution for RNA extraction (pH ≤4.5) (Carl Roth).
5. Trichloromethane/chloroform (Carl Roth).
6. 20 mg/mL Glycogene (Roche).
7. 100% ethanol, ice-cold. Store at −20 °C.
8. RNase-free water.

2.3 Radioactive 5′-End Labeling and Detection of Co-Immunoprecipitated piRNAs

1. Calf intestine phosphatase, CIP (NEB, M0290).
2. 10× CutSmart Buffer (NEB).
3. TRIzol reagent (also known as TRI Reagent or TriFast) (Peqlab).
4. 100% isopropanol, ice-cold solution. Store the solution at −20 °C.
5. RNA Clean & Concentrator-5 kit (Zymo Research).
6. T4 polynucleotide kinase (PNK) (NEB, M0201).
7. 10× PNK buffer (NEB).
8. 10 μCi/μL [γ-^{32}P]-ATP, 3000 Ci/mmol (PerkinElmer).
9. Illustra Microspin G25 columns (GE Healthcare).
10. 2× formamide loading dye: 95% formamide (v/v), 18 mM EDTA (pH 8.0), 0.025% SDS, 0.025% bromophenol blue (w/v), 0.025% xylene cyanol (w/v).
11. 25 mL denaturing 10% urea PAGE gel solution (SequaGel Urea Gel System, National Diagnostics): 10 mL UreaGel concentrate, 12.5 mL UreaGel diluent, 2.5 mL UreaGel buffer, 200 μL 10% ammonium persulfate, 10 μL TEMED.
12. 10× TBE buffer: 108 g Tris base, 55 g boric acid, 40 mL 0.5 M EDTA (pH 8.0), distilled water up to 1 L of total volume.
13. Low-molecular-weight 10–100 nt ladder (Affymetrix) (*see* **Note 1**). Label 1 μL low-molecular-weight ladder with [γ-^{32}P]-ATP and T4 PNK (NEB) following the instructions

given in Subheading 3.4, **step 9**. Optional: Purify the ladder from unincorporated radionucleotides using Illustra Microspin G25 columns (GE Healthcare).

14. Air-cooled vertical electrophoresis unit (Hoefer, SE410).
15. Phosphorimaging plate (Elysia-Raytest GmbH).
16. Phosphorimaging reader (CR 35 Bio phosphorimage scanner, Elysia-Raytest GmbH).

2.4 Adenylation of 3′-Adapter

1. 5′ DNA adenylation Kit (NEB, E2610).
2. 3 ′- Adapter, including a UMI sequence: 5 ′- (P) **NNN**GTC**NNN**TGGAATTCTCGGGTGCCAAGG/ddC/-3′. The adapter has the following modifications: 5′-phosphate (P) and dideoxycytosine (ddC). The 3′-adapter is PAGE purified (*see* **Note 2**).
3. Phenol-chloroform-isoamyl alcohol (P/C/I) solution for extraction of nucleic acids (pH 7.5–8.0) (Carl Roth).
4. 3 M sodium acetate (pH 5.2).

2.5 piRNA Library Construction

2.5.1 Ligation of Pre-Adenylated 3′-Adapter

1. T4 RNA ligase 2, trunc. K227Q (NEB, M0351).
2. 10× T4 RNA Ligase Reaction Buffer (NEB).
3. 50% PEG 8000 (NEB).

2.5.2 Ligation of 5′-RNA Adapter

1. T4 RNA ligase 1 (ssRNA ligase) (NEB, M0204).
2. 10× T4 RNA Ligase Reaction Buffer (NEB).
3. 100 mM ATP (NEB).
4. 100% DMSO (NEB).
5. 5′-RNA adapter(a) (with UMI): 5 ′-GUUCAGAGUUCUAC AGUCCGACGAUC**NNN**CGA**NNN**TAC**NNN**-3′.
6. 5′-RNA adapter(b) (with UMI): 5′-GUUCAGAGUUCUAC AGUCCGACGAUC**NNN**ATC**NNN**AGT**NNN**-3′.

Both adapters are PAGE-purified. After solubilization, store 5′-RNA adapters at −80 °C. For library preparation, dilute adapters to 10 μM and mix in an equimolar amount.

2.5.3 Reverse Transcription

1. RT primer (RTP primer): 5 ′-GCCTTGGCACCCGAGAA TTCCA-3′. The primer is PAGE purified.
2. SuperScript III reverse transcriptase (Invitrogen).
3. 5× first strand buffer (Invitrogen).
4. 100 mM DTT (Invitrogen).
5. 10 mM dNTP mix (Invitrogen).

2.5.4 Small-Scale PCR Amplification

1. Phusion High-Fidelity DNA Polymerase (NEB).
2. 5× Phusion HF buffer (NEB).
3. Universal Primer 1 (RP1):

   5′-AATGATACGGCGACCACCGAGATCTACACGTTCA GAGTTCTACAGTCCGA-3′
4. Specific PCR primers:

   PCR Index Primer 2, Index ACAGTG (P2.1) (*see* **Note 3**):

   5′-CAAGCAGAAGACGGCATACGAGAT**CACTGT**GTGA CTGGAGTTCCTTGGCACCCGAGAATTCCA-3′

   PCR Index Primer 2, Index GTGAAA (P2.2)

   5′-CAAGCAGAAGACGGCATACGAGAT**TTTCAC**GTGA CTGGAGTTCCTTGGCACCCGAGAATTCCA-3′

   PCR Index Primer 2, Index GCCAAT (P2.3)

   5′-CAAGCAGAAGACGGCATACGAGAT**ATTGGC**GTGA CTGGAGTTCCTTGGCACCCGAGAATTCCA-3′

   PCR Index Primer 2, Index CTTGTA (P2.4)

   5′-CAAGCAGAAGACGGCATACGAGAT**TACAAG**GTGA CTGGAGTTCCTTGGCACCCGAGAATTCCA-3′

   PCR Index Primer 2, Index CAGATC (P2.5)

   5′-CAAGCAGAAGACGGCATACGAGAT**GATCTG**GTGA CTGGAGTTCCTTGGCACCCGAGAATTCCA-3′

   PCR Index Primer 2, Index GTAGAG (P2.6)

   5′-CAAGCAGAAGACGGCATACGAGAT**CTCTAC**GTGA CTGGAGTTCCTTGGCACCCGAGAATTCCA-3′

   PCR Index Primer 2, Index GTCCGC (P2.7)

   5′-CAAGCAGAAGACGGCATACGAGAT**GCGGAC**GTGA CTGGAGTTCCTTGGCACCCGAGAATTCCA-3′

   PCR Index Primer 2, Index ACTGAT (P2.8)

   5′-CAAGCAGAAGACGGCATACGAGAT**ATCAGT**GTGA CTGGAGTTCCTTGGCACCCGAGAATTCCA-3′
5. 25 mL **Native** 6% PAGE: 3.75 mL 40% **acrylamide/bis solution** (19:1) (Carl Roth), 18.5 mL deionized water, 2.5 mL 10× TBE buffer, 200 μL 10% ammonium persulfate, 10 μL TEMED.
6. TE buffer: 10 mM Tris–HCl (pH 8.0), 1 mM EDTA.
7. 6× DNA loading dye: 60% glycerol in TE buffer, 0.25% bromophenol blue (w/v), 0.25% xylene cyanol (w/v).
8. O'RangeRuler 20-bp DNA Ladder (Thermo Scientific).
9. SYBR Gold nucleic acid gel stain (Invitrogen).
10. UV transilluminator.

2.5.5 Final PCR Amplification

1. Syringe gauge needle, 0.8 mm.
2. 0.5 mL flat cap tubes (Thermo Scientific).
3. Gel elution buffer: 300 mM NaCl, 2 mM EDTA (pH 8.0).
4. SpinX centrifuge tube filter (Costar).
5. Fragment Analyzer system (Agilent).

3 Methods

3.1 Preparation of Whole-Cell Lysate from Planarian Flatworms

The outlined procedure and lysis buffer are optimized for immunoprecipitation of planarian SMEDWI proteins and co-bound piRNAs. During the preparation of whole-cell lysate, keep the samples on ice at all times to prevent proteolytic degradation. Moreover, add protease inhibitors freshly to the lysis buffer solution before use.

1. Collect 80–100 planarian flatworms (7–10 mm long) in a microcentrifuge tube (*see* **Note 4**).
2. Remove all liquid from the tube and snap-freeze the collected worms in liquid nitrogen. Store samples at −80 °C for later use or proceed to immediate homogenization (**step 3**).
3. Transfer the frozen sample into a precooled glass/Teflon Potter-Elvehjem homogenizer and add 4 mL of freshly prepared ice-cold lysis buffer (*see* **Note 5**). Triturate the sample up and down for a total of 15 strokes or continue homogenizing until the cell lysate becomes homogeneous. Keep the solution on ice throughout the procedure.
4. Transfer the whole-cell lysate into DNA low-binding microcentrifuge tubes and clear the lysate by centrifugation at 21,100 × *g* for 30 min at 4 °C.
5. Collect the supernatant and filter it through a 0.20 μm cellulose acetate syringe filter to remove lipids and other debris (*see* **Note 6**). Discard the pellet.
6. Measure the total protein concentration of the prepared lysate using a Bradford assay. The optimal protein concentration for the subsequent immunoprecipitation procedure is 1–3 mg/mL. If necessary, adjust the protein concentration of the cell lysate to the desired value using lysis buffer.
7. Save 20 μL of whole-cell lysate as input control. Store it at 4 °C until Subheading 3.3.2, and then proceed according to the protocol.

3.2 Immunoprecipitation of SMEDWI Proteins and Co-Bound piRNAs

In this section, the whole-cell lysate prepared in Subheading 3.1 will be used to isolate SMEDWI proteins and co-bound piRNAs using protein-specific antibodies.

1. Divide cleared lysate into four aliquots (~3 mg of total protein each) and transfer them into DNA low-binding microcentrifuge tubes.
2. To each lysate aliquot, add the appropriate amount of antibody and incubate the sample for 2 h at 4 °C under gentle agitation or rotation:

   10 μg of purified polyclonal anti-SMEDWI-1 or anti-SMEDWI-3 antibody (*see* **Note 7**)

   30 μL of anti-SMEDWI-2 antiserum

   30 μL of pre-immune serum. Pre-immune serum is used as a negative control (*see* **Note 8**)
3. To prepare protein A magnetic beads (*see* **Note 9**), transfer 200 μL of bead slurry to a new DNA low-binding tube, place the beads in a magnetic stand for 1 min, and remove the supernatant. Add 500 μL of lysis buffer to wash the beads and mix by pipetting five times. Then, place the beads in the magnetic stand for 1 min until the supernatant appears clear. Remove the supernatant and repeat the wash step. Split the washed beads into four new DNA low-binding tubes (50 μL of beads slurry per sample).
4. Remove the supernatant from the washed beads and immediately transfer the whole-cell lysate that was preincubated with antibodies to the magnetic beads. Do not let the beads run dry. Continue incubation for another 2 h at 4 °C under gentle agitation or rotation.
5. To wash the immunoprecipitated proteins captured on the magnetic beads, spin down the tubes briefly and place them in the magnet stand for 2 min until the supernatant appears clear.
6. Discard the supernatant, remove a tube from the magnet stand and resuspend the beads in 1 mL ice-cold low-salt wash buffer. Mix the solution by pipetting 5–7 times. Let it stand on ice for 2 min and then place the tube back in a magnetic stand for 2 min.
7. Carefully remove the supernatant and repeat the wash step one more time.
8. Discard the supernatant and wash the beads twice with 1 mL high-salt wash buffer following the above procedure.

9. After the second wash step, transfer immunoprecipitated complexes coupled to magnetic beads to a new 1.5 mL DNA low-binding tube (*see* **Note 10**).
10. Continue with an extraction of co-immunoprecipitated RNA (Subheading 3.3.2).

3.3 Extraction of piRNAs from Planarian Lysate

3.3.1 Enrichment of piRNAs from Planarian Lysate by β-Elimination

In case immuno-isolation techniques of PIWI-bound piRNAs are not desired, one can take advantage of the 2′-O-methylated 3′-ends of piRNAs and enrich for piRNAs by oxidizing all small RNAs using sodium periodate ($NaIO_4$) [22–24]. Due to their 2′-O-methylated 3′-ends, piRNAs are resistant to $NaIO_4$-mediated oxidization, while unmodified small RNAs undergo β-elimination, which renders their 3′-ends inaccessible for subsequent library construction [25].

1. Dounce or freeze (*see* **Note 11**) 1–3 worms in 500 μL TRIzol reagent (also known as TRI Reagent or TriFast), a mixture of guanidine thiocyanate and phenol in a monophasic solution that is used for the isolation of DNA, RNA, and protein. Mix thoroughly until most of the sample is dissolved and incubate the solution for 5 min at room temperature (RT).
2. Centrifuge at 12,000 × *g* for 15 min at 4 °C to precipitate most genomic DNA from the sample.
3. Transfer supernatant into a new 1.5 mL DNA low-binding tube.
4. Add 100 μL chloroform to the supernatant and vortex thoroughly for 1 min.
5. Incubate the solution for 3 min at RT.
6. Separate the aqueous layer containing RNA by centrifugation at high speed (>16,000 × *g*) for 10 min at 4 °C.
7. Transfer the aqueous layer to a new 1.5 mL DNA low-binding tube.
8. Add 1 μL glycogen (20 μg) to facilitate the precipitation of low RNA yields. Mix the solution by vortexing.
9. Add 250 μL ice-cold isopropanol to the aqueous phase. Mix the solution thoroughly by inverting a tube several times.
10. Incubate for 10 min at RT and then pellet RNA at >16,000 × *g* for 15 min at 4 °C.
11. Discard the supernatant. Wash the RNA pellet with 1 mL of 75% ice-cold ethanol at high speed (>16,000 × *g*) for 5 min.
12. Remove the supernatant and air-dry the pellet for 2 min.

13. Resuspend RNA pellet in 44 μL RNase-free water, and add 5 μL 10× DNase I buffer and 1 μL RNase-free DNase I (Roche).
14. Incubate at 37 °C for 30 min in a thermomixer at 800 rpm.
15. Purify small RNAs using the RNA Clean & Concentrator-5 kit from Zymo Research following the manufacturer's instructions. Elute in 27 μL H_2O.
16. To oxidize RNAs that do not carry 2′-O-methyl at their 3′ ends, add the following reagents in a 1.5 mL DNA low-binding tube:

Small RNA	27 μL
5× borax/boric acid buffer (pH 8.6)	8 μL
200 mM sodium periodate	5 μL

17. Incubate for 30 min at 25 °C.
18. To precipitate RNA, add 230 μL H_2O and 30 μL 3 M sodium acetate (pH 5.2) to a final concentration of 300 mM and 1 μL glycogen (20 μg).
19. Next, add 3 volumes 100% ethanol, mix the sample by inverting, and precipitate at −20 °C overnight.
20. Centrifuge the samples the following day at 21,000 × *g* for 30 min at 4 °C. Discard the supernatant.
21. Wash the precipitated RNA pellet once with ice-cold 75% ethanol at high speed (>16,000 × *g*) for 5 min at 4 °C. Discard the supernatant and air-dry the pellet at RT for 3 min.
22. Dissolve the RNA pellets with RNase-free water (*see* **Note 12**) as follows:
    (a) To perform radioactive 5′-end labeling of the entire sample, add 17 μL of RNase-free water and proceed to Subheading 3.4.
    (b) To prepare a piRNA sequencing library from the entire sample, add 11 μL of RNase-free water and proceed to Subheading 3.6.
    (c) To prepare piRNA libraries and save an aliquot for the detection of piRNAs, dissolve the pellet in 13 μL of RNase-free water. Use 2 μL sample for radioactive 5′-end labeling.

From this point on, either proceed to Subheading 3.6 (piRNA library construction) or to Subheading 3.4 (radioactive 5′-end labelling).

3.3.2 Extraction of Co-Immunoprecipitated piRNAs

For all procedures detailed below, use RNase-free filter pipette tips, DNA low-binding tubes, and RNase-free solutions.

1. Resuspend washed immunoprecipitated complexes coupled to magnetic beads (Subheading 3.2, **step 10**) in 200 μL Proteinase K buffer and add Proteinase K to a final concentration of 120 μg/mL.
2. Incubate the sample for 20 min at 42 °C in a thermomixer at 1000 rpm. Spin briefly and proceed with phenol-chloroform extraction and ethanol precipitation of isolated RNAs.
3. Add 400 μL (2 volumes) of acid phenol-chloroform-isoamyl alcohol (P/C/I) solution for RNA extraction (pH ≤ 4.5) and vortex the sample for 1 min. Incubate the mixture for 5 min at RT for complete dissociation of nucleoprotein complexes.
4. To separate the aqueous layer containing RNA, spin tubes in a tabletop centrifuge at high speed (>16,000 × *g*) for 5 min at RT.
5. Carefully collect the upper aqueous phase in a new DNA low-binding tube and add 200 μL (1 volume) of 100% chloroform solution to the sample, vortex thoroughly for 1–2 min. Spin the tube at high speed (>16,000 × *g*) for 5 min at RT. This step removes traces of phenol from the RNA-containing aqueous layer.
6. Collect the top aqueous solution and transfer into a new DNA low-binding tube.
7. Add 1 μL of glycogen (20 μg) to each sample and vortex to mix the solution. Glycogen facilitates the precipitation of small quantities of RNA. Next, add 2.5 volumes of ice-cold 100% ethanol. Mix well by inverting 8–10 times. Precipitate RNA overnight at −20 °C.
8. Centrifuge the sample at maximum speed (21,100 × *g*) in a tabletop centrifuge for 30 min at 4 °C.
9. Carefully discard the supernatant without disturbing the pellet. Wash RNA pellet with 1 mL of ice-cold 70% ethanol and centrifuge the sample at maximum speed (21,100 × *g*) for 5 min at 4 °C.
10. Remove supernatant and repeat spin briefly to collect residual traces of ethanol. Let the tube air-dry for 2 min or until all traces of ethanol have evaporated.
11. Dissolve the RNA pellets with RNase-free water (*see* **Note 12**) as follows:
    (a) To perform radioactive 5′-end labeling of the entire sample of immunoprecipitated RNAs, add 17 μL of RNase-free water and proceed to Subheading 3.4.
    (b) To prepare a piRNA sequencing library from the entire sample, add 11 μL of RNase-free water and proceed to Subheading 3.6.

(c) To prepare piRNA libraries and save an aliquot for the detection of co-immunoprecipitated piRNAs, dissolve the pellet in 13 μL of RNase-free water. Use 2 μL immunoprecipitated piRNAs for radioactive 5′-end labeling.

3.4 Radioactive 5′-End Labeling and Detection of Co-Immunoprecipitated piRNAs

Small RNAs associated with Argonaute family members, such as microRNAs and piRNAs, carry 5′-monophosphates. These phosphate groups anchor the small RNAs in the MID-PIWI domain of the respective Argonaute protein, thus enabling proper positioning of mRNA targets in the active site of the endonucleolytic PIWI domain [26–28]. The monophosphate at the 5′-ends of piRNAs can be exchanged for a radioactive terminal phosphate using [γ-^{32}P]-ATP. This exchange allows the detection and visualization of trace amounts of immunoprecipitated piRNAs. However, to ensure efficient 5′-end labeling, piRNAs need to first be dephosphorylated.

1. To remove the piRNA 5′-phosphate, combine the following components in a 1.5 mL DNA low-binding tube:

Extracted RNA (Subheading 3.3.1, **step 22a** or **c**; Subheading 3.3.2, **step 11a** or **c**)	17 μL
10× CutSmart buffer (NEB, B7204)	2 μL
Calf intestine phosphatase, CIP (NEB, M0290)	1 μL

2. Incubate the mixture for 30 min at 37 °C in a thermomixer at 850 rpm.
3. Purify dephosphorylated RNAs using TRIzol. Add 500 μL of TRIzol to the sample and mix thoroughly for 1 min. Incubate the solution for 5 min at RT.
4. Add 100 μL of 100% chloroform and vortex vigorously for 1 min. Separate the aqueous layer containing RNA by centrifugation at high speed (>16,000 × *g*) for 5 min at RT.
5. Collect the top aqueous phase into a new 1.5 mL DNA low-binding tube (*see* **Note 13**). Add 1 μL of glycogen (20 μg) to facilitate RNA precipitation and vortex to mix the solution.
6. Add 250 μL of 100% isopropanol. Invert the tube for 8–10 times to mix the sample and incubate at RT for 10 min.
7. Centrifuge the sample at high speed (>16,000 × *g*) for 15 min at 4 °C. Carefully discard the supernatant and wash RNA pellet with 1 mL of 70% ethanol at high speed (>16,000 × *g*) for 5 min at 4 °C.

8. Remove supernatant, and repeat short spin to remove trace amounts of ethanol. Let the tube air-dry for 2 min. Resuspend RNA pellet in 16 μL of RNase-free water.
9. To label the RNA 5′-end with [γ-^{32}P]- ATP, prepare the following reaction:

Dephosphorylated RNA	16 μL
10× PNK buffer (NEB, B0201S)	2 μL
10 μCi/μL [γ-32P]-ATP	1 μL
T4 polynucleotide kinase (NEB, M0201)	1 μL

10. Incubate the mixture for 35 min at 37 °C in a thermomixer at 850 rpm.
11. Shortly spin the sample and add 20 μL of 2× formamide loading dye. Heat the sample for 2 min at 95 °C. The sample is now ready to be loaded on a denaturing 10% urea PAGE (*see* **Note 14**).
12. Before loading samples, pre-run a denaturing 10% urea PAGE gel in 1× TBE buffer for 20 min at 160 V (10 V per 1 cm of gel). This procedure warms up the gel and equilibrates it with running buffer.
13. Using a syringe, remove residual urea from all wells of the gel with 1× TBE. Flush each well with 1× TBE again just before loading. This allows the sample to sink to the bottom of the well, which increases gel resolution and delivers clear bands without smearing.
14. Load 10 μL of immunoprecipitated RNA per gel lane. For the whole RNA extract control and radioactively labeled ladder, load ≤1 Bq/cm^2 of radioactivity per lane. Run the gel at a constant current of 21 mA until the bromophenol blue band reaches three quarters of the gel (~1.5 h) (*see* **Note 15**).
15. Visualize the resolved RNA species by exposing the gel for several hours or overnight to a phosphorimaging plate at RT. As phosphorimaging plates are sensitive to moisture, wrap the gel in clear plastic foil to protect the plate from contact with wet surfaces. Read the plate using a phosphorimager (Fig. 3a).

3.5 Adenylation of the 3′-Adapter

As mentioned before (Subheading 3.3.1), planarian piRNAs, like those of other animals, carry 2′-O-methyl groups on their 3′--termini, a modification that protects them from decay [19, 22, 23] . However, this modification also strongly reduces the efficiency of 3′-adapter ligation during sequencing library preparation. To at least partially remedy this problem, we use truncated T4 RNA ligase 2 that utilizes a pre-adenylated adapter along with an optimal concentration of crowding agents such as polyethylene glycol. Taken together, these two modifications significantly improve the ligation efficiency of piRNAs [29, 30] . In the following section, we describe the adenylation of the 3′-adapter using thermostable RNA ligase from *Methanobacterium thermoautotrophicum* (MthRnl) (*see* **Note 16**).

1. Prepare the following mix using the 5′-DNA adenylation Kit (NEB, E2610S):

Phosphorylated 3′-adapter (100 μM)	2 μL
Nuclease-free water	26 μL
5′ DNA adenylation reaction buffer	4 μL
1 mM ATP	4 μL
Mth RNA ligase	4 μL

   Incubate the mixture for 1 h at 65 °C followed by enzyme inactivation for 5 min at 85 °C.

2. Purify the adenylated adapter by phenol-chloroform extraction.
3. Add 160 μL of nuclease-free water to the mixture followed by 400 μL of basic phenol-chloroform-isoamyl alcohol (P/C/I) solution for extraction of nucleic acids (pH 7.5–8.0). Make sure to use the P/C/I/ solution from the lower phase; do not shake the bottle before use.
4. Vortex the mixture vigorously for 1 min, and incubate for 5 min at RT.
5. To separate the aqueous layer containing DNA, centrifuge the mixture at high speed (>16,000 × *g*) for 5 min at RT.
6. Collect the upper aqueous phase in a new DNA low-binding tube. Add 200 μL (1 volume) of 100% chloroform solution to the sample, vortex thoroughly for 1–2 min and centrifuge the mixture at high speed for 5 min at RT.
7. Collect the top aqueous solution and place into a new DNA low-binding tube. Add 3 M sodium acetate (pH 5.2) to a final concentration of 0.3 M and 1 μL of glycogen (20 μg); vortex to mix the sample.
8. Next, add 2.5 volumes 100% ethanol. Mix solution well and precipitate overnight at −20 °C.
9. Pellet precipitated oligonucleotides at maximum speed (21,100 × *g*) in a tabletop centrifuge for 30 min at 4 °C. Wash the pellet with 1 mL of 70% ethanol as described in Subheading 3.3.2, **steps 9** and **10**.

10. Resuspend DNA pellet in 15 μL of nuclease-free water. Measure the concentration of adenylated 3′-adapter with UV spectrophotometer. Adjust the total concentration of the adapter to 5 μM with nuclease-free water.

3.6 piRNA Library Construction

The protocol below is a modified version of the small RNA library preparation protocol from [31] . It was optimized for piRNA library construction.

3.6.1 Ligation of Pre-Adenylated 3′-Adapter

1. Prepare the following reaction mix and pipette it into the PCR tube (*see* **Note 17**) (Fig. 3b):

Immunoprecipitated RNA from Subheading 3.3, **step (b)**	11 μL
10× T4 RNA ligase reaction buffer	2 μL
50% PEG 8000	4 μL
5 μM pre-adenylated 3′-adapter (Fig. 2)	1 μL

Mix the sample by pipetting 5–7 times.

Heat the sample for 30 s at 90 °C in a preheated thermal cycler.

Place the tube on ice for 1 min.

Add to the sample:

T4 RNA ligase 2, trunc. K227Q (NEB, M0351S)	2 μL

2. Incubate the mixture overnight at 16 °C in a thermal cycler.
3. Inactivate the enzyme by heating to 65 °C for 10 min. Spin the reaction mix and place on ice.

3.6.2 Ligation of 5′-RNA Adapter

1. Add the following components to the ligated RNA-3′-adapter mixture:

3′-adapter-ligated RNA from Subheading 3.6.1	20 μL
10× T4 RNA ligase reaction buffer	2 μL
100 mM ATP	0.4 μL
RNase-free water	11.6 μL
10 μM 5′ RNA adapter mix	1 μL
100% DMSO	3 μL

Mix the sample by pipetting 5–7 times.

Heat the sample for 30 s at 90 °C in the preheated thermal cycler.

Place the tube on ice for 1 min.
Add to the sample:

T4 RNA ligase 1 (NEB, M0204S)	2 μL

2. Incubate the mixture for 1 h at 37 °C in a thermal cycler.
3. Spin briefly and place on ice. Proceed to reverse transcription step or store the sample at −80 °C for up to 1 week.
4. To precipitate RNA, add 230 μL H_2O and 30 μL 3 M sodium acetate (pH 5.2) to a final concentration of 300 mM and 1 μL glycogen (20 μg).
5. Add 3 volumes 100% ethanol, mix by inverting, and precipitate at −20 °C overnight.
6. Centrifuge the samples the following day at 21,000 × *g* for 30 min at 4 °C. Discard the supernatant.
7. Wash pellet once with ice-cold 75% ethanol at high speed (>16,000 × *g*) for 5 min at 4 °C. Discard the supernatant and air-dry pellets at RT for 3 min.
8. Dissolve the DNA-RNA hybrid pellet in 10 μL RNase-free water.

3.6.3 Reverse Transcription

1. To reverse-transcribe RNA, combine the following components in a PCR tube:

5′-3′ adapter-ligated RNA from Subheading 3.6.2	10 μL
10 μM RT primer (RTP primer)	1 μL

Mix the sample by pipetting 5–7 times.

Incubate the tube for 5 min at 65 °C in a preheated thermal cycler.

Place the tube on ice for 1 min.

Premix the following and add it to the sample:

5× first strand buffer	4 μL
100 mM DTT	2 μL
10 mM dNTP mix	1 μL
SuperScript III RT (Invitrogen, 18080093)	2 μL

2. Briefly spin the mixture. Place the tube in a thermal cycler and run the following program with a heated lid:
   (a) 50 °C for 50 min
   (b) 85 °C for 5 min
   (c) 4 °C to store the sample

3. Store the cDNA reaction at −20 °C or proceed directly to PCR amplification.

3.6.4 Small-Scale PCR Amplification

1. To determine the minimal number of required PCR amplification cycles, assemble at least three separate reactions (10 μL each) per sample on ice, using the master mix below. Use 0.2-mL PCR tubes.

Nuclease-free water	6.25 μL
5× Phusion HF buffer	2 μL
10 μM universal primer 1 (RP1)	0.2 μL
10 μM PCR index primer 2	0.2 μL
10 mM dNTP mix	0.25 μL
Phusion high-fidelity DNA polymerase	0.1 μL

Mix the sample by pipetting for 5–7 times.
Add to each 9 μL reaction:

cDNA from Subheading 3.6.3	1 μL

2. Place the tube in a thermal cycler. Incubate the tube using the PCR cycling conditions below and different numbers of cycles (12, 14, and 16 cycles):
   (a) Preheat lid to 100 °C
   (b) 98 °C for 30 s
   (c)

12 (14 or 16) cycles at	98 °C for 10 s
	60 °C for 30 s
	72 °C for 30 s

   (d) 72 °C for 3 min
   (e) hold at 4 °C.
3. Analyze the amplified products on a native 6% PAGE (18 × 16 × 0.75 cm) (for recipe, *see* Materials Subheading 2.5.4, **step 5**). Pre-run the gel for 10 min at a constant voltage of 160 V.
4. Mix each sample with 2 μL of 6× DNA loading dye. To prepare a gel ladder, combine 2 μL O'RangeRuler 20-bp DNA ladder (Thermo Scientific, SM1323) with 6 μL of TE buffer, 2 μL 5× Phusion HF buffer (to adjust salt concentration), and 2 μL of 6× DNA loading dye.

5. Flush each well with 1× TBE just before loading and load samples on the native gel. Allow the sample to enter the gel at 80 V, then increase the voltage to 180 V, and separate the amplified products in 1× TBE buffer for 3 h.
6. Stain the gel with SYBR Gold nucleic acid gel stain in 1× TBE for 5 min.
7. Visualize bands with UV transilluminator. Choose the minimal number of cycles that gives a clear band at 172–177 bp size, which corresponds to planarian piRNAs (Fig. 3c, 14 cycles). The presence of fuzzy bands that appear at about twice the expected size or even bigger (labeled "overamplified" in Fig. 3C) is a sign of overamplification of the libraries (Fig. 3c, 16 and 18 cycles). The data quality of such libraries may be reduced by various artifacts such as amplification bias, PCR duplicates, and low complexity.

3.6.5 Final PCR Amplification

1. Set up the following PCR using specific index primers for each sample (Fig. 3d):

Nuclease-free water	26.25 μL
5× Phusion HF buffer	10 μL
10 μM universal primer 1 (RP1)	1 μL
10 μM PCR index primer 2 (*see* **Note 18**)	1 μL
10 mM dNTP mix	1.25 μL
Phusion high-fidelity DNA polymerase	0.5 μL

Mix the sample by pipetting 5–7 times.
Add to each reaction:

cDNA from Subheading 3.6.3	10 μL

2. Place the tube in a thermal cycler. Incubate the tube using PCR cycling conditions below and the minimal number of cycles determined in Subheading 3.6.4:
   (a) Preheat lid to 100 °C.
   (b) 98 °C for 30 s
   (c)

N cycles	98 °C for 10 s
	60 °C for 30 s
	72 °C for 30 s

   (d) 72 °C for 3 min
   (e) Hold at 4 °C

3. Separate amplified products on a native 6% PAGE (length 16 cm, thickness 0.75 mm) (for recipe, *see* Materials Subheading 2.5.4, **step 5**). To each 50 μL reaction, add 10 μL of 6× DNA loading dye. Prepare ladder by mixing 10 μL of O'RangeRuler 20-bp DNA ladder, 10 μL of TE buffer with 5 μL 5× Phusion HF buffer, and 5 μL of 6× DNA loading dye.
4. Load samples onto a gel while splitting one reaction into two lanes (30 μL of sample per lane). Leave one lane empty between different replicates to avoid cross-contamination. Libraries from different immunoprecipitation experiments should be run on separate gels to avoid cross-contamination. Allow the sample to enter the gel at 80 V, increase voltage, and separate amplified products in 1× TBE buffer at 180 V for 3 h.
5. Stain the gel with SYBR Gold nucleic acid gel stain in 1× TBE for 5 min.
6. Visualize and cut the appropriate bands (Fig. 3c). Place gel pieces into 0.5 mL tubes punctured 2–3 times at the bottom with a gauge needle (0.8 mm).
7. Place 0.5 mL tubes with gel pieces into 1.5 mL DNA low-binding tubes and spin for 2 min at high speed (>16,000 × *g*) to shred gel into tiny pieces. Discard 0.5 mL tubes.
8. Add 300 μL of gel elution buffer to each tube and incubate overnight at 25 °C and 1000 rpm in a thermomixer or for 2 h at 37 °C and 1000 rpm.
9. To separate eluate from gel pieces, pipette the solution and gel pieces into SpinX centrifuge tube filter. Spin the tubes at high speed (>16,000 × *g*) for 2 min.
10. Transfer eluate into 1.5 mL DNA low-binding tubes, add 1 μL of glycogen, and mix the solution by vortexing. To precipitate DNA, add 2.5 volumes of ice-cold 100% ethanol and incubate overnight at −20 °C.
11. Spin the solution at high speed (>16,000 × *g*) for 30 min at 4 °C.
12. Remove the supernatant and wash DNA pellet with 1 mL of 70% ethanol at high speed (>16,000 × *g*) for 5 min at 4 °C.
13. Carefully remove the supernatant and shortly spin the tube once again to collect and remove residual ethanol. Leave the tube to air-dry with an open lid for 2 min.
14. Resuspend DNA pellet in 12 μL of nuclease-free water. Analyze and quantify small RNA-Seq libraries using a Fragment Analyzer system (Agilent) (Fig. 3d).

4 Notes

1. The low-molecular weight standard from Affymetrix cannot be obtained anymore. Alternatively, we use the oligo length standard 20/100 ladder from Integrated DNA Technologies (IDT).
2. Please note that Fig. 3b illustrates library preparation using standard 3′-(5′-(P)TGGAATTCTCGGGTGCCAAGG/ddC/-3′) and 5′-adapters (5′-GUUCAGAGUUCUACAG UCCGACGAUC-3′) that do not include unique molecular identifiers (UMIs). The use of UMIs adds an additional 24 nt to the bands corresponding to piRNAs, miRNA, and empty adapters (*see* Fig. 3c).
3. The index sequence is read as its reverse-complement during sequencing on Illumina platforms.
4. The whole-cell extract from 80 to 100 animals homogenized in 4 mL of lysis buffer is sufficient to carry out four independent immunoprecipitation assays (SMEDWI-1, SMEDWI-2, SMEDWI-3 proteins and one negative control sample). The volume of lysis buffer must be adjusted to the amount of starting tissue material to obtain a whole cell extract with an optimal total protein concentration of 1–3 mg/mL.
5. Planarian SMEDWI proteins are stabilized in the presence of at least 1 mM DTT, which maintains their free sulfhydryl groups in a reduced state. The absence of DTT in the lysis buffer greatly reduces the amount of recovered SMEDWI proteins from whole cell extract. Always add DTT solution freshly to the lysis buffer before use. In addition, we observed that HEPES hemisodium salt (Sigma-Aldrich, **H9897**) as a buffer agent of the lysis solution favorably affects SMEDWI protein recovery and signal-to-noise ratio of co-immunoprecipitated RNA. While other buffers, including TRIS titrated with hydrochloric acid, HEPES free acid (Merck, 391340) titrated with NaOH or KOH, or HEPES sodium salt (Sigma-Aldrich, H7006) titrated with hydrochloric acid, cause aggregation and precipitation of SMEDWI proteins at least in concentrated cell extracts (3–4 mg/mL protein concentration). We speculate that an increase in ionic strength during titration leads to a significant change in the osmolality of the solution and affects protein stability.
6. The presence of lipids in the whole-cell lysate does not increase the background noise of immunoprecipitated proteins and co-bound piRNAs but it interferes with the correct estimation of protein concentration.

7. The antibody used for the immunoprecipitation assay must be able to recognize the native protein conformation. Therefore, not all antibodies that are approved for immunostaining and western blot can be applicable to immunoprecipitation procedure. In addition, the amounts of antibody depend on their affinity to the specific protein and must be optimized in each case. However, for purified antibodies, 5–10 μg is a good starting point.
8. The use of a negative control is very important to assess the background of the experiment. We used rabbit serum extracted prior to immunization (pre-immune serum) as a negative control for anti-SMEDWI-2 antiserum and other polyclonal antibodies raised in a rabbit. In cases where a pre-immune serum is not available, isotypic control antibodies should be used. Thus, a suitable isotope control instead of rabbit pre-immune serum would be rabbit polyclonal IgG immunoglobulin. A "beads only" negative control will account for unspecific binding to magnetic beads and protein A or G but it cannot be used as the only negative control in an experiment. It is highly recommended to always include a pre-immune serum or immunoglobulin control.
9. Protein A and protein G have different binding affinities to immunoglobulins of different species. For optimal and consistent results, check the compatibility of protein A or G with your antibodies of interest.
10. It is crucial to transfer the washed beads to a new DNA low-binding tube for subsequent RNA extraction step. This removes the nonspecific RNA signal that otherwise sticks to the surface of a DNA low-binding tube during its exposed contact with the whole-cell lysate. The use of DNA low-binding tubes also increases the recovery of co-immunoprecipitated RNAs.
11. Collect worms in 500 μL TRIzol solution and flash-freeze in liquid nitrogen. Store the sample in TRIzol at −80 °C before processing. We find that freezing worms and homogenizing them by vortexing is sufficient.
12. We do not recommend using DEPC-treated water as DEPC, if not completely removed, might interfere with enzymatic reactions during library preparation.
13. Alternatively, piRNAs can be purified from aqueous phase using the RNA Clean & Concentrator-5 kit from Zymo Research following the manufacturer's instructions. piRNAs are eluted from the columns using 16 μL of RNase-free water. The protocol can then be continued with **step 9**.

14. Labeled RNA can be purified from unincorporated radionucleotides using Illustra Microspin G25 columns (GE Healthcare) before adding the formamide dye, following the manufacturer's instructions. G25 spin columns will remove any oligonucleotides shorter than 10 nt in length. This step is optional.
15. Formamide loading dye contains bromophenol blue and xylene cyanol that migrate at 12 and 55 nt, respectively, in 10% denaturing urea PAGE.
16. Alternatively, a PAGE-purified pre-adenylated 3′-adapter can be used. In this case, skip Subheading 3.5 and proceed directly to small RNA library preparation (Subheading 3.6).
17. The addition of 10% PEG 8000 to the reaction mixture significantly increases the ligation rate of the 3′-adapter to the 2′-O-methylated 3′-ends of piRNAs.
18. It is important to optimize the color balance of the final pooled library by using appropriate indexed adapters. This is particularly critical when sequencing low-plex pools (2–4 pooled libraries). During DNA sequencing, DNA polymerase adds fluorescently labeled terminator nucleotides to the sequenced DNA clusters attached on a flow cell. After each individual base is added, the flow cell is imaged, and clusters emitting fluorescence signal are counted. Illumina NextSeq platforms use a red laser/LED to sequence A/C and a green laser/LED to sequence G/T. For each cycle, both the red and green channel need to be read to ensure proper cluster detection. Hence, A/C (A or C) and G/T (G or T) must be present in index sequences in each cycle. The optimal primer combination that will result in efficient pool demultiplexing would be P2.1 (ACAGTG) + P2.2 (GTGAAA), P2.3 (GCCAAT) + P2.4 (CTTGTA), P2.5 (CAGATC) + P2.6 (GTAGAG), P2.7 (GTCCGC) + P2.8 (ACTGAT).

Acknowledgements

This work was supported by the Elite Network of Bavaria, the German Research Foundation (DFG, grant no. KU 3514/5-1), and the University of Bayreuth.

References

1. Aravin A, Gaidatzis D, Pfeffer S et al (2006) A novel class of small RNAs bind to MILI protein in mouse testes. Nature 442:203–207. https://doi.org/10.1038/nature04916
2. Girard A, Sachidanandam R, Hannon GJ et al (2006) A germline-specific class of small RNAs binds mammalian Piwi proteins. Nature 442:199–202. https://doi.org/10.1038/nature04917
3. Tolia NH, Joshua-Tor L (2007) Slicer and the argonautes. Nat Chem Biol 3:36–43. https://doi.org/10.1038/nchembio848

4. Brennecke J, Aravin AA, Stark A et al (2007) Discrete small RNA-generating loci as master regulators of transposon activity in drosophila. Cell 128:1089–1103. https://doi.org/10.1016/j.cell.2007.01.043
5. Watanabe T, Takeda A, Tsukiyama T et al (2006) Identification and characterization of two novel classes of small RNAs in the mouse germline: retrotransposon-derived siRNAs in oocytes and germline small RNAs in testes. Genes Dev 20:1732–1743. https://doi.org/10.1101/gad.1425706
6. Rojas-Ríos P, Simonelig M (2018) piRNAs and PIWI proteins: regulators of gene expression in development and stem cells. Development 145. https://doi.org/10.1242/dev.161786
7. Lee H-C, Gu W, Shirayama M et al (2012) C. elegans piRNAs mediate the genome-wide surveillance of germline transcripts. Cell 150: 78–87. https://doi.org/10.1016/j.cell.2012.06.016
8. Shirayama M, Seth M, Lee H-C et al (2012) piRNAs initiate an epigenetic memory of nonself RNA in the C. elegans germline. Cell 150: 65–77. https://doi.org/10.1016/j.cell.2012.06.015
9. Seth M, Shirayama M, Gu W et al (2013) The C. elegans CSR-1 argonaute pathway counteracts epigenetic silencing to promote germline gene expression. Dev Cell 27:656–663. https://doi.org/10.1016/j.devcel.2013.11.014
10. Miesen P, Girardi E, van Rij RP (2015) Distinct sets of PIWI proteins produce arbovirus and transposon-derived piRNAs in Aedes aegypti mosquito cells. Nucleic Acids Res 43:6545–6556. https://doi.org/10.1093/nar/gkv590
11. Reddien PW, Oviedo NJ, Jennings JR et al (2005) SMEDWI-2 is a PIWI-like protein that regulates planarian stem cells. Science 310:1327–1330. https://doi.org/10.1126/science.1116110
12. Palakodeti D, Smielewska M, Lu Y-C et al (2008) The PIWI proteins SMEDWI-2 and SMEDWI-3 are required for stem cell function and piRNA expression in planarians. RNA 14: 1174–1186. https://doi.org/10.1261/rna.1085008
13. Friedländer MR, Adamidi C, Han T et al (2009) High-resolution profiling and discovery of planarian small RNAs. Proc Natl Acad Sci U S A 106:11546–11551. https://doi.org/10.1073/PNAS.0905222106
14. Kim IV, Duncan EM, Ross EJ et al (2019) Planarians recruit piRNAs for mRNA turnover in adult stem cells. Genes Dev 33:1575–1590. https://doi.org/10.1101/gad.322776.118
15. Nakagawa H, Ishizu H, Hasegawa R et al (2012) Drpiwi-1 is essential for germline cell formation during sexualization of the planarian Dugesia ryukyuensis. Dev Biol 361:167–176. https://doi.org/10.1016/J.YDBIO.2011.10.014
16. Shibata N, Kashima M, Ishiko T et al (2016) Inheritance of a nuclear PIWI from pluripotent stem cells by somatic descendants ensures differentiation by silencing transposons in planarian. Dev Cell 37:226–237. https://doi.org/10.1016/j.devcel.2016.04.009
17. Rouhana L, Weiss JA, King RS et al (2014) PIWI homologs mediate histone H4 mRNA localization to planarian chromatoid bodies. Development 141:2592–2601. https://doi.org/10.1242/dev.101618
18. Kim IV, Ross EJ, Dietrich S et al (2019) Efficient depletion of ribosomal RNA for RNA sequencing in planarians. BMC Genom 20: 909. https://doi.org/10.1186/s12864-019-6292-y
19. Kim IV, Riedelbauch S, Kuhn C-D (2020) The piRNA pathway in planarian flatworms: new model, new insights. Biol Chem 401:1123–1141. https://doi.org/10.1515/hsz-2019-0445
20. Kashima M, Agata K, Shibata N (2020) What is the role of PIWI family proteins in adult pluripotent stem cells? Insights from asexually reproducing animals, planarians. Develop Growth Differ 62:407–422. https://doi.org/10.1111/dgd.12688
21. Guo T, Peters AHFM, Newmark PA (2006) A Bruno-like gene is required for stem cell maintenance in planarians. Dev Cell 11:159–169. https://doi.org/10.1016/j.devcel.2006.06.004
22. Gainetdinov I, Colpan C, Cecchini K et al (2020) Terminal modification, sequence, and length determine small RNA stability in animals. Mol Cell. https://doi.org/10.1016/j.molcel.2021.09.012
23. Ji L, Chen X (2012) Regulation of small RNA stability: methylation and beyond. Cell Res 22: 624–636. https://doi.org/10.1038/cr.2012.36
24. Wu P-H, Fu Y, Cecchini K et al (2020) The evolutionarily conserved piRNA-producing locus pi6 is required for male mouse fertility. Nat Genet 52:728–739. https://doi.org/10.1038/s41588-020-0657-7
25. Ohara T, Sakaguchi Y, Suzuki T et al (2007) The 3′ termini of mouse Piwi-interacting RNAs are 2'-O-methylated. Nat Struct Mol Biol 14:349–350. https://doi.org/10.1038/nsmb1220

26. Martinez J, Tuschl T (2004) RISC is a 5′ phosphomonoester-producing RNA endonuclease. Genes Dev 18:975–980. https://doi.org/10.1101/gad.1187904
27. Cora E, Pandey RR, Xiol J et al (2014) The MID-PIWI module of Piwi proteins specifies nucleotide- and strand-biases of piRNAs. RNA 20:773–781. https://doi.org/10.1261/rna.044701.114
28. Matsumoto N, Nishimasu H, Sakakibara K et al (2016) Crystal structure of silkworm PIWI-clade Argonaute Siwi bound to piRNA. Cell 167:484–497.e9. https://doi.org/10.1016/J.CELL.2016.09.002
29. Munafó DB, Robb GB (2010) Optimization of enzymatic reaction conditions for generating representative pools of cDNA from small RNA. RNA 16:2537–2552. https://doi.org/10.1261/rna.2242610
30. Zhuang F, Fuchs RT, Robb GB (2012) Small RNA expression profiling by high-throughput sequencing: implications of enzymatic manipulation. J Nucleic Acids 2012:360358. https://doi.org/10.1155/2012/360358
31. Hauptmann J, Schraivogel D, Bruckmann A et al (2015) Biochemical isolation of Argonaute protein complexes by Ago-APP. Proc Natl Acad Sci U S A 112:11841–11845. https://doi.org/10.1073/pnas.1506116112

Chapter 3

Genome-Wide Analysis of Planarian piRNAs

Andreas Pittroff, Iana V. Kim, Tim Demtröder, and Claus-D. Kuhn

Abstract

In planarian flatworms, the piRNA pathway is operated by three PIWI proteins, termed SMEDWI-1, SMEDWI-2, and SMEDWI-3 (SMEDWI = *Schmidtea mediterranea* PIWI). The interplay between these three PIWI proteins and their associated small noncoding RNAs, termed piRNAs, fuels the outstanding regenerative abilities of planarians, enables tissue homeostasis, and, ultimately, ensures animal survival. As the molecular targets of PIWI proteins are determined by the sequences of their co-bound piRNAs, it is imperative to identify these sequences by next-generation sequencing applications. Following sequencing, the genomic targets and the regulatory potential of the isolated piRNA populations need to be uncovered. To that end, here we present a bioinformatics analysis pipeline for processing and systematic characterization of planarian piRNAs. The pipeline includes steps for the removal of PCR duplicates based on unique molecular identifier (UMI) sequences, and it accounts for piRNA multimapping to different loci in the genome. Importantly, our protocol also includes a fully automated pipeline that is freely available at GitHub. Together with the piRNA isolation and library preparation protocol (see accompanying chapter), the presented computational pipeline enables researchers to explore the functional role of the piRNA pathway in flatworm biology.

Key words piRNAs, Small RNAs, Small RNA sequencing, Bioinformatics, NGS analysis, Automated pipeline, piRNAnalyzer, Planarians

1 Introduction

PIWI-interacting RNAs (piRNAs) are a class of small RNAs that is expressed in the animal germline. There, piRNAs guide PIWI proteins to transposable elements to silence them at both transcriptional and posttranscriptional levels [1, 2]. In planarian flatworms, PIWI proteins and their co-bound piRNAs are not restricted to the germline, yet they are abundantly expressed in the stem cells of these animals, where they play an essential role in animal regeneration and homeostasis. In addition to their role in suppressing transposable elements, piRNA-bound PIWI proteins were also found to regulate transcript abundance in planarian stem cells, underlining the biological relevance of the piRNA pathway for

Luca Gentile (ed.), *Schmidtea Mediterranea: Methods and Protocols*, Methods in Molecular Biology, vol. 2680,
https://doi.org/10.1007/978-1-0716-3275-8_3,
© The Author(s), under exclusive license to Springer Science+Business Media, LLC, part of Springer Nature 2023

planarians [3–5]. To characterize their biological function, piRNAs need to be isolated and prepared for next-generation sequencing applications, as described in the accompanying chapter. Following sequencing, the resulting large amounts of data need to be analyzed using computational methods. To enable researchers to carry out such analyses, we describe a computational pipeline that covers the initial steps of piRNA library analysis in this chapter (Fig. 1). The described computational pipeline includes preprocessing steps to remove sequencing adapters and to identify unique molecular identifiers (UMIs). The latter is important for the removal of PCR duplicates and for a quantitative analysis of the sequenced piRNAs. Next, the processed sequences are mapped to the planarian genome accounting for mapping to multiple loci, an inherent feature of all piRNA libraries. Finally, to explore piRNA biogenesis as well as to determine the molecular targets of the analyzed piRNA pool, mapped and normalized piRNAs are annotated to genomic features. The presented workflow is available as a fully automated pipeline, termed piRNAnalyzer, on GitHub (https://github.com/kuhnlab-bayreuth/piRNAnalyzer). piRNAnalyzer includes the most recent version of the planarian genome assembly [6] as well as reference fasta files for rRNAs [7] and tRNAs (http://gtrnadb.ucsc.edu). In addition to the described protocol, the automated version of the pipeline produces figures to visually inspect the preprocessing steps, the length distribution of the small RNA population, and the genomic distribution of mapped piRNAs.

2 Materials

All third-party tools and software used in the pipeline are listed in Table 1. Throughout the entire pipeline, the number of processor cores is set to 8. For improved performance this parameter can be adjusted to the number of processors in your computational setup. A reduced sample dataset for training is available for download under GSE192524. The sample dataset includes three biological replicates of a *S. mediterranea* piRNA library that was isolated from the epidermis and $NaIO_4$ treated to eliminate small RNA populations that do not carry 2′-O-methyl modification at their 3′ ends.

3 Methods

3.1 Preprocessing of Sequencing Reads

The first step of any next-generation sequencing data analysis is demultiplexing or sorting sequenced reads into different files according to their index sequences that were added during library preparation. In most cases sequencing data will already be demultiplexed by the sequencing facility. Otherwise, please refer to the reference manual of a suiting demultiplexing tool (e.g., Illumina's

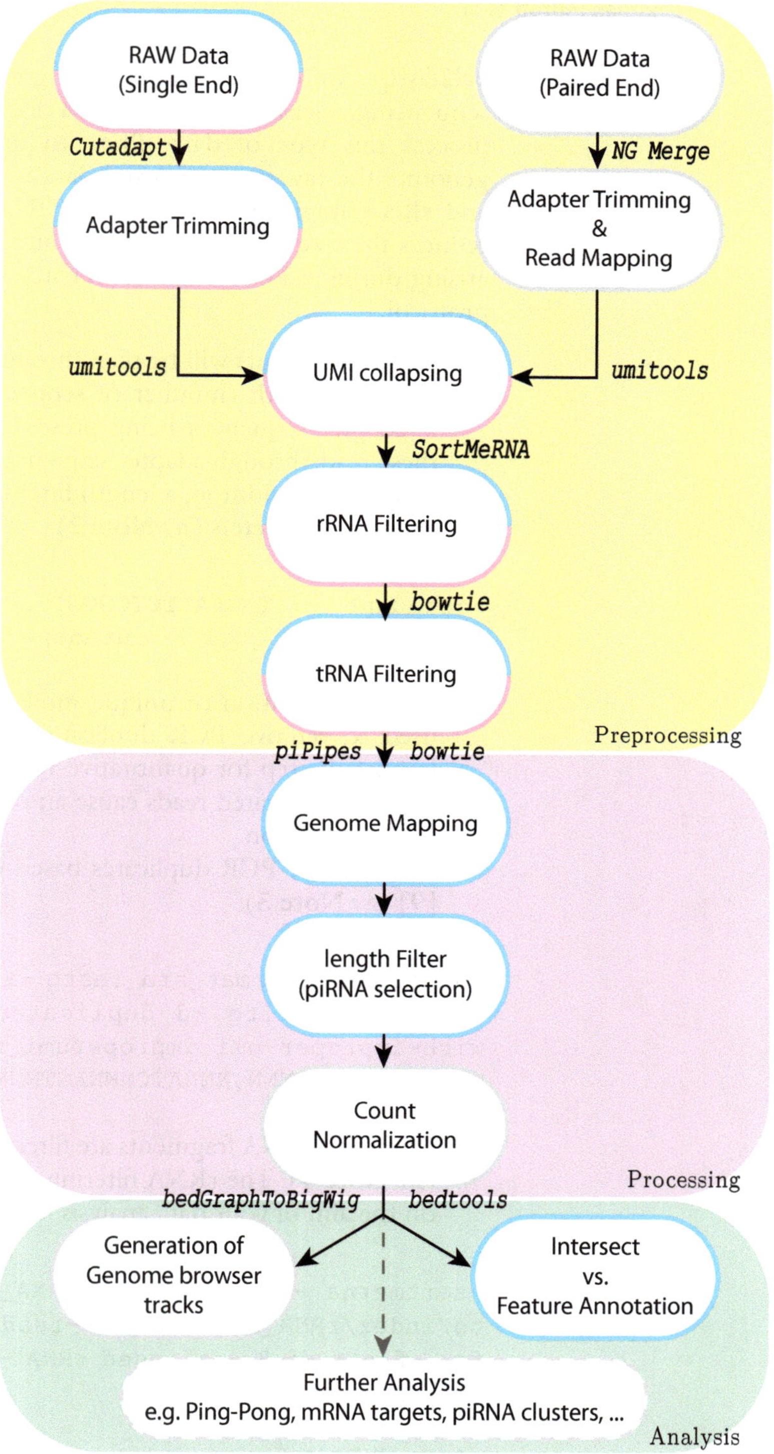

Fig. 1 piRNAnalyzer workflow. Steps that are part of the automated quality assessment in piRNAnalyzer are marked by colored contour lines (rose, fastQC; blue, pipeline generated plots; gray, no quality assessment). As indicated, further analysis steps are to be considered downstream of the automated pipeline depending on the biological question. The required software for each step is listed

bcl2fastq). In case a dataset was generated using paired-end sequencing, please refer to **Note 1** for a description of how to process this type of data. Prior to mapping to the reference genome, the raw demultiplexed data is filtered to remove rRNA and tRNA fragments as well as PCR duplicates. Together, this reduces the size of the working file and eliminates possible biases arising during a PCR amplification step of the library preparation protocol.

1. cDNA fragments will be read through if they are shorter than the read length (number of sequencing cycles), resulting in 3′-adapter sequences being present in the read. In this step these read-through adapter sequences are trimmed using cutadapt [8]. In addition, a length filter is applied by using the –m and –M parameters (*see* **Note 2**).

```
>cutadapt -a TGGAATTCTCGGGTGCCAAGG -j 8 -m 42 -M
64 raw_reads.fq.gz 1> cutadapt.fq 2> log.txt
```

2. The identification of unique molecular identifiers (UMIs) is used to remove PCR duplicates from the data. This is an important step for quantitative RNA sequencing experiments as PCR-duplicated reads cause an overestimation of measured RNA expression.

   Collapse PCR duplicates based on UMIs using umitools [9] (*see* **Note 3**):

```
>umitools reformat_sra_fastq -i cutadapt.fq -o col-
lapsed_output.fq -d duplicated_reads.fq --reads-
with-improper-umi improperumi_reads.fq -e 1 -v -5
NNNCGANNNTACNNN,NNNATCNNNAGTNNN -3 NNNGTCNNN
```

3. Ribosomal RNA fragments are filtered using SortMeRNA [10] (*see* **Note 4**). The rRNA filtering step is optional and depends on the aim of your data analysis (*see* **Note 5**).

```
>sortmerna --ref path/to/rRNA_sequences.fa','path/
to/index/rRNA_sequences' --reads collapsed_output.
fq --fastx -a 8 --aligned rRNA --other rRNAfiltered
-v --log
```

4. Small RNA fragments that map to tRNAs are removed using bowtie [11]. This step is optional (*see* **Note 6**):

```
>bowtie -q -p 8 -v 1 -a path/to/tRNA_BowtieIndex/
Smed_tRNA rRNAfiltered.fq -S tRNA_hits.sam --un
filtered_reads.fq 2>> log.txt
```

Table 1
Required third-party software

Software	Minimum version	Link	Reference
Cutadapt	1.15	https://cutadapt.readthedocs.io/en/stable/	[8]
umitools	0.2.0	https://github.com/weng-lab/umitools	[9]
SortMeRNA	2.1	https://bioinfo.lifl.fr/RNA/sortmerna/	[10]
bowtie	1.2.2	http://bowtie-bio.sourceforge.net/index.shtml	[11]
piPipes	1.5.0	https://github.com/bowhan/piPipes	[12]
samtools	1.7	http://www.htslib.org/	[13]
bedtools	2.26.0	https://bedtools.readthedocs.io/en/latest/	[14]
bedGraphtoBigWig	–	https://genome.ucsc.edu/goldenpath/help/bigWig.html	[15]

It is good practice to assess the quality of data after each preprocessing step. The automated version of the pipeline will generate plots that assist in evaluating the composition and quality of your sequencing libraries (*see* **Note 7**, Fig.2).

3.2 Genome Alignment and Annotation of Sequenced piRNAs

1. To map and quantify piRNAs, we use the piPipes smallRNA workflow [12].
   (a) Convert fastq files from the previous step to the "insert" file format using fastq_to_insert [12]:

```
>piPipes_fastq_to_insert filtered_reads.fq filter-
ed_reads.insert
```

   (b) Map filtered reads to the reference genome using bowtie [11] allowing one mismatch and an indefinite number of multimapping positions with the best alignment score (*see* **Note 8**):

```
>bowtie -r -p 8 -v 1 -a --best --strata path/to/
bowtie-index/ filtered_reads.insert -S mapped_-
reads.sam 2 >> bowtie_Log.txt
```

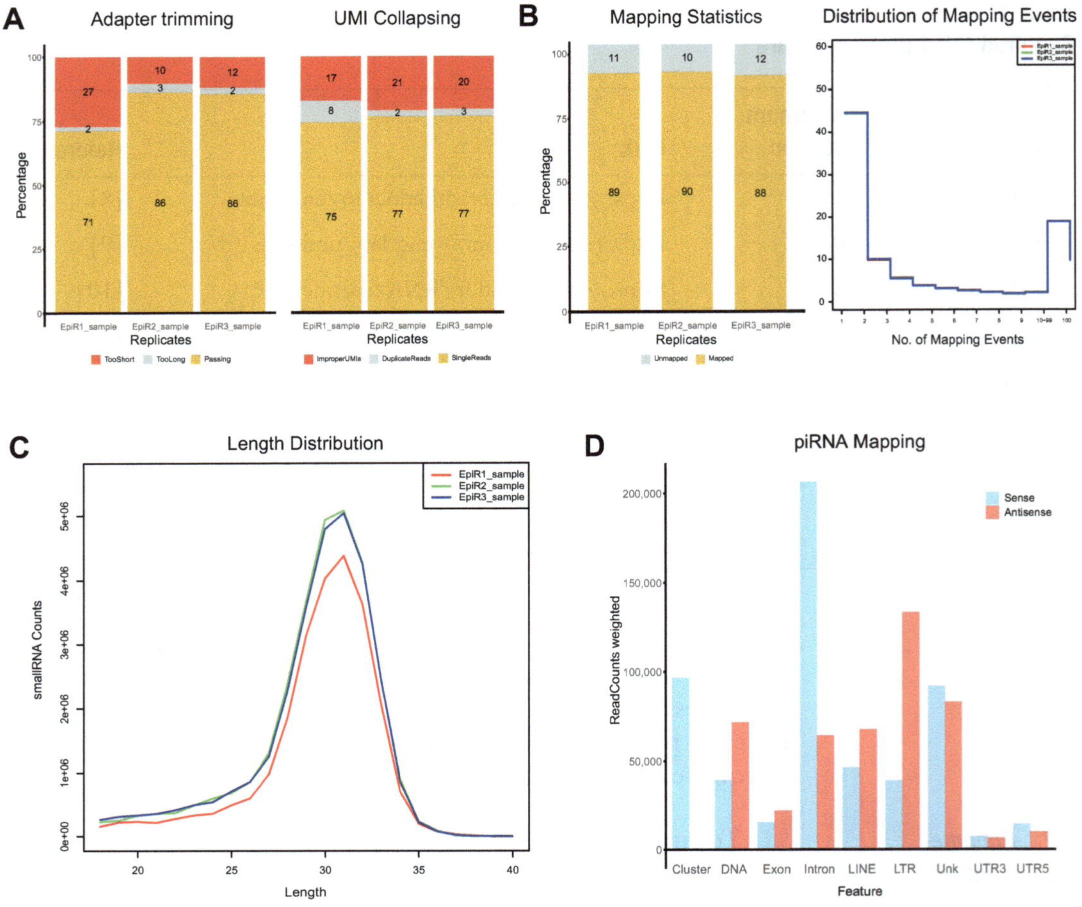

Fig. 2 piRNAnalyzer output. (**a**) Quality control plots for preprocessing steps, e.g., adapter trimming, UMI collapsing. (**b**) Visualization of mapping statistics, e.g., number of genome mapping reads and distribution of multimapping events. (**c**) Length distribution of the sequenced small RNA population. (**d**) Strand-specific annotation of sequenced piRNAs to genomic features using weighted piRNA counts

2. Aligned sam files are next converted to bam and bed format using samtools [13] and bedtools [14]:

```
>samtools view -bS -@ 8 mapped_reads.sam | samtools
sort -@ 8 - > mapped_reads.bam
>bedtools bamtobed -i mapped_reads.bam > mapped_-
reads.bed
```

3. Bed files are then extended to bed2 format, which adds sequence information, count data, and the number of reported alignments, using piPipes [12]:

```
>piPipes_insertBed_to_bed2 mapped_reads.insert
mapped_reads.bed > mapped_reads.bed2
```

4. To exclude miRNAs and siRNAs from the downstream analysis, we apply an additional length filtering step that only retains reads longer than 25 nt:

```
>awk 'length($7) > 25' mapped_reads.bed2 > piRNAs.
bed2
```

5. To account for piRNAs that map to multiple loci in the genome, piRNA counts are weighted by dividing read counts by the number of reported possible alignment positions:

```
>awk '{print $0"\t"$4/$5}' piRNAs.bed2 > piRNAs_-
weighted.bed2
```

6. The genomic coverage of mapped piRNAs can be visually inspected in a genome browser such as IGV (https://igv.org/app/) [16, 17]. To that end, using genomeCoverageBed [14] and UCSC's bedGraphtoBigWig utility [15], we generate two separate genome coverage tracks for the piRNAs, which are mapped in sense and antisense orientation. The BigWig format is recommended when working with large files due to its indexed binary nature [15].

```
>genomeCoverageBed -bg -strand + -i piRNAs_-
weighted.bed2 -g genome.sizes | sort -k1,1 -k2,2n
- > piRNAs_forward.bedgraph
>genomeCoverageBed -bg -strand - -scale -1 -i
piRNAs_weighted.bed2 -g genome.sizes | sort -k1,1
-k2,2n - > piRNAs_reverse.bedgraph
>bedGraphToBigWig piRNAs_forward.bedgraph genome.
sizes piRNAs_forward.bw
>bedGraphToBigWig piRNAs_reverse.bedgraph genome.
sizes piRNAs_reverse.bw
```

7. Mapped piRNAs are annotated to genomic features using bedtools [14]. To determine genomic locations of sequenced and mapped piRNAs, the pipeline overlaps them with transcriptomic features, such as exons, introns, 5′- and 3′-untranslated regions (UTRs), and repeat-masked elements, e.g. long interspersed nuclear elements (LINE), DNA transposons (DNA), long terminal repeat elements (LTR) and transposable elements of unknown origin (Unk), or annotated piRNA clusters [18]. By restricting the intersection step to a certain genomic strand, we can distinguish between sense and antisense mapped piRNAs.

The automated pipeline (https://github.com/kuhnlab-bayreuth/piRNAnalyzer) uses an extended planarian annotation file based on the most recent transcriptome assembly [6] in this step. In addition to exon and CDS structures of the predicted transcripts, we annotated and included their intron and UTR regions.

```
>bedtools intersect -split -wo -s -f 0.5 -a piR-
NAs_weighted.bed2 -b TranscriptAnnotation.gtf >
piRNAs_Transcripts_sense_mapping.bed
>bedtools intersect -split -wo -S -f 0.5 -a piR-
NAs_weighted.bed2 -b TranscriptAnnotation.gtf >
piRNAs_Transcripts_antisense_mapping.bed
```

3.3 Further Analysis

The final bed2 files can be used for a variety of different downstream analyses, such as characterizing ping-pong signatures, length distributions, piRNA cluster assembly, and mRNA target analysis [19].

4 Notes

1. For paired-end sequencing data adapter trimming should be performed in paired-end mode. Overlapping paired-end reads can be merged using utilities such as NGmerge [20] or usearch's fastq_mergepairs command [21].

   Merging read pairs and trimming adapter sequences using NGmerge [20]:

```
>NGmerge -1 forward_reads.fq.gz -2 reverse_reads.
fq.gz -d -e 42 -m 42 -n 8 -c Log.txt -j alignmen-
t_output.fq -o merged_readpairs.fq
```

   Prepare fastq header for processing by umitools:

```
>sed -e 's/\(@.*\):\(.*\)/\1 e\2/' merged_read-
pairs.fq > merged_reads.fq
```

   This step will replace adapter trimming. Thus, following this step the protocol can be used as indicated from **step 2** of preprocessing (Subheading 3.1).

2. The -m and -M parameters set the lower and upper limits of the read length after adapter removal. By retaining fragments of 18–42 nt in length, we are covering the entire small RNA population including siRNAs, miRNAs, and piRNAs. We recommend a minimum length of 18 nt for nonambiguous

alignments. UMI sequences are still present in the reads at this point and have to be added to the desired fragment length. For the library preparation approach described in the accompanying chapter, UMI sequences sum up to 24 nt (3′-UMI = 9 nt + 5-′-UMI = 15 nt). This is why reads are filtered for a length between 42 and 64 nt.

3. Reads are collapsed based on the UMIs and on the read sequence. If the combination of both is unique, the read will be retained; else it will be printed to the duplicated reads file. We are allowing for one mismatch to account for possible sequencing errors.
4. The automated pipeline from GitHub includes a pre-built reference database for planarian rRNAs.
5. The accumulation of ribosomal RNA-derived piRNAs was detected independently in diverse organisms in normal conditions [22–25] and upon mutation of the piRNA pathway-associated factors [26]. Therefore, include the rRNA filtering step according to your research question.
6. This step is optional. Please note that SMEDWI-bound small RNAs were recently shown to be partially derived from tRNAs in planarians [27]. Similar to rRNA removal, tRNAs are filtered by aligning reads to a tRNA reference fasta file allowing one mismatch (http://gtrnadb.ucsc.edu). tRNA contamination should be rather low due to the size selection step in library preparation. The automated pipeline from GitHub includes a pre-built bowtie index for planarian tRNAs.
7. To visually assist in quality control and troubleshooting of the sequenced data, piRNAnalyzer generates the following plots:
    (a) Percentages of sequencing reads that pass the applied filtering criteria are plotted for all preprocessing steps. This helps to assess technical problems that occurred during library construction (e.g., adapter ligation, library complexity, or rRNA contamination). Additional information can be obtained from automatically generated fastQC plots [28].
    (b) The length distribution plot gives an overview over all small RNA populations that were sequenced. In planarians, piRNAs are 30–35 nt long [19, 29, 30], whereas miRNAs and siRNAs are 21–22 nt in length [29, 30]. A successfully prepared piRNA library shows a peak in the length distribution of small RNAs between 30 and 35 nt. The populations of miRNA and siRNA are significantly reduced in small RNA libraries prepared using $NaIO_4$-mediated oxidization and β-elimination protocol.

(c) The piRNA mapping plot visualizes the genomic distribution of sequenced piRNAs. This analysis provides an immediate genome-wide assessment of a given piRNA population.

8. The pipeline does allow for one mismatch in alignments to account for sequencing errors and high heterozygosity of the planarian genome. All alignments with the best alignment score are reported. If multiple alignments share the best score, all of them will be written to the final output file. The automated pipeline from GitHub includes a pre-built bowtie index for the latest version of the planarian genome.

Acknowledgements

This work was supported by the German Research Foundation (DFG, grant No. KU 3514/5-1), the Elite Network of Bavaria, and the University of Bayreuth.

References

1. Aravin A, Gaidatzis D, Pfeffer S et al (2006) A novel class of small RNAs bind to MILI protein in mouse testes. Nature 442:203–207. https://doi.org/10.1038/nature04916
2. Girard A, Sachidanandam R, Hannon GJ et al (2006) A germline-specific class of small RNAs binds mammalian Piwi proteins. Nature 442:199–202. https://doi.org/10.1038/nature04917
3. Rojas-Ríos P, Simonelig M (2018) piRNAs and PIWI proteins: regulators of gene expression in development and stem cells. Development 145. https://doi.org/10.1242/dev.161786
4. Shirayama M, Seth M, Lee H-C et al (2012) piRNAs initiate an epigenetic memory of nonself RNA in the C. elegans germline. Cell 150:65–77. https://doi.org/10.1016/j.cell.2012.06.015
5. Seth M, Shirayama M, Gu W et al (2013) The C. elegans CSR-1 argonaute pathway counteracts epigenetic silencing to promote germline gene expression. Dev Cell 27:656–663. https://doi.org/10.1016/j.devcel.2013.11.014
6. Rozanski A, Moon H, Brandl H et al (2019) PlanMine 3.0-improvements to a mineable resource of flatworm biology and biodiversity. Nucleic Acids Res 47:D812–D820. https://doi.org/10.1093/nar/gky1070
7. Kim IV, Ross EJ, Dietrich S et al (2019) Efficient depletion of ribosomal RNA for RNA sequencing in planarians. BMC Genomics 20:909. https://doi.org/10.1186/s12864-019-6292-y
8. Martin M (2011) Cutadapt removes adapter sequences from high-throughput sequencing reads. EMBnet j 17:10. https://doi.org/10.14806/ej.17.1.200
9. Fu Y, Wu P-H, Beane T et al (2018) Elimination of PCR duplicates in RNA-seq and small RNA-seq using unique molecular identifiers. BMC Genomics 19:531. https://doi.org/10.1186/s12864-018-4933-1
10. Kopylova E, Noé L, Touzet H (2012) SortMeRNA: fast and accurate filtering of ribosomal RNAs in metatranscriptomic data. Bioinformatics 28:3211–3217. https://doi.org/10.1093/bioinformatics/bts611
11. Langmead B, Trapnell C, Pop M et al (2009) Ultrafast and memory-efficient alignment of short DNA sequences to the human genome. Genome Biol 10:R25. https://doi.org/10.1186/gb-2009-10-3-r25
12. Han BW, Wang W, Zamore PD et al (2015) piPipes: a set of pipelines for piRNA and transposon analysis via small RNA-seq, RNA-seq, degradome- and CAGE-seq, ChIP-seq and genomic DNA sequencing. Bioinformatics 31:593–595. https://doi.org/10.1093/bioinformatics/btu647
13. Li H, Handsaker B, Wysoker A et al (2009) The sequence alignment/map format and

SAMtools. Bioinformatics 25:2078–2079. https://doi.org/10.1093/bioinformatics/btp352
14. Quinlan AR, Hall IM (2010) BEDTools: a flexible suite of utilities for comparing genomic features. Bioinformatics 26:841–842. https://doi.org/10.1093/bioinformatics/btq033
15. Kent WJ, Zweig AS, Barber G et al (2010) BigWig and BigBed: enabling browsing of large distributed datasets. Bioinformatics 26: 2204–2207. https://doi.org/10.1093/bioinformatics/btq351
16. Robinson JT, Thorvaldsdóttir H, Winckler W et al (2011) Integrative genomics viewer. Nat Biotechnol 29:24–26. https://doi.org/10.1038/nbt.1754
17. Robinson JT, Thorvaldsdóttir H, Turner D et al (2020) igv.js: an embeddable JavaScript implementation of the Integrative Genomics Viewer (IGV). Bioinformatics 15. https://doi.org/10.1101/2020.05.03.075499
18. Rosenkranz D, Zischler H (2012) proTRAC--a software for probabilistic piRNA cluster detection, visualization and analysis. BMC Bioinform 13:5. https://doi.org/10.1186/1471-2105-13-5
19. Kim IV, Duncan EM, Ross EJ et al (2019) Planarians recruit piRNAs for mRNA turnover in adult stem cells. Genes Dev 33:1575–1590. https://doi.org/10.1101/gad.322776.118
20. Gaspar JM (2018) NGmerge: merging paired-end reads via novel empirically-derived models of sequencing errors. BMC Bioinform 19:536. https://doi.org/10.1186/s12859-018-2579-2
21. Edgar RC (2010) Search and clustering orders of magnitude faster than BLAST. Bioinformatics 26:2460–2461. https://doi.org/10.1093/bioinformatics/btq461
22. Brennecke J, Aravin AA, Stark A et al (2007) Discrete small RNA-generating loci as master regulators of transposon activity in Drosophila. Cell 128:1089–1103. https://doi.org/10.1016/j.cell.2007.01.043
23. Houwing S, Kamminga LM, Berezikov E et al (2007) A role for Piwi and piRNAs in germ cell maintenance and transposon silencing in Zebrafish. Cell 129:69–82. https://doi.org/10.1016/j.cell.2007.03.026
24. Wahba L, Hansen L, Fire AZ (2021) An essential role for the piRNA pathway in regulating the ribosomal RNA pool in C. elegans. Dev Cell 56:2295–2312.e6. https://doi.org/10.1016/j.devcel.2021.07.014
25. García-López J, Alonso L, Cárdenas DB et al (2015) Diversity and functional convergence of small noncoding RNAs in male germ cell differentiation and fertilization. RNA 21:946–962. https://doi.org/10.1261/rna.048215.114
26. Reuter M, Chuma S, Tanaka T et al (2009) Loss of the Mili-interacting Tudor domain-containing protein-1 activates transposons and alters the Mili-associated small RNA profile. Nat Struct Mol Biol 16:639–646. https://doi.org/10.1038/nsmb.1615
27. Lakshmanan V, Sujith TN, Bansal D et al (2021) Comprehensive annotation and characterization of planarian tRNA and tRNA-derived fragments (tRFs). RNA 27:477–495. https://doi.org/10.1261/rna.077701.120
28. Andrews S (2019) FastQC. Babraham Bioinformatics
29. Friedländer MR, Adamidi C, Han T et al (2009) High-resolution profiling and discovery of planarian small RNAs. Proc Natl Acad Sci U S A 106:11546–11551. https://doi.org/10.1073/pnas.0905222106
30. Palakodeti D, Smielewska M, Lu Y-C et al (2008) The PIWI proteins SMEDWI-2 and SMEDWI-3 are required for stem cell function and piRNA expression in planarians. RNA 14: 1174–1186. https://doi.org/10.1261/rna.1085008

Chapter 4

Combining Fluorescent In Situ Hybridization with Immunofluorescence and Lectin Staining in Planarians

Daniel Font-Martín, Eudald Pascual-Carreras, and Emili Saló

Abstract

The capability to simultaneously apply different molecular tools to visualize a wide variety of changes in genetic expression and tissue composition in *Schmidtea mediterranea* has always been of great interest. The most commonly used techniques are fluorescent in situ hybridization (FISH) and immunofluorescence (IF) detection. Here, we describe a novel way to perform both protocols together adding the possibility to combine them with fluorescent-conjugated lectin staining to further broaden the detection of tissues. We also present a novel lectin fixation protocol to enhance the signal, which could be useful when single-cell resolution is required.

Key words Fluorescent ISH, Immunofluorescence, Antibodies, Lectin, Fluorescent-conjugated, Planarian

1 Introduction

1.1 Molecular Tools in the Planarian Field

Originally from the occidental Mediterranean area, planarian *Schmidtea mediterranea* is a model organism for regeneration and developmental studies. Planarians can regenerate an entire organism from any fragment of their body by rebuilding lost tissue with the correct pattern. Moreover, they are also constantly resizing according to food availability and temperature through a process that involves repatterning of existing tissues [1]. The high regenerative capability of these animals resides in their population of adult stem cells, called neoblasts [2]. During homeostasis, neoblasts are the source of the constant cell turnover required to maintain the body size according to food availability [3]. All these characteristics make these animals ideal models to study changes in genetic

Daniel Font-Martín and Eudald Pascual-Carreras contributed equally with all other contributors.

Supplementary Information The online version contains supplementary material available at https://doi.org/10.1007/978-1-0716-3275-8_4.

Luca Gentile (ed.), *Schmidtea Mediterranea: Methods and Protocols*, Methods in Molecular Biology, vol. 2680,
https://doi.org/10.1007/978-1-0716-3275-8_4,

© The Author(s), under exclusive license to Springer Science+Business Media, LLC, part of Springer Nature 2023

expression in vivo under a wide umbrella of experimental conditions. Therefore, the capability to simultaneously apply different molecular tools to visualize these changes is of great interest. The most commonly used techniques are fluorescent in situ hybridization (FISH), detecting simultaneously more than one mRNA per cell [4–7], and immunofluorescence (IF) detection, unveiling the cellular localization of different protein epitopes [8–10]. Interestingly, laboratories and private companies have generated antibodies that specifically label tissues and cell types in planarians, such as neural tissue [11–13], photoreceptors and optic chiasm [14], muscle fibers [8, 15, 16], digestive system and epidermis [17], stem cells [18, 19] and proliferating cells [15, 20, 21]. Along the same lines, fluorescent-conjugated lectins have been shown to bind with high affinity to several glycoproteins. These sugar-binding proteins are ideal to stain a broad diversity of glycan conjugates, detecting a variety of planarian tissue types. In planarians, they have been reported to especially stain subpopulations of secretory cells [22].

1.2 Method Overview

In molecular and cellular techniques, tissue fixation is one of the most critical steps. In planarians, prior to fixation, the body-recovering mucus is removed, normally with HCl or N-acetyl cysteine [4, 5, 8]. Then, formaldehyde is used as fixative in FISH and IF protocols. After FISH fixation, samples must be dehydrated with methanol. At this point, they can be stored for several weeks or months. When samples are required, they are hydrated, the pigmentation is removed by bleaching, and they are permeabilized to allow probe incubation. Specific nucleotide-labelled probes are hybridized overnight. After several washes, samples are incubated overnight with anti-hapten antibodies. Finally, probes are serially developed. To proceed with IF and/or lectin staining, samples are washed, and pre-block treated. Primary antibodies and/or lectins are incubated overnight. After several washes, the tissue is blocked and incubated with secondary antibodies. Finally, samples need to be cleared and mounted for their analysis and image acquisition.

1.3 Combining Different Molecular Techniques

Recently, authors have tried to combine different molecular techniques, being able to perform FISH followed by immunofluorescence, detecting PIWI protein (stem cells) [23, 24], the optic chiasm [25], or muscle fibers [16]. However, the combination of both techniques is difficult since the methanol dehydration and/or the permeabilization in the FISH protocol might damage protein epitopes, impeding its consecutive detection. In this chapter, we propose to combine both techniques and present new working antibodies and lectin staining visible after FISH.

2 Materials

Prepare solution with H_2O filtered (ultrapure), and perform the DEPC treatment when it is indicated. Prepare and store all the reagents at room temperature (otherwise will be indicated). Diligently follow all waste disposal regulations when disposing waste materials. Calculations for most of the following solutions are added in Supplementary Tables 1, 2, and 3.

2.1 Stock Solutions

See Supplementary Table 1 of stock solutions for their preparation at desired volumes. *See* Table 1 for the adapted concentrations of antibodies and lectins tested.

1. 10× PBS: 1.37 M NaCl, 27 mM KCl, 100 mM Na_2HPO_4, and 20 mM KH_2PO_4 diluted in autoclaved ultrapure H_2O. pH adjusted to 7.4 (adding HCl). Treat with DEPC and autoclave.
2. Tx: 10% Triton-X-100 in ultrapure H_2O.
3. PBT: 3% Tx in 1× PBS.
4. 50% methanol: methanol (v/v) diluted in PBT.
5. 20× SSC: 3 M sodium chloride and 0.3 M sodium citrate diluted in ultrapure H_2O. Treat with DEPC and autoclave.
6. 1× SSC: 20× SSC stock solution diluted to 1× with ultrapure H_2O.
7. 2× SSCTx: 20× SSC stock solution diluted to 2× with ultrapure H_2O + Tx diluted to 0.1%.
8. 0.2× SSCTx: 20× SSC stock solution diluted to 0.2× with ultrapure H_2O + Tx diluted to 0.1%.
9. Pre-hybe buffer: 50% non-deionized formamide, 5× SSC, 0.1 mg/mL yeast RNA, 1% Tween-20, diluted in DEPC-treated water. Store at −20 °C.
10. Hybe buffer: pre-hybe buffer with 5% dextran sulphate, stored at −20 °C.
11. TNTx: 0.1 M Tris–HCl, 0.15 M NaCl, and 0.3% Tx. pH adjusted to 7.5 and filtered.
12. 4-IPBA: 20 mg/mL 4-iodophenylboronic acid in dimethylformamide (DFM), stored at −20 °C.
13. TSA buffer: 2 M NaCl and 0.1 M boric acid–HCl. pH adjusted to 8.5. Filter, autoclave, and store at 4 °C.
14. Glycerol 80%: glycerol (v/v) diluted in 1× PBS.
15. DAPI solution: 0.02 μg/mL diluted in 1× PBS.
16. TO-PRO®-3 solution: 0.33 μM diluted in 1× PBS.
17. Lectin kit: stored at 4 °C.
18. DEPC treatment: DEPC 1:1000 dilution in required solution. Stir and incubate O/N in the hood. Autoclave.

Table 1
Summarized table of all the antibodies and lectins tested in combination with FISH

Source	1ry antibody	Host animal	Working dilution	Fixation	After FISH
Orii Lab	VC1 (arrestin)	Mouse (monoclonal)	1:15000	HCl—FA	**Yes**
Hybridoma Bank (DSHB)	3C11 (synapsin)	Mouse (monoclonal)	1:50	HCl—FA/ HCl—FA—Met	No
Sigma	FMRF	Rabbit (polyclonal)	1:500	HCl—FA	No
Sigma	NPF	Rabbit (polyclonal)	1:1000	HCl—FA/ HCl—FA—Met	**Yes**
Sigma	A2066 (actin)	Rabbit (polyclonal)	1:100	HCl—FA	No
Hybridoma Bank (DSHB)	JLA20 (actin)	Mouse (monoclonal)	1:100	HCl—FA	No
Sigma	T9028 (tubulin tyrosine)	Mouse (monoclonal)	1:400	HCl—FA	No
Upstate Biotechnology	Phospo-histone3 ser10	Rabbit (polyclonal)	1:500	HCl—FA/ HCl—FA—Met	No
Millipore	Phospo-histone3 ser10	Rat (monoclonal)	1:1000	HCl—FA	**Yes**
Salo & Adell Lab	SMED-β catenin2	Rabbit (polyclonal)	1:2000	HCl—FA	No
Bartscherer Lab	SMED-PIWI	Rabbit (polyclonal)	1:1000	HCl—FA	**Yes**
Hybridoma Bank (DSHB)	22C10 (Futsch)	Mouse (monoclonal)	1:250	HCl—FA	No
Hybridoma Bank (DSHB)	6G10	Mouse (monoclonal)	1:1000	HCl—FA	**Yes**
Reddien Lab	V5277	Rabbit (polyclonal)	1:500	HCl—FA	**Yes**
Source	**2ry antibody**	**Host animal**	**Working dilution**		
Molecular Probes	α-Mouse Alexa 488	Goat	1:400		
Molecular Probes	α-Rabbit Alexa 568	Goat	1:1000		

(continued)

Table 1 (continued)

Molecular Probes	α-Mouse Alexa 568	Goat	1:1000		
Molecular Probes	α-Rabbit Alexa 488	Goat	1:400		
Molecular Probes	α-Rat Alexa 488	Goat	1:400		
Source	**Lectin**	**Conjugation**	**Working dilution**	**Fixation**	**After FISH**
Vector Laboratories, Inc.	ConA	FITC	1:2000	HCl-FA/HCl—FA—Met	**Yes**
Vector Laboratories, Inc.	SBA	FITC	1:2000	HCl-FA/HCl—FA—Met	**Yes**
Vector Laboratories, Inc.	WGA	FITC	1:2000	HCl-FA/HCl—FA—Met	**Yes** (no pharynx)
Vector Laboratories, Inc.	DBA	FITC	1:400	HCl-FA/HCl—FA—Met	**Yes** (no pharynx)
Vector Laboratories, Inc.	PNA	FITC	1:400	HCl-FA/HCl—FA—Met	Dimmer (no pharynx)
Vector Laboratories, Inc.	RCAI	FITC	1:400	HCl-FA/HCl—FA—Met	**Yes**
Vector Laboratories, Inc.	GSLI	Rhodamine	1:2000	HCl-FA/HCl—FA—Met	**Yes**

2.2 FISH Solutions

See Supplementary Table 2 of FISH solutions for their preparation at desired volumes.

2.2.1 NAC-FA, Fixation, and Dehydration

1. 7.5% NAC solution: 7.5% (w/v) N-acetyl-L-cysteine diluted in 1× PBS.
2. 4% fixative solution: 4% formaldehyde (v/v) in PBT.

2.2.2 Bleaching, Permeabilization, and Hybridization

1. Formamide bleaching solution: 5% non-deionized formamide, 0.5× SSC, and 1.2% H_2O_2 (*see* **Note 1**).
2. Proteinase K solution: 10% (w/v) SDS and 5 μL 20 mg/mL Proteinase K diluted in PBT.

3. Riboprobe solution: dilute riboprobes in hybe buffer (final concentration of 0.1–5 ng/μL), denature the solution at 80 °C for 5 min and chill on ice for 2 min (*see* **Note 1**).

2.2.3 Post-Hybe Washes and Antibody Incubation

1. Blocking solution: 5% horse serum, 0.5% Western Block Reagent in TNTx.

2.2.4 Antibody Washes and TSA Reaction

1. Tyramide solution: For FAM and TMR fluor-tyramide, use 1: 500 dilution (1:250 for DYLight633), 4-IPBA (1:1000), and H_2O_2 (0.003%) in the TSA buffer. Prepare the tyramide working solution stocks as stated in the commercial kits (*see* **Note 1**).

2.2.5 Double FISH

1. Azide solution: 100 mM sodium azide in PBT.

2.3 Immunofluorescence Solutions

2.3.1 Blocking and Primary Antibody Incubation

1. HCl solution: 2% (v/v) HCl dissolved in ultrapure H_2O on ice from 32% HCl stock solution (*see* **Note 2**).
2. Fixative solution: 4% formaldehyde solution dissolved in PBT (v/v) from 37% formaldehyde stock solution (*see* **Note 2**).
3. Bleaching solution: 6% H_2O_2 solution in PBT (v/v) from 30% H_2O_2 stock (*see* **Note 2**).
4. Blocking solution: 1% BSA (w/v) in PBT.

Check Supplementary Table 3 of immunochemistry solutions for their preparation at desired volumes. Check Table 1 for the tested antibodies and lectin dilutions as well as for their appropriate fixation and FISH protocol compatibility (*see* **Note 2** for fixation details).

3 Methods

All procedures are carried out at room temperature, unless otherwise specified.

3.1 Fluorescent In Situ Hybridization

3.1.1 NAC-FA Fixation and Dehydration

1. Collect and transfer planarians to the chosen container. A small number of animals (10 intact or 15 regenerating planarians) could be fixed in 1.5 mL tubes, using 1 mL per step. Until 30 planarians, a 15 mL tube should be used, adding between 10 and 12 mL per step. For more than 30, a 50 mL tube will be used, adding 40 mL per step (*see* **Note 3**).
2. Remove planarian water and add 7.5% NAC solution. Gently invert and swirl the tubes. Then place them in a rocking shaker for 7–10 min (*see* **Note 3**).
3. Remove 7.5% NAC solution and rapidly add 4% fixative solution. Gently invert and swirl tubes, placing them in a rocking shaker for 15 min. During the fixative incubation, ensure that the animals do not stick together. If so, gently invert.

4. Remove the fixative solution and wash two times with PBT. Finally, dehydrate with 50% methanol for 5 min while rotating, and keep them in 100% methanol at the freezer (−20 °C) (*see* **Note 3**).

3.1.2 Bleaching, Permeabilization, and Hybridization

1. Rehydrate samples by incubating in 50% methanol for 5 min and then in PBT for 5 min.
2. Wash in 1× SCC for 5 min.
3. Bleach animals by incubating in freshly prepared formamide bleaching solution for 2 h under bright LED light (cold light, 450 lumen).
4. Wash for 5 min in 1× SCC.
5. Wash in PBT two times for 5 min each.
6. Incubate in Proteinase K solution for 10–15 min with gentle agitation according to animal size since larger animals (>5 mm) need more time for proper permeabilization (*see* **Note 4**).
7. Postfix with 4% fixative for 10 min.
8. Wash in PBT two times for 5 min each.
9. Transfer animals to small in situ baskets in a 24-well plate and remove PBT. Incubate in 1:1 pre-hybe/PBT for 10 min (*see* **Note 5**).
10. Remove the pre-hybe/PBT and incubate with pre-hybe buffer for 2 h at 56 °C with gentle agitation.
11. Remove the pre-hybe buffer and hybridize in 500 μL denatured Riboprobe solution for >16 h at 56 °C with gentle agitation (*see* **Note 5**).

3.1.3 Post-Hybe Washes and Antibody Incubation

1. Save the Riboprobe solution and/or replace it with 2× SCCTx for 20 min twice at 56 °C. Follow it by four washes with 0.2× SCCTx four times for 20 min each (*see* **Note 6**).
2. Return samples to room temperature and wash them twice with TNTx 10 min each.
3. Incubate animals in blocking solution for 2 h (or overnight at 4 °C).
4. Remove blocking solution and incubate with antibody solution overnight at 4 °C with gentle agitation. Use a dilution of 1: 2000 for anti-DIG-POD and anti-FITC-POD and 1:300 for anti-DNP-POD in blocking solution.

3.1.4 Antibody Washes and TSA Reaction

1. Remove the antibody solution and wash animals for 5 min, 10 min, and then six times for 20 min each with TNTx.
2. Incubate in freshly made tyramide solution (TSA reaction) for 10 min.

3. Wash with TNTx two times for 5 min each.
4. For a single FISH, wash four more times with PBT for 10 min (Fig. 1a for single FISH detection). For double FISH, proceed to **step 5** in Subheading 3.1. For immunofluorescence detection and/or lectin staining, proceed to Subheading 3.2.

3.1.5 Double FISH

1. Inactivate peroxidase activity by incubating in azide solution for 45 min (*see* **Note** 7).
2. Wash peroxidase inactivated samples in TNTx four times for 10 min each (*see* **Note 8**).
3. Block in blocking solution for 1 h.
4. Incubate in antibody solution overnight at 4 °C. Then repeat all the steps in Subheading 3.1.4.

3.2 Immunofluorescence Detection/Lectin Staining

If it is of one interest to start the immunofluorescence or lectin staining from scratch (*see* **Note 9**).

3.2.1 Blocking and Primary Antibody Incubation

1. Transfer samples from the baskets to a new 48-well plate (*see* **Note 10**).
2. Wash in PBT three times for 5 min.
3. Replace the PBT with the blocking solution. Let it incubate for 2 h on a rocker (*see* **Note 11**).
4. Replace it with the primary antibody and/or conjugated lectins diluted in blocking solution (see Table 1 for concentration). Incubate overnight at 4 °C on a rocker shaker (*see* **Note 11**).

3.2.2 Primary Antibody Washes, Blocking, and Secondary Antibody Incubation

1. Rinse thrice with PBT followed by six washes with PBT for 6–8 h (*see* **Note 12**).
2. Replace the PBT, adding the blocking solution. Let it incubate for 2 h on a rocker (*see* **Note 11**).
3. Replace it with the secondary antibody diluted in blocking solution (*see* Table 1 for concentration). Incubate overnight at 4 °C on a rocker shaker. For lectin stainings, only this step is skipped since there is no need for a fluorescent-conjugated secondary antibody.

3.2.3 Secondary Antibody Washes

1. Rinse three times with PBT followed by six washes with PBT for 2–3 h (*see* **Note 12**).

3.2.4 Clearing and Mounting

1. Finally add DAPI/TO-PRO®-3 incubation for 2 h at room temperature or 4 °C overnight.

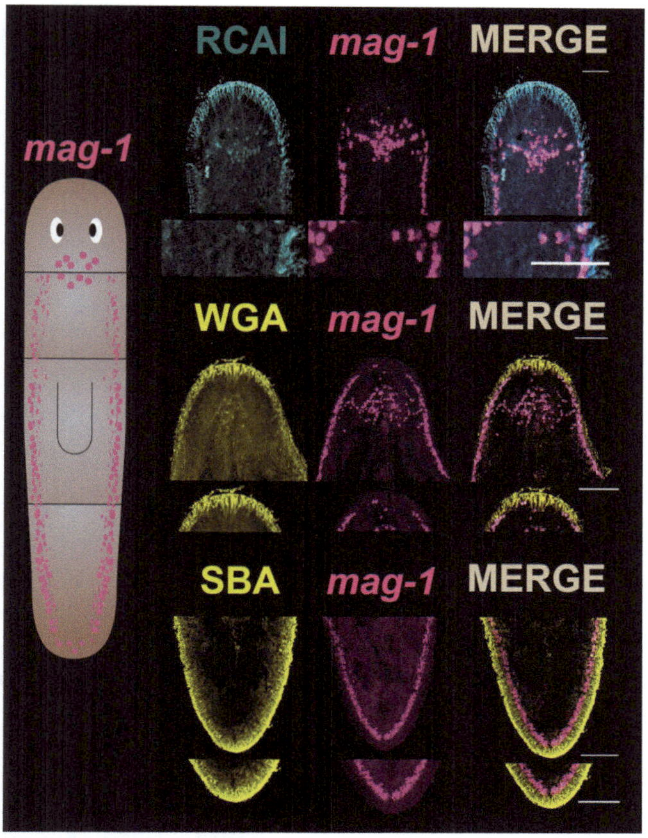

Fig. 1 Summary of lectin staining coupled with FISH. Schematic illustration of *mag-1* expression pattern was added. Fluorescent in situ hybridization (FISH) of *mag-1* (marginal adhesive gland cells) (magenta) followed by either RCAI (cian) or WGA and SBA (yellow) lectin stainings. Magnifications were added to better visualize their colocalizations and/or tissue disposition. Scale bars in all the panel: 100 μm

2. Remove DAPI/TO-PRO®-3 solution and add 80% glycerol solution for tissue clearance. Proceed to mounting on microscope slides and top it with a coverslip sealed with nail polish topcoat (*see* **Note 13**).

See Fig. 1 for some combination examples of FISH and lectin stainings and Fig. 2 for more IF and lectin combinations.

4 Notes

1. Diluting high formamide concentrations in H_2O_2 produce a violent reaction. Always dilute these reagents into the water before mixing. For probe concentration testing, start with 1 ng/μL. For the tyramide solution, prepare a fresh stock of 0.5% H_2O_2.

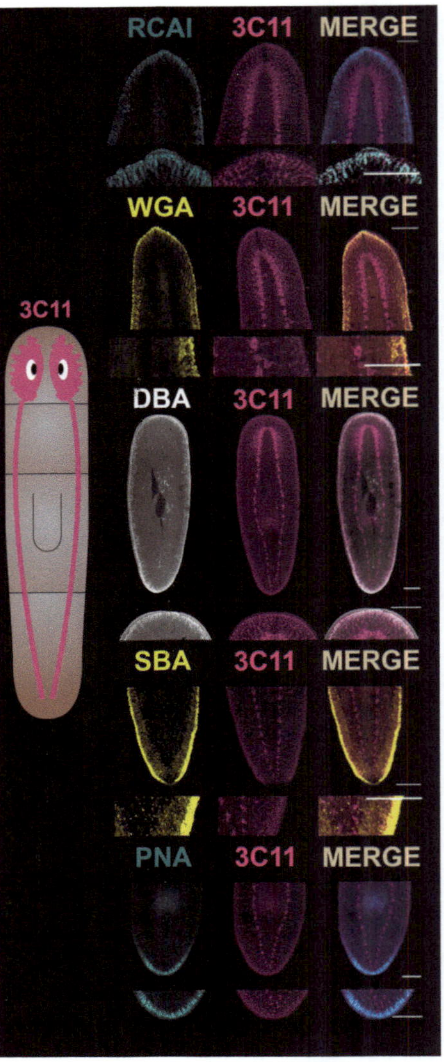

Fig. 2 Summary of lectin staining coupled with IF. Schematic illustration of the nervous system was added. Immunofluorescence (IF) staining with 3C11 labeling synapsin protein (magenta) coupled with either RCAI and PNA (cian), WGA and SBA (yellow), or DBA (white) lectin stainings. Magnifications were added to better visualize their colocalizations and/or tissue disposition. Scale bars in all the panel: 100 μm

2. For immunofluorescence and/or lectin fixation, transfer planarians to the appropriate vial for fixation. Replace planarian water with pre-chilled 2% HCl (v/v). Incubate for 5 min, alternating 1 min on ice, 1 min inverting. Remove 2% HCl solution and rinse once with PBT. Remove the PBT and add 4% fixative solution. Gently agitate the tubes and put them on the rocking shaker for 15 min. Remove the fixative and rinse the animals twice with PBT. Fixed animals may last up to 1 week in

PBS 1× at 4 °C. For longer periods of time (months) proceed to perform methanol dehydration. Replace PBT with 50% methanol solution for 5–10 min on the shaker. Afterwards, replace it with 100% methanol for 5–10 more min. Store them at −20 °C. For methanol stored fixations, rehydrate first the vials replacing the 100% methanol with 50% methanol for 5–10 min followed by a replacement with PBT for the same amount of time. Replace PBT and add the bleaching solution. Vials are placed under direct cold light in an aluminum foil closed box overnight (16 h). If using 1.5 mL tubes, seal properly the vials with parafilm if they seem loose. Gases from the reaction may open them and leak the planarians outwards ruining the experiment. Remove the bleaching solution and rinse the specimens twice with PBT. Transfer planarians of each condition to a 48-well plate and proceed with Subheading 3.2.1.

3. In all fixative steps, try to remove as much solution as possible. During the methanol step, check that the animals do not clump together. They should stay 1 h minimum at −80 °C before the next step. On the other hand, they can be stored at −20 °C from days to months.
4. Proteinase K activity might change in the aliquots following storage at −20 °C, over months. The mentioned time and concentration are good starting point but concentrations and incubation times might need adjustments depending on enzyme activity. If the same animals need to be tested with different probes, use the same tube during all the steps, until the animals are transferred to the baskets.
5. In 24-well plates, 500 μL is enough to cover the samples. After placing animals in the baskets, washes can be accelerated with a tip-attached small tube connected to a vacuum pump. During probe hybridization, seal the plate with parafilm to avoid evaporation in the oven.
6. Remove the Riboprobe solution carefully with a transfer pipette to store, not vacuum. Alternatively, add 500 μL per well to dilute the probe.
7. Azide solution should be disposed of in a special container.
8. For better staining, develop first the digoxigenin (DIG)-labeled riboprobe with red tyramide followed the next day by the FITC-labeled riboprobe with green tyramide.
9. Use a 200 μL tip with the bottom sectioned to transfer planarians with more ease.
10. Add 400 μL of each solution per well. Prepare double the volume of blocking solution in advance to use it with primary and secondary antibodies. Store it in the freezer.

11. FITC-conjugated lectins ConA, SBA, WGA, and rhodamine-conjugated GSLI should be diluted to 1 μg/mL using blocking solution. On the other hand, FITC-conjugated lectins DBA, PNA, and RCAI should be diluted to 5 μg/mL. Some lectin kits might provide other lectins such as FITC-conjugated UEAI and rhodamine-conjugated PHA-E, PHA-L, and SJA which do not present any specific planarian staining. It is advisable to dispense of them if the opportunity is given. Although methanol fixation is allowed for lectin staining, it is recommended not to do so if a stronger staining is required. One might find it especially useful when single-cell resolution images are requested. The same statement shall be emphasized after in situ since the lectin signal appears dimmer and the in situ protocol compromises the visualization of internal structures of the animals such as the pharynx (as in the case of DBA, WGA, and PNA staining). In these conditions, a simple lectin staining, or an immunofluorescence approach, is more advisable as seen in Fig. S1.
12. More washes correlate with less washing time, although at least 3 h is required.
13. Optionally, to avoid photobleaching, embed the animals in a SlowFade mounting medium before mounting.

References

1. Saló E (2006) The power of regeneration and the stem-cell kingdom: freshwater planarians (Platyhelminthes). BioEssays 28:546–559. https://doi.org/10.1002/bies.20416
2. Reddien PW, Alvarado AS (2004) Fundamentals of planarian regeneration. Annu Rev Cell Dev Biol 20:725–757. https://doi.org/10.1146/annurev.cellbio.20.010403.095114
3. Pellettieri J, Fitzgerald P, Watanabe S et al (2010) Cell death and tissue remodeling in planarian regeneration. Dev Biol 338:76–85. https://doi.org/10.1016/j.ydbio.2009.09.015
4. King RS, Newmark PA (2013) In situ hybridization protocol for enhanced detection of gene expression in the planarian Schmidtea mediterranea. BMC Dev Biol 13:8. https://doi.org/10.1186/1471-213X-13-8
5. Pearson BJ, Eisenhoffer GT, Gurley KA et al (2009) Formaldehyde-based whole-mount in situ hybridization method for planarians. Dev Dyn 238:443–450. https://doi.org/10.1002/dvdy.21849
6. Currie KW, Brown DDR, Zhu S et al (2016) HOX gene complement and expression in the planarian Schmidtea mediterranea. EvoDevo 7: 2663–2670. https://doi.org/10.1186/s13227-016-0044-8
7. Umesono Y, Watanabe K, Agata K (1997) A planarian orthopedia homolog is specifically expressed in the branch region of both the mature and regenerating brain. Develop Growth Differ 39:723–727
8. Ross KG, Omuro KC, Taylor MR et al (2015) Novel monoclonal antibodies to study tissue regeneration in planarians. BMC Dev Biol 15: 2. https://doi.org/10.1186/s12861-014-0050-9
9. Forsthoefel DJ, Waters F a, Newmark P a (2014) Generation of cell type-specific monoclonal antibodies for the planarian and optimization of sample processing for immunolabeling. BMC Dev Biol 14:1–22. https://doi.org/10.1186/s12861-014-0045-6
10. Cebrià F, Newmark PA (2005) Planarian homologs of netrin and netrin receptor are required for proper regeneration of the central nervous system and the maintenance of nervous system architecture. Development 132: 3691–3703. https://doi.org/10.1242/dev.01941

11. Cebrià F (2008) Organization of the nervous system in the model planarian Schmidtea mediterranea: an immunocytochemical study. Neurosci Res 61:375–384. https://doi.org/10.1016/j.neures.2008.04.005
12. Fraguas S, Barberán S, Ibarra B et al (2012) Regeneration of neuronal cell types in Schmidtea mediterranea: an immunohistochemical and expression study. Int J Dev Biol 56:143–153. https://doi.org/10.1387/ijdb.113428sf
13. Monjo F, Romero R (2015) Embryonic development of the nervous system in the planarian Schmidtea polychroa. Dev Biol 397:305–319. https://doi.org/10.1016/j.ydbio.2014.10.021
14. Sakai F, Agata K, Orii H, Watanabe K (2000) Organization and regeneration ability of spontaneous supernumerary eyes in planarians — eye regeneration field and pathway selection by optic nerves—. Zool Sci 17:375–381. https://doi.org/10.2108/jsz.17.375
15. Cebrià F, Vispo M, Newmark P et al (1997) Myocyte differentiation and body wall muscle regeneration in the planarian Girardia tigrina. Dev Genes Evol 207:306–316. https://doi.org/10.1007/s004270050118
16. Scimone ML, Cote LE, Reddien PW (2017) Orthogonal muscle fibres have different instructive roles in planarian regeneration. Nature 551:623–628. https://doi.org/10.1038/nature24660
17. Chai G, Ma C, Bao K et al (2010) Complete functional segregation of planarian β-catenin-1 and -2 in mediating Wnt signaling and cell adhesion. J Biol Chem 285:24120–24130. https://doi.org/10.1074/jbc.M110.113662
18. Guo T, Peters AHFM, Newmark PA (2006) A Bruno-like gene is required for stem cell maintenance in planarians. Dev Cell 11:159–169. https://doi.org/10.1016/j.devcel.2006.06.004
19. Zeng A, Li H, Guo L et al (2018) Prospectively isolated Tetraspanin + Neoblasts are adult pluripotent stem cells underlying planaria regeneration. Cell 173:1593–1608.e20. https://doi.org/10.1016/j.cell.2018.05.006
20. Yamamoto H, Agata K (2011) Optic chiasm formation in planarian I: cooperative netrin- and robo-mediated signals are required for the early stage of optic chiasm formation. Develop Growth Differ 53:300–311. https://doi.org/10.1111/j.1440-169X.2010.01234.x
21. Pascual-Carreras E, Marin-Barba M, Herrera-Úbeda C et al (2020) Planarian cell number depends on blitzschnell, a novel gene family that balances cell proliferation and cell death. Development 147. https://doi.org/10.1242/dev.184044
22. Zayas RM, Cebrià F, Guo T et al (2010) The use of lectins as markers for differentiated secretory cells in planarians. Dev Dyn 239:2888–2897. https://doi.org/10.1002/dvdy.22427
23. Seebeck F, März M, Meyer AW et al (2017) Integrins are required for tissue organization and restriction of neurogenesis in regenerating planarians. Development 144:795–807. https://doi.org/10.1242/dev.139774
24. de Sousa N, Rodríguez-Esteban G, Rojo-Laguna JI et al (2018) Hippo signaling controls cell cycle and restricts cell plasticity in planarians. PLoS Biol 16:1–28. https://doi.org/10.1371/journal.pbio.2002399
25. Scimone ML, Atabay KD, Fincher CT et al (2020) Muscle and neuronal guidepost-like cells facilitate planarian visual system regeneration. Science (80-) 368:eaba3203. https://doi.org/10.1126/science.aba3203

Chapter 5

Colorimetric Whole-Mount In Situ Hybridization in Planarians

Susanna Fraguas, Mª. Dolores Molina, and Francesc Cebrià

Abstract

Whole-mount in situ hybridization (WISH) is an extremely useful technique for visualizing specific mRNA targets and solving many biological questions. In planarians, this method is really valuable, for example, for determining gene expression profiles during whole-body regeneration and analyzing the effects of silencing any gene to determine their functions. In this chapter, we present in detail the WISH protocol routinely used in our lab, using a digoxigenin-labelled RNA probe and developing with NBT-BCIP. This protocol is basically that already described in Currie et al. (EvoDevo 7:7, 2016), which put together several modifications developed from several laboratories in recent years that improved the original protocol developed in the laboratory of Kiyokazu Agata in 1997. Although this protocol, or slight modifications of it, is the most common protocol in the planarian field for NBT-BCIP WISH, our results show that key steps such as the use and time of NAC treatment to remove the mucus need to be taken into account depending on the nature of the gene analyzed, especially for the epidermal markers.

Key words Whole-mount in situ hybridization, WISH, *Schmidtea mediterranea*, Planarian, N-acetyl-cysteine, Riboprobe, DIG-labelled probe, NBT-BCIP

1 Introduction

In situ hybridization on tissues and whole animals allows us to detect and localize gene expression in many model organisms [1–6], when an antisense RNA probe detects its complementary mRNA target. This powerful tool has been described in a number of protocols for a variety of animal models [7–10], which have been slightly modified and improved over time with the use of different fixatives, permeabilization reagents, and more sensitive detection tools.

Freshwater planarians are an important model in which to study stem cell-based regeneration [11]. Although a classical model for regeneration, the application of modern molecular tools is relatively

Luca Gentile (ed.), *Schmidtea Mediterranea: Methods and Protocols*, Methods in Molecular Biology, vol. 2680, https://doi.org/10.1007/978-1-0716-3275-8_5,
© The Author(s), under exclusive license to Springer Science+Business Media, LLC, part of Springer Nature 2023

recent [12]. At present, whole-mount in situ hybridizations and immunostainings, BrdU labelling, and RNAi are techniques routinely used in any laboratory [13–18]. In 1997 the laboratory of Kiyokazu Agata provided to the community the first consistent protocol for whole-mount in situ hybridization in these animals [18].

In planarians, the first step to get rid of the external mucus layer is critical for a correct fixation and proper penetration of the probe. In the original protocol this was achieved by treating the animals with 2% HCl for 5 min on ice (being this a critical step which also depended on the size of the animals), followed by a fixation with Carnoy's solution [18]. More recently, these initial steps have been modified and the classical hydrochloric acid treatment and alcohol-/acid-based fixation involving methanol were replaced by a N-acetyl-L-cysteine (NAC) treatment to remove the mucus followed by a fixation with formaldehyde [16]. Also, a reduction step can be added to facilitate probe penetration in the pre-pharyngeal region [16]. Finally, body pigmentation in planarians impedes successful gene expression detection by WISH. Freshwater planarians show different pigment types in their body and eyes (melanin, porphyrin, and ommochrome) [19], which need to be bleached in order to increase the sensitivity of the signal during the WISH development. In this sense, the original bleaching step in hydrogen peroxide in methanol overnight [18] has been optimized by a shorter treatment of hydrogen peroxide in formamide solution that not only eliminates pigmentation, but also increases the staining intensity and specificity [13].

These modifications were included in the protocol reported in [20], which constitutes the main base of the protocol described here. However, we have included some slight modifications that we carry out in our laboratory. Briefly, the whole protocol is 4 days long and the main steps each day are as follows: day 1, mucus removal, fixation, and permeabilization; day 2, bleaching, pre-hybridization, and riboprobe hybridization; day 3, washes and antibody incubation; and day 4, washes and enzymatic colorimetric development. In addition, we show that the final concentration of NAC is critical for a robust result and, in general, is not recommended for genes expressed in the epidermis. The integrity of the mature epidermis composed by a single-layered epithelial sheet [21] is really sensitive to NAC treatment and, accordingly, the classical fixation in HCl and Carnoy's would be better for epidermal genes. We show that variations in the final concentration of NAC is essential for a favorable and robust WISH result, above all to optimize the detection of epidermal markers.

2 Materials

All solutions are prepared using DEPC (diethylpyrocarbonate)-treated water and at room temperature, unless indicated otherwise. It follows a list of stock solutions and general buffers and material.

- DEPC (diethylpyrocarbonate)-treated water: Add 1:1000 volume of DEPC into ultrapure water in a hood with swirling, let it stay overnight, and autoclave.
- 10× PBS stock solution: 1.37 M NaCl, 27 mM KCl, 20 mM KH_2PO_4, 100 mM Na_2HPO_4 in DEPC-treated water and autoclaved, pH 7.4.
- 1× PBS stock: 10× PBS stock diluted to 1× with DEPC water.
- PBSTx: 0.5% (v/v) Triton X-100 in 1× PBS.
- 1× planarian artificial medium (PAM) stock solution: 1.6 mM NaCl, 1.0 mM $MgSO_4{\cdot}7H_2O$, 1.2 mM $NaHCO_3$, 0.1 mM KCl, 0.1 mM $MgCl_2{\cdot}6H_2O$, 1.0 mM $CaCl_2$ in ultrapure water [22].
- In situ baskets. Alternatively, homemade baskets can be prepared, using Eppendorf tubes and nylon mesh. Cut the bottom of the tube, melt the plastic at the bottom, and attach a piece of the nylon mesh.
- Tissue culture 24-well plates.

2.1 Day 1: Mucus Removal, Fixation, and Permeabilization

- 5%, 7,5%, or 10% (w/v) N-acetyl-L-cysteine dissolved in 1× PBS. To dissolve NAC completely in PBS, use a rocker or vortex but do not heat. Made fresh.
- Fixative solution: 4% formaldehyde from 37% stock in PBSTx. Made fresh.
- Reduction solution: 50 mM DTT (1,4-dithiothreitol), 1% NP-40, 0.5% SDS in 1× PBS.
- 50% methanol in PBSTx.

2.2 Day 2: Bleaching and Incubation with the Riboprobes

- Formamide bleaching solution: 5% non-deionized formamide, 0.5× SSC, 1.2% H_2O_2, in DEPC water. Always dilute these reagents into water before mixing to avoid a violent reaction. In any case the formamide should be directly added to concentrated H_2O_2. Made fresh.
- Pre-hybe solution: 50% deionized formamide, 5× SSC (from 20× SSC stock), 1× Denhardt's solution, 100 ug/mL heparin from porcine intestine, 1 mg/mL ribonucleic acid from torula yeast, 1% (v/v) Tween-20 (from a 10% stock). Bring to volume with DEPC water. Store at −20 °C.

- Hybe solution: Pre-hybe solution with 5% dextran sulfate. Heat at 60 °C and vortex the solution for several times until the dextran sulfate is completely dissolved. Store at −20 °C.

2.3 Day 3: Washing and Antibody Incubation

- 20× SSC: 3 M NaCl, 0.3 M sodium citrate, pH 7.0, DEPC-treated and autoclaved.
- 0.2×, 0.05×, and 2× SSCx: 20× SSC stock diluted to 0.2×, 0.05×, and 2× with DEPC water, 0.1% Triton X-100.
- MABT solution: 100 mM maleic acid, 150 mM NaCl, 0.1% Tween-20, in ultrapure water, pH to 7.5.
- Blocking solution: 5% horse serum and 0.5% Western blocking reagent from 10% stock, in ultrapure water.
- Antibody solution: anti-DIG-AP in blocking solution.

2.4 Day 4: Antibody Washes and Development

- AP buffer: 100 mM Tris–HCl pH 9.5, 100 mM NaCl, 50 mM $MgCl_2$, 0.1% Tween-20. Bring to volume with ultrapure water. Made fresh.
- 10% polyvinyl alcohol (PVA) solution: Dilute PVA in ultrapure water and stir on a hot plate at 80 °C until dissolved. Filter and store at 4 °C for several months.
- Developing buffer: AP buffer but in that case bring to volume with 10% PVA solution and add 20 μL/mL of NBT/ BCIP stock solution. Made fresh.
- 80% glycerol in PBS.

3 Methods

3.1 Day 1: Mucus Removal, Fixation, and Permeabilization

During the first day of the protocol, the animals are killed and their mucus is removed with N-acetyl-L-cysteine; then they are fixed with formaldehyde treatment, and the tissues are permeabilized with Triton X-100 and dehydrated in methanol for a long conservation, if needed. Finally, they are bleached before continuing with the WISH protocol.

The following treatments and washing steps are performed at room temperature and under gentle shaking unless otherwise indicated:

1. Transfer at least 7-day starved asexual planarians to a 1.5-mL microfuge tube (*see* **Note 1**).
2. Rinse the worms with fresh PAM water twice before fixation to remove dirty.
3. Remove planarian water and gently treat the animals with 5% NAC solution for 5 min (*see* **Notes 2** and **3**).

4. Incubate the animals in the fixative solution for 15 min.
5. Rinse the samples twice with PBSTx for 5 min.
6. Replace PBSTx with Reduction Solution and incubate 5 min at 37 °C without shaking (*see* **Note 4**).
7. Rinse samples with PBSTx for 5 min.
8. Incubate the worms in 50% methanol/PBSTx for 10 min.
9. Dehydrate the worms completely with 100% methanol for 10 min.
10. Replace with new 100% methanol and store the samples at − 20 °C for at least 1 h up to several months until use.

3.2 Day 2: Bleaching and Incubation with the Riboprobes

Planarians are rehydrated and bleached (in order to clear the pigment and improve the signal) and incubated with the specific riboprobe. All washing steps are performed at room temperature and under gentle shaking unless otherwise indicated.

1. Bring animals back to RT, replace 100% methanol with 50% methanol/PBSTx, and incubate the samples for 10 min.
2. Rinse with PBSTx for 5 min.
3. Incubate the planarians in bleaching solution for 2 h under direct light without shaking (*see* **Note 5**).
4. Bleached worms are rinsed 5 min twice with PBSTx.
5. Transfer the worms to baskets placed into a 24-well plate. Replace PBSTx with 50% pre-hybridization solution/PBSTx for 10 min (*see* **Note 6**).
6. Incubate worms in the pre-hybridization solution for 2 h rotating in a hybridizer oven at 56 °C (*see* **Note** 7).
7. Dilute the riboprobe in hybridization solution, heat at 80 °C for 5 min, and leave it at 56 °C until use (*see* **Note 8**).
8. Replace the pre-hybridization solution with riboprobe mix and incubate the samples at least 16 h at 56 °C in a hybridizer oven without rotation (*see* **Note** 7).

3.3 Day 3: Washing and Antibody Incubation

During the following steps, keep the animals at 56 °C under rotation in a hybridizer oven (*see* **Note 8**):

1. Remove the riboprobe mix and wash 5 min with 2× SSCTx at 56 °C.
2. Rinse with 2× SSCTx twice 30 min at 56 °C.
3. Rinse with 0.2× SSCTx twice 30 min at 56 °C.
4. Rinse with 0.05× SSCTx twice 30 min at 56 °C.

Get the animals back to room temperature

5. Wash with MABTw twice 10 min, with shaking.
6. Replace MABTw with blocking solution and incubate at least 2 h with shaking.
7. Incubate the samples with the antibody solution, under rotation overnight at room temperature (*see* **Note 9**).

3.4 Day 4: Antibody Washes and Development

All steps are carried out at RT and gentle shaking unless otherwise indicated.

1. Rinse the antibody solution with MABTw twice for 5 min.
2. Wash the animals with MABTw at least for 2 h. Perform at least 6–8 washes to wash out completely the unbound antibody and improve the signal/background staining.
3. Before proceeding to develop the probes, take the animals out of the baskets and transfer them in MABTw directly into the 24-well plate.
4. Wash the samples in AP buffer for 10 min (*see* **Note 10**).
5. Replace AP buffer with development buffer. From this step, animals are kept in the dark and without shaking (*see* **Note 10**).
6. Check the intensity of the signal every 20–30 min until optimal staining/background is obtained (*see* **Note 11**).
7. To stop the NBT/BCIP reaction, replace the development buffer with PBS and wash twice for 5 min (*see* **Note 12**).
8. Fix the animals with the fixative solution for 15 min.
9. Rinse with PBSTx twice for 5 min.
10. Wash with 100% ethanol for 15 min (*see* **Note 13**).
11. Rinse the samples with PBSTx for 5 min.
12. Store the animals with 80% glycerol solution at 4 °C until mounting.
13. Transfer the animals to a slide with 80% glycerol solution and cover them under a coverslip.
14. Keep the mounted slides at RT until the images are taken. Samples can be maintained for >6 months.

4 Notes

1. Use a 1,5-mL microfuge tube for processing up to 20 animals or a 15-mL Falcon tube for processing up to 50 worms. To prevent the animals from drying out, use 1 mL of solution volume when planarians are fixed in a microfuge tube and 10 mL of solution volume in a 15-mL tube.

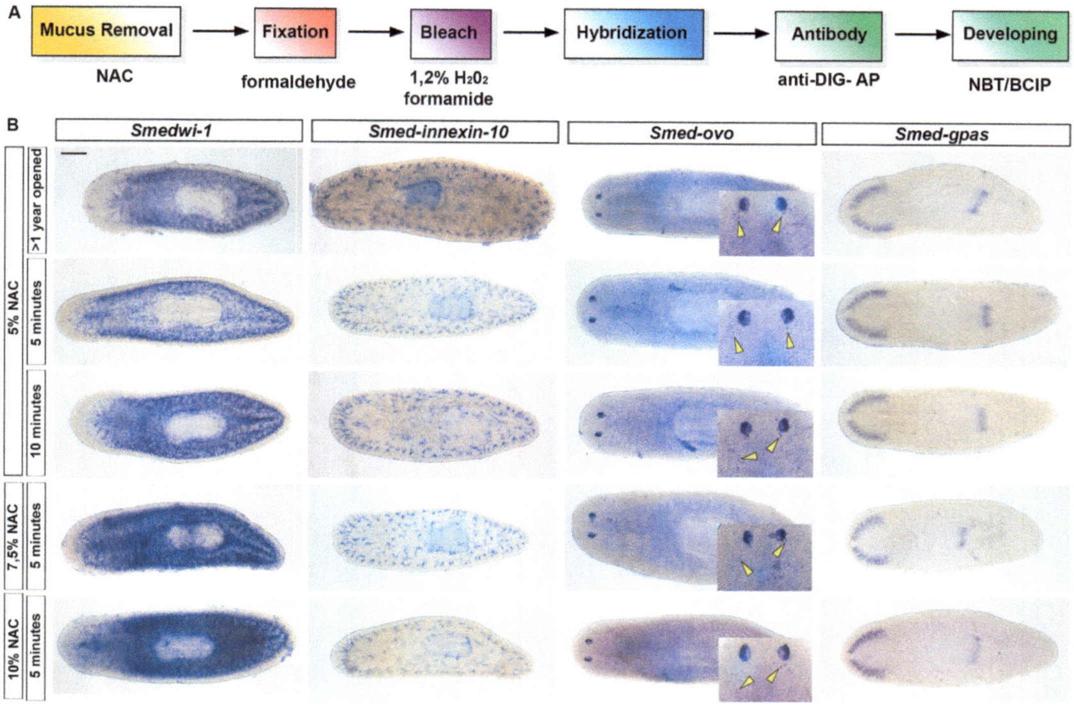

Fig. 1 Effects of different concentrations of NAC solution on the output of whole-mount in situ hybridization for non-epidermal markers. (**a**) Workflow of the main steps in the whole-mount in situ hybridization (WISH) method for planarians. (**b**B) NAC solution prepared with NAC powder older than 1 year can compromise the integrity of the epidermis even after a short 5-min treatment before fixation. When using the new NAC powder in general, it is better to use it at a concentration of 5% for 5–10 min. Even though for some genes as *Smed-gpas* and *Smed-ovo* no obvious differences are seen when using higher concentrations of NAC, for genes such as *Smedwi-1* and *Smed-innexin-10*, the quality of the signal decreases at higher concentrations of NAC. Yellow arrowheads point to *Smed-ovo+* eye progenitor cells. In all panels anterior to the left. Scale bar: 200 μm

2. Prepare a fresh NAC solution [15] for each experiment. NAC reagent appears to be stable for 1 year and its activity drops after that time (Fig. 1). The time of the NAC treatment seems to be crucial for the success of the protocol and depends on the size of the animals and expression pattern of the probes used (Figs. 1 and 2): for 1–4 mm planarians, more than 5 min may damage the epithelial integrity. Moreover, higher concentrations of NAC solution also affect the epidermis (Fig. 2). When epithelial probes are used, we recommend carefully testing different times and concentrations of NAC treatment or use HCl treatment and Carnoy's fixative solution instead (*see* Fig. 3 and **Note 3**).
3. Treating the animals with 2% HCl (from 32% stock in ultrapure water) for 5 min on ice (this being a critical step which also depended on the size of the animals) followed by a fixation with Carnoy's solution (ethanol/chloroform/acetic acid anhydride

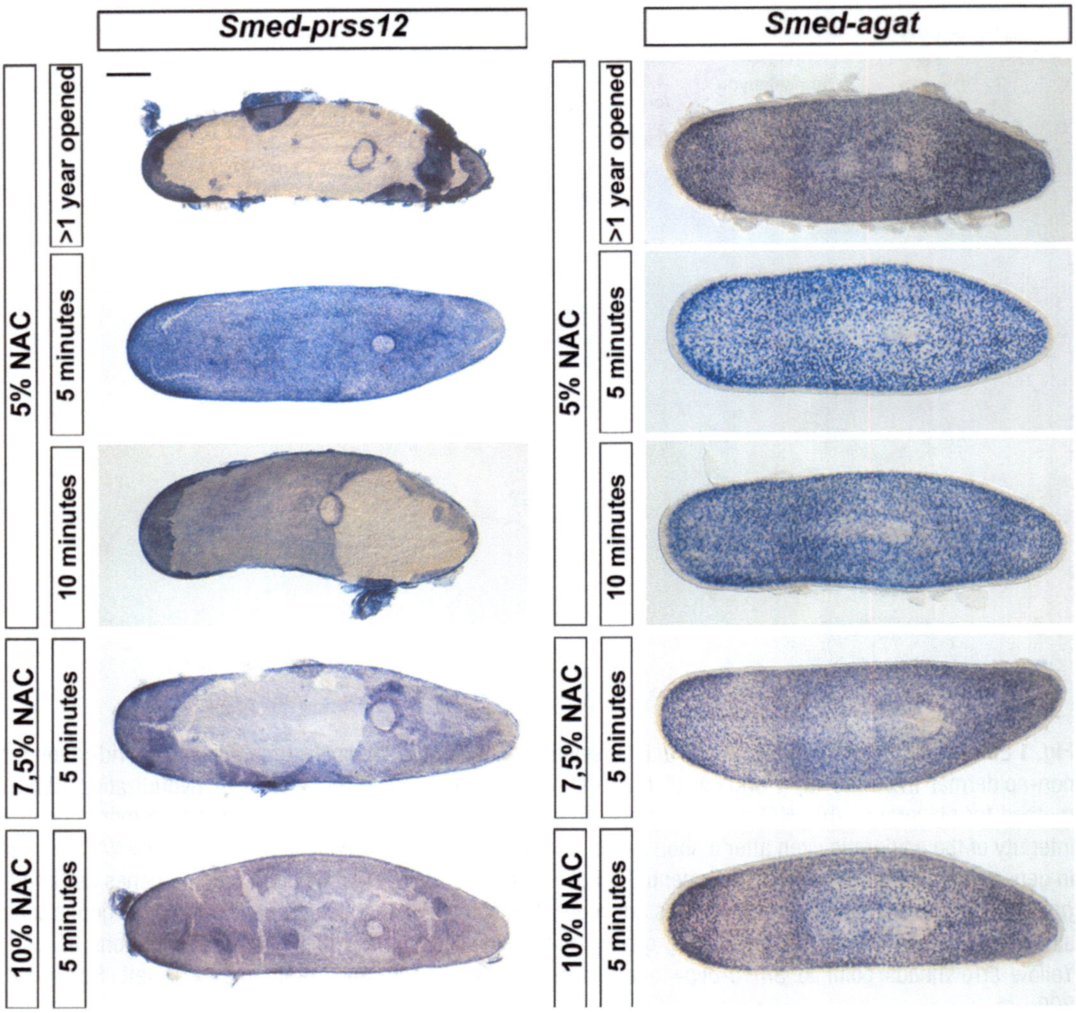

Fig. 2 Effects of different concentrations of NAC solution on the output of whole-mount in situ hybridization for epidermal markers. The integrity of the epidermis is severely affected when using either old NAC powder (for 5 min) or higher concentrations of new one, which compromises the in situ results for epidermal markers. As described for non-epidermal markers, in general 5–10 min of treatment in 5% NAC is better compared to 5 min in higher concentrations of NAC. Scale bar: 200 μm

in a proportion of 6:3:1; [17]) seems to be a good alternative to NAC treatment when riboprobes labelling the planarian epidermis are used (such as *Smed-prss12).* When treating the animals with Carnoy's, the short bleaching with the formamide solution is not recommended. In this case, bleaching of the animals should be performed with a 6% H_2O_2 in methanol solution overnight (Fig. 3).

4. Reduction solution is not always necessary but it helps to get a more clear and consistent signal in the pre-pharyngeal region

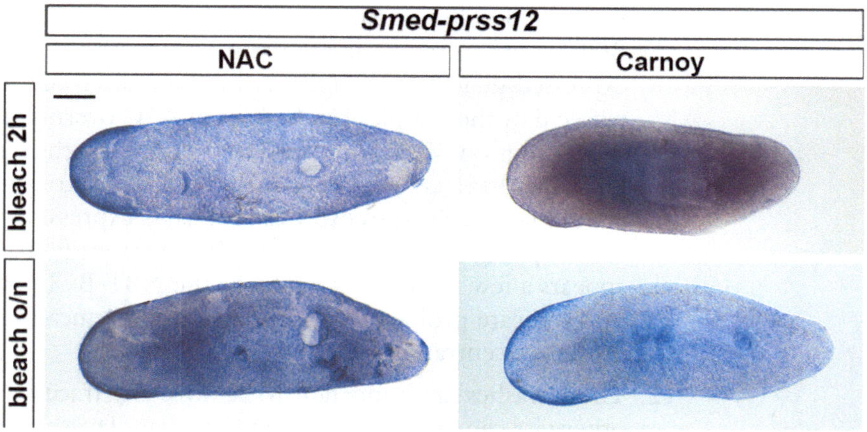

Fig. 3 Comparison between NAC-formaldehyde and HCl-Carnoy's treatment for epidermal markers. For epidermal markers such as *Smed-prss12* HCl and Carnoy's, fixation seems more appropriate as it keeps a better integrity of the epidermis. When using Carnoy's, a long bleaching time is also recommended. Scale bar: 200 μm

[15], which secretes more mucus [19], or in big animals, more resilient to permeabilization. A 5-min step in the reduction solution at 37 °C should be enough.

5. In our hands, larger animals may require longer bleachings. For proper bleaching, a polystyrene tray covered with aluminum foil is constructed, where the tubes are laid down under the light of a LED lamp. To avoid insufficient unpigmented animals, cover the tray with a cardboard box.
6. We use in situ baskets in a 24-well plate instead of conical tubes or glass vials in all hybridization and washing steps to avoid damaging the animals when washing. All solutions can be rapidly changed from the wells by vacuum aspiration. Use at least 300–400 μL of each wash solution to be sure that the bottom of the baskets is always immersed in the solution.
7. To avoid evaporation during the prehybridization, hybridization, and washing steps at 56 °C, we construct a humidity chamber adding a beaker full of water in the oven.
8. It is recommended to test different concentrations of the probes to determine the optimal concentration and obtain good signal/noise staining. In our hands, concentrations ranging from 0.05 ng/μL up to 0.5 ng/μL work well in planarians.
9. For digoxigenin-labelled probes, we use an anti-DIG-AP antibody diluted 1:4000 in blocking solution.
10. Prepare fresh for each experiment.
11. Developing time will vary depending on the level of gene expression. For some genes, a strong specific signal will appear in some minutes, whereas for others several hours are needed.

For long developing times, we recommend replacing the NBT-BCIP solution to accelerate the reaction. Also, for very lowly expressed genes, if the signal is still faint after developing the whole day, the samples can be kept at 4 °C overnight in NBT-BCIP solution. We recommend to overdevelop the samples the first time that a gene is being analyzed in order to have a better picture of all cell types expressing it as its expression level may vary depending on the cell type. If a nonspecific background appears a few minutes after adding the NBT-BCIP solution, it may indicate problems with the fixation, permeabilization, or probe concentration.

12. These washes are important to remove any trace of PVA that otherwise can precipitate during the ethanol washes and posterior storage at 4 °C in PBS.
13. To remove nonspecific staining and turn the NBT/BCIP signal blue, animals are washed with 100% ethanol usually for 15 min. However, the exact time may depend on the intensity of both the background and the signal. If the specific signal is low, too much time in ethanol may make this signal faint significantly.

Acknowledgements

We thank all members of the E. Saló and T. Adell laboratory for discussions and comments. Maria Dolores Molina has received funding from the postdoctoral fellowship program Beatriu de Pinós, funded by the Secretary of Universities and Research (Government of Catalonia) and by the European Union Horizon 2020 research and innovation program under Marie Sklodowska-Curie grant agreement no. 801370. Francesc Cebrià was supported by grants PGC2018-100747-B-100 from the Ministerio de Ciencia, Innovación y Universidades (Spain), and grant 2017 SGR 1455 from AGAUR, Generalitat de Catalunya.

References

1. Alonso C (1973) Improved conditions for *in situ* RNA-DNA hybridization. FEBS Lett 31(1):85–88
2. Harland RM (1991) *In situ* hybridization: an improved whole-mount method for Xenopus embryos. Methods Cell Biol 36:685–695
3. Hennig W (1973) Molecular hybridization of DNA and RNA *in situ*. Int Rev Cytol 36:1–44
4. Jin L, Lloyd RV (1997) In situ hybridization: methods and applications. J Clin Lab Anal 11(1):2–95. Panoskaltsis-Mortari A, Bucy RP (1995) *In situ* hybridization with digoxigenin-labeled RNA probes: facts and artifacts. BioTechniques 18(2):300–7
5. Tautz D, Pfeifle C (1989) A non-radioactive *in situ* hybridization method for the localization of specific RNAs in Drosophila embryos reveals translational control of the segmentation gene hunchback. Chromosoma 98(2):81–85
6. Broihier H (2012) Whole-mount fluorescence in situ hybridization and antibody staining of Drosophila embryos. Cold Spring Harb Protoc, p 8

7. Lécuyer E, Parthasarathy N, Krause HM (2008) Fluorescent *in situ* hybridization protocols in Drosophila embryos and tissues. Methods Mol Biol 420:289–302
8. Saint-Jeannet JP (2017) Whole-mount *in situ* hybridization of *Xenopus* embryos. Cold Spring Harb Protoc 2017(12):pdb. prot097287
9. Thisse C, Thisse B (2008) High-resolution *in situ* hybridization to whole-mount zebrafish embryos. Nat Protoc 3:59–69
10. Reddien PW (2018) The cellular and molecular basis for planarian regeneration. Cell 175(2): 327–345
11. Newmark PA, Sánchez Alvarado A (2002) Not your father's planarian: a classic model enters the era of functional genomics. Nat Rev Genet 3(3):210–219
12. King RS, Newmark PA (2013) *In situ* hybridization protocol for enhanced detection of gene expression in the planarian *Schmidtea mediterranea*. BMC Dev Biol 13:8
13. King RS, Newmark PA (2018) Whole-mount *in situ* hybridization of planarians. Methods Mol Biol 1774:379–392
14. Newmark PA, Sánchez Alvarado A (2000) Bromodeoxyuridine specifically labels the regenerative stem cells of planarians. Dev Biol 15;220 (2):142–153
15. Pearson BJ, Eisenhoffer GT, Gurley KA, Rink JC, Miller DE, Sánchez Alvarado A (2009) Formaldehyde-based whole-mount *in situ* hybridization method for planarians. Dev Dyn 238:443–450
16. Sánchez Alvarado A, Newmark PA (1999) Double-stranded RNA specifically disrupts gene expression during planarian regeneration. Proc Natl Acad Sci U S A 96:5049–5054
17. Umesono Y, Watanabe K, Agata K (1997) A planarian orthopedia homolog is specifically expressed in the branch region of both the mature and regenerating brain. Develop Growth Differ 39:723–727
18. Lindsay-Mosher N, Pearson BJ (2019) The true colours of the flatworm: mechanisms of pigment biosynthesis and pigment cell lineage development in planarians. Semin Cell Dev Biol 87:37–44
19. Currie KW, Brown DD, Zhu S, Xu C, Voisin V, Bader GD, Pearson BJ (2016) HOX gene complement and expression in the planarian Schmidtea mediterranea. EvoDevo 7:7
20. Tu KC, Cheng LC, Vu HTK, Lange JJ, McKinney SA, Seidel CW, Sánchez Alvarado A (2015) Egr-5 is a post-mitotic regulator of planarian epidermal differentiation. eLife 4:e10501
21. Cebrià F, Newmark PA (2005) Planarian homologs of netrin and netrin receptors are required for proper regeneration of the central nervous system and the maintenance of nervous system architecture. Development 132(16):3691–3703

Chapter 6

Single-Molecule Fluorescent In Situ Hybridization (smFISH) on Whole-Mount Planarians

Elke F. Roovers and Kerstin Bartscherer

Abstract

Since the establishment of planarian species as laboratory models, investigation of molecular pathways has relied heavily on visualization of transcripts using in situ hybridization (ISH). ISH has revealed various aspects ranging from anatomical details of different organs to distribution of planarian stem cell populations and signaling pathways involved in their unique regenerative response. High-throughput sequencing techniques including single-cell approaches have allowed us to investigate gene expression and cell lineages in more detail. One application that could provide important new insights into more subtle intercellular transcriptional differences and intracellular mRNA localization is single-molecule fluorescent in situ hybridization (smFISH). In addition to obtaining an overview of the expression pattern, this technique allows for single-molecule resolution and hence quantification of a transcript population. This is achieved by hybridization of individual oligonucleotides antisense to a transcript of interest, all carrying a single fluorescent label. This way, a signal is produced only when the combination of labelled oligonucleotides, targeting the same transcript, are hybridized, minimizing background and off-target effects. Moreover, it requires only a few steps compared to the conventional ISH protocol and thus saves time. Here we describe a protocol for the tissue preparation, probe synthesis, and smFISH, combined with immunohistochemistry, for whole-mount *Schmidtea mediterranea* samples.

Key words Double smFISH, Immunohistochemistry, *smedwi-1*, *smedwi-2*, Whole-mount *Schmidtea mediterranea*

1 Introduction

Over the last years, many high-throughput techniques have been applied to planarians, such as RNA sequencing, proteomics approaches, and single-cell RNA sequencing [1–3]. As a first step in analyzing a gene's biological relevance and function, validation of selected candidates extracted from large datasets is key. Robust molecular biology techniques, such as (fluorescent) in situ hybridization ([F]ISH), have been developed successfully for applications in planarian samples including *Schmidtea mediterranea* *(S. mediterranea)* [4, 5]. Because only few antibodies are available

Luca Gentile (ed.), *Schmidtea Mediterranea: Methods and Protocols*, Methods in Molecular Biology, vol. 2680,
https://doi.org/10.1007/978-1-0716-3275-8_6,
© The Author(s), under exclusive license to Springer Science+Business Media, LLC, part of Springer Nature 2023

for planarians, this technique has been indispensable to the field in order to analyze expression patterns of genes of interest using labelled RNA probes, which every lab can synthesize.

Since single-cell techniques have been applied to planarians, even more heterogeneity has been found within cell types. This way, different subclasses of neoblasts were characterized in detail for the first time [3]. By now, extensive single-cell sequencing datasets are publicly available, allowing us to follow subtle expression dynamics, for instance, throughout differentiation steps during tissue homeostasis, or following amputation events, including transcriptomes of stem cell subpopulations for virtually all cell types [6, 7]. In order to validate these types of datasets, single-molecule fluorescent in situ hybridization (smFISH) approaches have been established that demonstrate intercellular differences and subcellular localization patterns at the resolution of a single transcript [8, 9]. This enables the researcher to identify (rare) cell types, complementing single-cell sequencing data [10]. Furthermore, smFISH also provides quantitative information about transcript *density* at certain (subcellular) areas [11, 12].

The concept of smFISH is based on using arrays of oligonucleotide probes designed against a target transcript, rather than one, long riboprobe. Each smFISH probe carries a single fluorescent label, producing a linearly amplified signal, which will only be detectable when multiple probes hybridize. This is the case when hybridized to a *specific* target, whereas potential nonspecific binding of few probes will remain undetected. Compared to signals from the long riboprobes used in standard ISH procedures, smFISH signals provide higher specificity and only little background. On the other hand, the signal intensity of smFISH is lower, requiring imaging with more sensitive detection.

Probe sets for smFISH can be synthesized commercially, which comes with relatively high costs. However, a great effort has been published recently by Gaspar et al [13], in which a protocol for the production of labelled probe sets is described at significantly reduced costs and successfully applied this to visualize transcripts in *Drosophila* egg chambers. This makes this technique suitable for similar throughput levels as conventional (F)ISH techniques, accessible to all laboratories.

Here, we describe a protocol for double smFISH, using self-labelled probe sets based on the protocol by Gaspar et al [13], on whole-mount *S. mediterranea* samples. We will describe how to make custom probe sets against your gene of interest, how to perform the hybridization, and how to combine this with immunohistochemistry (IHC).

2 Materials

2.1 Sample Preparation

1. Small (up to 5 mM) planarians (here: asexual *Schmidtea mediterranea*).
2. 1× PBS, pH 7.4.
3. 1× PBSTx: dissolve 0.5% (v/v) Triton X-100 in 1× PBS, pH 7.4.
4. 5% NAC solution: dissolve 5% (w/v) N-acetyl-L-cysteine (NAC) powder in 1× PBS.
5. 4% formaldehyde solution: dissolve 1 mL 37% formaldehyde in 8 mL PBSTx.
6. 100% MeOH.
7. 50% MeOH: 50% (v/v) methanol in PBSTx.
8. Bleaching solution: 1.2% H_2O_2 and 5% formamide, in 0.5× SSC (from a stock solution of 20× SSC, pH 7.0).
9. Petri dish or 6-well plate.
10. Lamp with bright light (we use a desk lamp by Maul, 82138, with an 11-W G23 halogen bulb).

2.2 Probe Design and Labelling

1. DNA oligos against the target sequence at a 200 μM concentration in nuclease-free water (see details for their design in **step 1** of Subheading 3.2).
2. Amino-11-ddUTP (Lumiprobe, 15040/25040).
3. Fluorescent dyes coupled to N-Hydroxysuccinimide (NHS) (used here: ATTO 565-NHS, ATTO-TEC, AD 565-31, and ATTO-633-NHS, ATTO-TEC, AD 633-31).
4. DMSO (anhydrous).
5. 1 M NaHCO3 at pH 8.4.
6. Terminal deoxynucleotidyl transferase (TdT) (Thermo Fisher Scientific, EP0162).

2.3 Purification of the Labelled Probes

1. 100% ethanol.
2. 80% ethanol: 80% (v/v) ethanol in nuclease-free water.
3. 3 M sodium acetate, pH 5.2.
4. Linear acrylamide 5 mg/mL.
5. 40% 29:1 acrylamide/bis stock solution.
6. Urea.
7. 10× Tris/borate/EDTA (TBE) buffer: 890 mM Tris base, pH 8.3, 890 mM boric acid, 20 mM EDTA.
8. Ammonium persulfate (APS).

9. Tetramethylethylenediamine (TEMED).
10. 2× PAGE loading buffer: 8 M urea in 2× TBE, 20% glycerol 0.5% bromophenol blue (indicates ~8 nt), and 0.5% xylene cyanol (indicates ~28 nt).
11. SYBR™ Gold nucleic acid gel stain.
12. Amersham ImageQuant 800 Fluor imager.
13. 1 M magnesium acetate tetrahydrate solution.
14. 5 M ammonium acetate solution.
15. Nuclease-free water.
16. Tris/EDTA (TE) buffer: 10 mM Tris–HCl, 1 mM disodium EDTA at pH 8.0.
17. Spectrophotometer.

2.4 Smedwi-1 Immunohistochemistry and smedwi-1/smedwi-2 Double smFISH on Whole-Mount Planarians

1. Block buffer: 1% BSA, 2% donkey serum in 1× PBSTx.
2. Primary antibody, e.g., rabbit anti-Smedwi-1 [14].
3. Secondary antibody, e.g., donkey-Anti-Rabbit Alexa-488.
4. smFISH hybridization buffer: 10% dextran sulfate, 10% deionized formamide, 1 mg/mL tRNA, 2× SSC (from a stock solution of 20× SSC, pH 7.0), 0.02% BSA, 2 mM vanadyl-ribonucleoside complex, and 0.05% Triton X-100, dissolved in nuclease-free water (*see* **Note 1**).
5. Hoechst 33342.
6. Wash buffer: 10% formamide, 2x SSC (from a stock solution of 20× SSC, pH 7.0) and 0.05% Triton X-100, in nuclease-free water.
7. ProLong™ Gold Antifade Mountant.

3 Methods

3.1 Sample Preparation

1. Collect worms that were starved for 7 or more days up to a maximum size of ~5 mm in a 50 mL tube.
2. Treat the worms for 10 min in 5% NAC solution, rotating at RT.
3. Replace NAC with 4% formaldehyde solution and fix for 15 min at RT (*see* **Note 2**).
4. Remove the fixative and rinse the worms 2 × 5 min with PBSTx at RT.
5. Replace the PBSTx with 50% MeOH and incubate for 5 min at RT.
6. Next, bring the worms to 100% MeOH and incubate another 5 min at RT.

7. Put the samples to −20 °C and leave for at least 1 h (*see* **Note 3**).
8. After incubation at −20 °C, bring the samples back to PBSTx via 50% MeOH, at RT.
9. Once in PBSTx, replace with bleaching solution at RT.
10. Move the samples to a petri dish or a well plate and place under the lamp.
11. Leave the worms over a couple of hours (up to overnight) at RT, until they are bleached (monitor the bleaching progression under a microscope).
12. Bring back to PBSTx. The samples are now ready for hybridization (Subheading 3.5). If the samples are used at a later timepoint, they can be stored in 100% MeOH at −20 °C.

3.2 Probe Design and Labelling

Subheadings 3.2 and 3.3 have been described previously in Gaspar et al [13]. Here, we describe an adaptation of their protocol.

1. Probes can be designed using the online probe designer tool by Stellaris®, which will provide an array of DNA probes, antisense against the input sequence that are between 18 and 22 nucleotides in length, separated by at least two nucleotides, with an optimal GC content between 40% and 60%. Make a free account and paste the target sequence of interest in the probe designer (5′-3′ orientation; the resulting probes will be reverse-complemented to the input sequence). The standard probe number provided is 48, but we generally obtain good results with 30–35 probes (*see* **Note 4**). Ideally, use the coding sequence of your gene of interest (*see* **Note 5**).
2. Copy the desired probe sequences and order them at a custom DNA oligo production service. It is important to use the option to dissolve them in nuclease-free water at a concentration of 200 μM, in a 96- or 384-well plate (*see* **Note 6**).
3. Pool a similar volume of each oligo targeting the same gene in a tube (*see* **Note 7**). Use at **step 8** and store the remainder of the oligo pool at −20 °C for future labeling reactions. The plates containing the remaining, individual oligonucleotide stocks can be stored at −20 °C as well.
4. Next, the terminal NH_2-11-ddUTP has to be coupled to the fluorescent dye (coupled to the NHS-ester) of choice (*see* **Note 8**). Dissolve the dye-NHS-ester to 40 mM in DMSO and the amino-11-ddUTP to 20 mM in nuclease-free water.
5. Combine the 20 mM amino-11-ddUTP with 40 mM dye-NHS in equal volumes, so that the dye is present in a molar excess of twofold, buffered with 0.1 M $NaHCO_3$ at

pH 8.3. For instance, pipette 20 μL of 20 mM amino-11-ddUTP, 4 μL 1 M of $NaHCO_3$ pH 8.3, mix well, and then add 20 μL 40 mM of dye-NHS-ester (add the dye-NHS at the end).

6. Leave the coupling reaction for ~3 h in the dark at RT.
7. Adjust the final concentration of the stock to 5 mM labelled ddUTP with nuclease-free water. This stock can be stored at − 20 °C and can be used for multiple labelling reactions.
8. For the labelling reaction, pipette in a PCR strip: 4 μL of nuclease-free water, 4 μL of oligo mixture of Subheading 3.2, **step 3**, 3 μL of 5× TdT buffer, 1 μL of labelled ddUTP (*see* **Note 9**), and 3 μL of TdT enzyme (*see* **Note 10**).
9. Mix well and incubate in a PCR machine at 37 °C overnight for 18 h, and set the lid at 40 °C.

3.3 Purification of the Labelled Probes

1. The next day, clean up the probes through ethanol precipitation. To a 15 μL reaction, add 20 μL of 3 M sodium acetate, 0.5 μL of 5 mg/mL linear acrylamide, and 165 μL of nuclease-free water.
2. Transfer to an Eppendorf tube and add 800 μL of 100% ethanol; mix well by vortexing.
3. Incubate for 30 min at −80 °C.
4. Spin at 16,000 × g at 4 °C for 20 min and remove supernatant.
5. Wash the pellet with 1 mL of pre-cooled 80% ethanol (~4 °C), briefly vortex (such that the pellet is displaced), and spin again at 16,000 × g at 4 °C, for 20 min.
6. Remove supernatant, add 1 mL of cold 80% ethanol, and transfer the pellet plus the ethanol to a new Eppendorf tube.
7. Spin at 16,000 × g at 4 °C for 20 min and wash two more times.
8. After the last wash, remove the supernatant and air-dry the pellet.
9. Resuspend the pellet in 50 μL of TE buffer or nuclease-free water.
10. Measure the absorption of the unlabeled (input) probe set (with a known concentration of 200 μM) and the labelled purified probes on a nanodrop spectrometer (using UV-Vis spectrometry) at 260 nM (A_{260}) and at the absorption maximum of the dye (A_{dye}, for instance, 565 or 633 nm) divided by the molar extinction coefficient of the probe set (*see* **Note 11**). To calculate the recovery rate, the concentration times the resuspension volume can be divided by the initial amount of the oligo mix. The concentration (c) degree of labelling (DOL)

and recovery fraction can be calculated using the following formulae (or use the probe calculator provided by [13]; *see* **Note 12**):

$$c = \frac{A_{260} - cf_{260} \times A_{dye}}{\varepsilon_{oligo}}$$

$$DOL = \frac{A_{dye} \times c^{-1}}{\varepsilon_{oligo}}$$

$$\%\text{of recovery} = \frac{c \times V}{n_{initial}}$$

11. If the probes have a sufficient degree of labelling (~95%≤), continue with the smFISH procedure (Subheading 3.5). If the labelling is insufficient or the smFISH or the recovery is suboptimal, the probe sets can be visualized using denaturing polyacrylamide-urea gel electrophoresis (Subheading 3.4) in order to decide for optimization strategies for the labelling or do a gel extraction (*see* **Note 13**).

3.3.1 Probe Validation Using Denaturing PAGE

1. Assemble a glass sandwich minigel (spacer plate of 0.75 or 1 mm + short glass plate).
2. Prepare a 15% denaturing polyacrylamide gel in 8 M urea in 1× TBE. Dissolve the urea by incubating the mixture in a 60 °C water bath, while inverting regularly.
3. Polymerize the gel by adding 25 μL of TEMED and 250 μL of 10% APS to a 20 mL volume; invert carefully to avoid air bubbles.
4. Pour directly and insert the comb. Allow polymerization for 30 min.
5. Pre-run the minigel at 200 V for 30 min in 1× TBE at RT.
6. Mix the sample 1:1 with 2× PAGE loading buffer.
7. Heat up the sample mixed with the loading dye to 90 °C for 5 min and directly load the sample on the pre-run gel.
8. Run the minigel at 200 V for 1 h and 15 min at RT.
9. Stain the gel by incubating the gel in 2× SYBR Gold in 1× TBE for 15 min at RT.
10. Image with Amersham ImageQuant 800 Fluor imager (or any other imager that can excite and detect the required wavelengths of SYBR Gold and the fluorescent dyes used for the labelling).

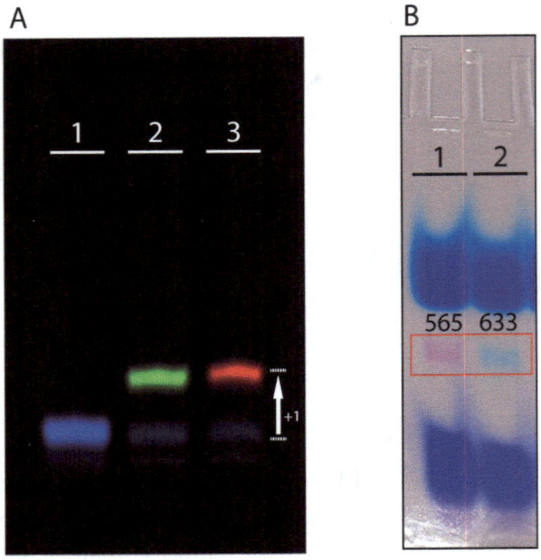

Fig. 1 Visualization of labelled probes on a 15% polyacrylamide/Urea gel. (**a**) unlabeled (lane 1), ATTO 565-labelled (green, lane 2), and ATTO 633-labelled (red, lane 3) probe sets. The labelled ATTO565 (green) and ATTO633 (red) probe sets are detected with a + 1-nucleotide shift. SYBR Gold visualizes the unlabeled fraction in lane 1 (blue). (**b**) One can detect the labelled fraction by eye as well in the case of both ATTO 565 (lane 1) and ATTO 633 (lane 2) as indicated with the red rectangle

11. Labelling plus a nucleotide shift, compared to the unlabeled oligo mixture in lane 1, can be detected on the gel (Fig. 1a) (*see* **Note 13**). Labeled probe sets can also be distinguished by eye (Fig. 1b).

3.4 Smedwi-1 Immunohistochemistry and smedwi-1/smedwi-2 Double smFISH on Whole-Mount Planarians

Below describes the procedure of smFISH combined with IHC, for which the primary antibody incubation is performed first in this case. One can also choose to perform smFISH without IHC, which involves only the overnight hybridization step and 2–3 washes the next day, followed by mounting (Subheading 3.5, **steps 5–8**).

1. Bring fixed and bleached whole-mount worms from Subheading 3, **step 1** from 100% MeOH to PBSTx, via 50% MeOH/PBSTx (incubate each **step 5** min at RT).
2. Block the worms for 1 h at RT in block buffer.
3. Replace block buffer with block buffer containing 1 in 500 Rb-anti-Smedwi-1 and incubate overnight at 4 °C.
4. The next day, wash the samples three times in 1× PBSTx for at least 30 min per wash at RT.
5. After the last wash, replace the PBSTx with hybridization buffer and incubate for 30 min at 30 °C in a heat-block or oven (*see* **Note 14**).

6. Replace the hybridization buffer with hybridization buffer containing 2 nM of the labelled probe set, 1 in 500 anti-Rb Alexa488 and 0.5 μg/mL Hoechst 33342, and mix gently (*see* **Note 15**). Incubate overnight at 30 °C (*see* **Note 16**).
7. The next day, wash three times for 30 min at 30 °C with wash buffer (*see* **Note 17**).
8. After the last wash, remove as much wash buffer as possible and replace with a few drops of ProLong Gold or another mounting medium. Cure the mounting medium for 24 h at RT (or when another medium is used, according to the manufacturer's instruction) and continue with imaging the sample.
9. Imaging can be performed using widefield, (spinning disk) confocal, and superresolution microscopy [8, 11, 12]. We performed imaging using a Zeiss LSM900, using a 40× water objective with an NA of 1.2 with a pixel size of 40 nM, in order to obtain a stitched, single-plane overview of stainings of Smedwi-1 protein combined with *smedwi-1* and *smedwi-2* transcripts, counterstained with Hoechst (Fig. 2a) (*see* **Notes 18–20**). Furthermore, we used Airyscan microscopy to obtain detailed 3D information (one plane depicted in Fig. 2b). We also compared our probe sets to commercially labeled probes for *smedwi-1* (Quasar 570) and *smedwi-2* (Quasar 670). We note that this generally provides a dimmer signal than we found for probes that were labelled with ATTO 565 and ATTO 633 (Fig. 2c).
10. In order to perform quantitative analysis, it is recommended to collect 3D data and analyze with available software such as the Airlocalize software [15], FISH-quant [16] (both written in MATLAB), or available ImageJ macros (such as described in [17]). Comprehensive step-by-step description of imaging and processing of your smFISH images using these softwares can be found in, for instance, (Trcek et al. 2017 and Wang 2019). Fig. 2d, d′ shows an example of *smedwi-1* ATTO 565 processed by the "smFISH-thresholdOptimization" ImageJ macro by [17].

4 Notes

1. Make aliquots of the deionized formamide and store at −20 ° C, in order to minimize freeze-thaw cycles. Prepare stocks of 20 mg/mL tRNA in 2x SSC and 1% BSA in 2x SSC. Aliquot the hybridization buffer according to the scale of your experiments, for instance, 1 or 10 mL aliquots, and store at −20 °C.
2. To prevent clumping during this step, do not wait for the worms to settle and quickly add the formaldehyde solution.

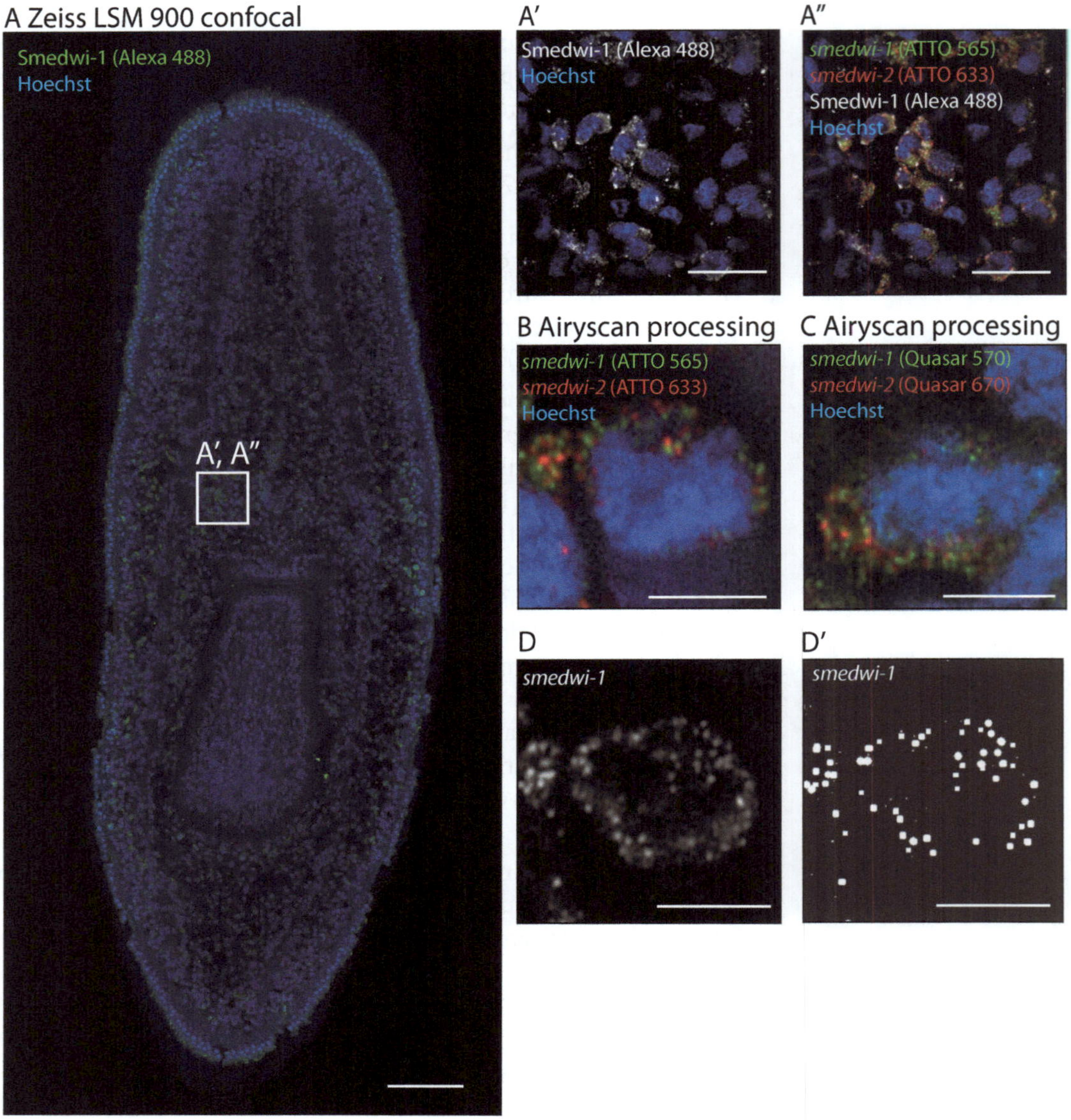

Fig. 2 Fluorescence imaging of whole-mount *S. mediterranea*. (**a**) Overview of whole-mount IHC against Smedwi-1 protein (white), with zoomed images that contain (**a**′) Smedwi-1 IHC only or (**a**″) Smedwi-1 IHC combined with double smFISH against *smedwi-1* (labelled with ATTO565, green) and *smedwi-2* (labelled with ATTO633, red), counterstained with Hoechst (blue). (**b**) Single plane of a Z-stack. Airyscan imaging of double smFISH labeling of *smedwi-1*-ATTO 565 and *smedwi-2*-ATTO 633. (**c**) Single plane of a Z-stack. Airyscan imaging of double smFISH labeling using commercially labeled *smedwi-1*-Quasar 570 and *smedwi-2*-Quasar 670. (**d**, **d**′) Example of spot detection (**d**′) of the *smedwi-1* channel (**d**) from Fig. 2b, using the "smFISH-thresholdOptimization" ImageJ macro published by [17]. Depicted circles indicate fluorescence maxima of the detected spot in Z. Other shapes indicate the maximum is in a different plane. Scalebar indicates 100 μm (**a**), 20 μm (**a**′, **a**″), and 5 μm (**b**, **c**, **d**, **d**′). Both **a**′ and **a**″ images were deconvolved using the Huygens deconvolution software

3. Worms can be stored in MeOH at −20 °C for longer periods until usage.
4. Overall, 35 probes per target provide a detectable, specific signal without introducing nonspecific background signals. The lower limit of the probe number in order to detect a transcript approximately lies between ~20 and 25 probes; even a signal with as few as 17 probes has been detected successfully [11]. Results can really differ per target and has to be tested per gene. Limiting the probe number below 25 is only recommended in case the target transcript is too short to cover with more oligos. More probes than 48 can be used as well, without noticeably increasing background signals, but this can also differ per target. Increasing the probe number could be of help in case a target is difficult to detect.
5. However, if the CDS is too short or has too few regions to obtain probes from with appropriate GC percentage to design a probe set, one can include UTR sequences as well. Since *S. mediterranea* has a relatively low GC content in coding regions (36.6%) [19], it could be that UTRs are required as well to obtain a sufficient number of probes.
6. Make sure to keep the spec sheet that contains the extinction coefficient of each oligo. This is required for measuring the degree of labelling (Subheading 3.3, **step 9**). Alternatively, extinction coefficients can also be calculated by the "Multiple Primer Analyzer" webtool by Thermo Scientific.
7. For instance, take 10 μL of each 200 μM oligo; this stock can be stored at −20 °C for future additional labelling experiments.
8. During this step, the dyes are coupled to the available amino group of the ddUTP. NHS esters are highly reactive at physiological conditions so it is critical to dissolve them in anhydrous DMSO and keep free from water during storage at −20 °C.
9. We tested this for the previously published ATTO 565-ddUTP and ATTO 633-ddUTP. We also tested this for ATTO 488 (ATTO TEC, AD 488-31), but in order to obtain a labeled probe set, purification as described in Subheading 3.3 was not sufficient and we required another gel extraction step as described in **Note 13**.
10. The standard amount of enzyme is not always sufficient. This depends on the probe set and the label [13] and also the enzyme batch. In our case, we obtained consistent results using 60 U of TdT (3 μL) in an overnight incubation reaction (18 h).
11. This can be found in the specification sheet of the oligos or can be calculated by the Thermo Scientific "Multiple Primer

Analyzer" webtool: average the extinction coefficient of all probes used in the probe set to obtain the approximate molar extinction coefficient of the pooled oligos in the set.

12. Gaspar et al., 2017, provide a very user-friendly "probe calculator" which provides extinction coefficients of the different dyes, the relative absorbance at 260 nm of the different dyes (cf_{260}), and calculates c, % loss, and DOL with your nanodrop values. Be sure to multiply your nanodrop values and extinction coefficient of your probe set by 10, since the probe calculator is adjusted for 10 mm path length measurements and not 1 mm that is used by the nanodrop.

13. If labelling is not complete and optimization strategies do not work, the pellet at Subheading 3.3, **step 8** can also be dissolved in 10 μL TE buffer and mixed with 10 μL 2× PAGE loading buffer and loaded entirely on a PAGE/urea gel. After running the gel, the labelled fraction will be clearly visible by eye (Fig. 1b) (while being separated from the unlabeled fraction because of the shift) and can be excised of the acrylamide gel, followed by DNA isolation of the gel as described in [20]. In short, collect the gel slab in an Eppendorf tube and add 300 μL of gel elution buffer (10 mM magnesium acetate tetrahydrate, 0.5 M ammonium acetate, 1 mM EDTA pH 8.0 in nuclease-free water). Incubate overnight at 37 °C. Centrifuge at 16,000 × g for 1 min at RT and collect the supernatant. (Avoid gel pieces to come along! For this reason, we also do not crush the gel slice but incubate the piece as a whole with gel elution buffer.) Add another 100 μL gel elution buffer to the gel piece, vortex and spin again at 16,000 × g for 1 min at RT, and pool the supernatants. Add 1 mL of 100% ethanol, mix, and incubate at −80 °C for 30 min. Centrifuge at 16,000 × g for 10 min at 4 °C and discard the supernatant. Take up the pellet in 200 μL TE buffer, add 25 μL 3 M sodium acetate pH 5.2, and precipitate with 1 mL of 100% ethanol. Wash the pellet with 80% ethanol, spin at 16,000 × g for 10 min at 4 °C, and discard the supernatant. Air-dry the pellet and take up in 20 μL TE buffer. Running the probe set on a PAGE gel will cause a loss of probe; therefore, it is recommended to use a smaller volume to reconstitute the pellet. The probe set can be used for the smFISH procedure as well.

14. The hybridization buffer is relatively viscous due to the dextran sulfate, so make sure the worms are mixed carefully well, without damaging them.

15. We typically use a 2 nM probe concentration; however, recommended working concentrations of probe sets vary and working concentrations are reported at, for instance, 12.5–25 nM [13], 1.33 ng/μL [18], and 1–2 nM [17]. Stellaris®

recommends a working concentration of 125 nM of their probe sets (consisting of 48 probes); for more information, see their website:

https://www.biosearchtech.com/support/resources/stellaris-protocols.

16. We have not experienced background issues at hybridization and wash temperatures at 30 °C. However, if this is the case for your gene of interest, hybridization temperature can be increased to 37 °C. This can be optimized per probe set, depending on GC content of the probes and their length, but also buffer adjustments.
17. Two washes are usually sufficient for the smFISH procedure alone; however, in order to get rid of remaining, unbound secondary antibody from whole-mount samples, more washes are typically required.
18. Alternatively, we have also successfully imaged sections using a Leica TCS Sp8 confocal microscope, using a 63× oil objective with an NA 1.4, using hybrid detectors (HyDs) for both smFISH probe sets (smFISH signals are much less bright than immunofluorescence signals and therefore require higher sensitivity detectors than PMT detectors). Images can be post-processed using deconvolution software, such as Huygens Essential or DeconvolutionLab [21].
19. Depending on the antibody, immunohistochemistry can also be performed simultaneously with the smFISH hybridization step. In this case, add the primary antibody during the overnight hybridization and the secondary antibody during the second wash step. Wash 3–4 more times in order to remove unbound secondary antibody and mount the sample. In the case of our Smedwi-1 staining, the primary antibody did not successfully stain Smedwi-1 during the overnight smFISH hybridization, which we therefore performed prior to the hybridization step. Once bound, it was not removed by the hybridization buffer and the secondary antibody was therefore added during the smFISH hybridization step as described.
20. Lastly, this technique can also be applied on sections.

Acknowledgments

We would like to thank Wessel van Leeuwen and Lucas Bruurs for help during troubleshooting the labelling procedure and sharing protocols. We would also like to thank members of the Bartscherer lab for helpful discussions.

References

1. Blythe MJ, Kao D, Malla S et al (2010) A dual platform approach to transcript discovery for the planarian schmidtea mediterranea to establish RNAseq for stem cell and regeneration biology. PLoS One. https://doi.org/10.1371/journal.pone.0015617
2. Böser A, Drexler HCA, Reuter H et al (2013) SILAC proteomics of planarians identifies Ncoa5 as a conserved component of pluripotent stem cells. Cell Rep. https://doi.org/10.1016/j.celrep.2013.10.035
3. Van Wolfswinkel JC, Wagner DE, Reddien PW (2014) Single-cell analysis reveals functionally distinct classes within the planarian stem cell compartment. Cell Stem Cell. https://doi.org/10.1016/j.stem.2014.06.007
4. King RS, Newmark PA (2018) Whole-mount in situ hybridization of planarians. In: Methods in molecular biology
5. Pearson BJ, Eisenhoffer GT, Gurley KA et al (2009) Formaldehyde-based whole-mount in situ hybridization method for planarians. Dev Dyn. https://doi.org/10.1002/dvdy.21849
6. Plass M, Solana J, Alexander Wolf F et al (2018) Cell type atlas and lineage tree of a whole complex animal by single-cell transcriptomics. Science 80. https://doi.org/10.1126/science.aaq1723
7. Fincher CT, Wurtzel O, de Hoog T et al (2018) Cell type transcriptome atlas for the planarian Schmidtea mediterranea. Science 80. https://doi.org/10.1126/science.aaq1736
8. Raj A, van den Bogaard P, Rifkin SA et al (2008) Imaging individual mRNA molecules using multiple singly labeled probes. Nat Methods. https://doi.org/10.1038/nmeth.1253
9. Femino AM, Fay FS, Fogarty K, Singer RH (1998) Visualization of single RNA transcripts in situ. Science 80. https://doi.org/10.1126/science.280.5363.585
10. Grün D, Lyubimova A, Kester L et al (2015) Single-cell messenger RNA sequencing reveals rare intestinal cell types. Nature 525:251–255. https://doi.org/10.1038/nature14966
11. Trcek T, Grosch M, York A et al (2015) Drosophila germ granules are structured and contain homotypic mRNA clusters. Nat Commun 6:7962. https://doi.org/10.1038/ncomms8962
12. Little SC, Sinsimer KS, Lee JJ et al (2015) Independent and coordinate trafficking of single drosophila germ plasm mRNAs. Nat Cell Biol 17:558–568. https://doi.org/10.1038/ncb3143
13. Gaspar I, Wippich F, Ephrussi A (2017) Enzymatic production of single-molecule FISH and RNA capture probes. RNA. https://doi.org/10.1261/rna.061184.117
14. März M, Seebeck F, Bartscherer K (2013) A pitx transcription factor controls the establishment and maintenance of the serotonergic lineage in planarians. Development. https://doi.org/10.1242/dev.100081
15. Lionnet T, Czaplinski K, Darzacq X et al (2011) A transgenic mouse for in vivo detection of endogenous labeled mRNA. Nat Methods. https://doi.org/10.1038/nmeth.1551
16. Mueller F, Senecal A, Tantale K et al (2013) FISH-quant: automatic counting of transcripts in 3D FISH images. Nat Methods
17. Wang S (2019) Single molecule RNA FISH (smFISH) in whole-mount mouse embryonic organs. Curr Protoc Cell Biol. https://doi.org/10.1002/cpcb.79
18. Trcek T, Lionnet T, Shroff H, Lehmann R (2017) mRNA quantification using single-molecule FISH in drosophila embryos. Nat Protoc. https://doi.org/10.1038/nprot.2017.030
19. Lamolle G, Fontenla S, Rijo G et al (2019) Compositional analysis of flatworm genomes shows strong codon usage biases across all classes. Front Genet. https://doi.org/10.3389/fgene.2019.00771
20. Green MR, Sambrook J (2019) Isolation of DNA fragments from polyacrylamide gels by the crush and soak method. Cold Spring Harb Protoc. https://doi.org/10.1101/pdb.prot100479
21. Sage D, Donati L, Soulez F, et al (2017) DeconvolutionLab2: An open-source software for deconvolution microscopy. Methods

Chapter 7

Whole-Mount In Situ Hybridization in Large Sexual *Schmidtea mediterranea*

Miquel Vila-Farré, Hanh Thi-Kim Vu, and Jochen C. Rink

Abstract

Whole-mount in situ hybridization (WISH), colorimetric or fluorescent (FISH), allows for the visualization of endogenous RNA. For planarians, robust WISH protocols exist for small-sized animals (>5 mm) of the model species *Schmidtea mediterranea* and *Dugesia japonica*. However, the sexual strain of *Schmidtea mediterranea* studied for germline development and function reaches much larger body sizes in excess of 2 cm. The existing whole-mount WISH protocols are not optimal for such large specimens, owing to insufficient tissue permeabilization. Here, we describe a robust WISH protocol for 12–16 mm long sexually mature *Schmidtea mediterranea* individuals that could serve as a starting point for adapting WISH to other large planarian species.

Key words Whole-mount in situ hybridization (WISH), Fluorescent in situ hybridization (FISH), Permeabilization, Size, Planarian, *Schmidtea mediterranea*, Sexual strain

1 Introduction

Whole-mount in situ hybridization (WISH) is a powerful approach for visualizing gene expression in intact tissue. The technique is based on hybridizing a single-stranded antisense RNA riboprobe with complementary endogenous RNA. Because of frequent use and a concerted community effort, the WISH protocols for *Schmidtea mediterranea* (*Smed* in the following) have continuously improved over the years [1–3]. However, they have been optimized mainly for small specimens of 3–5 mm in body length that are most commonly used for experiments [4, 5]. Moreover, smaller specimens are easier to work with, owing to easier permeabilization, pigment removal, and availability in the lab. However, the planarian model species can grow to significantly larger body sizes. For example, the sexual biotype of *Schmidtea mediterranea* (*Smes* in the following) can grow to a size >2.5 cm [5–7]. Instead of asexually reproducing by fissioning like the common asexual

Luca Gentile (ed.), *Schmidtea Mediterranea: Methods and Protocols*, Methods in Molecular Biology, vol. 2680,
https://doi.org/10.1007/978-1-0716-3275-8_7,
© The Author(s), under exclusive license to Springer Science+Business Media, LLC, part of Springer Nature 2023

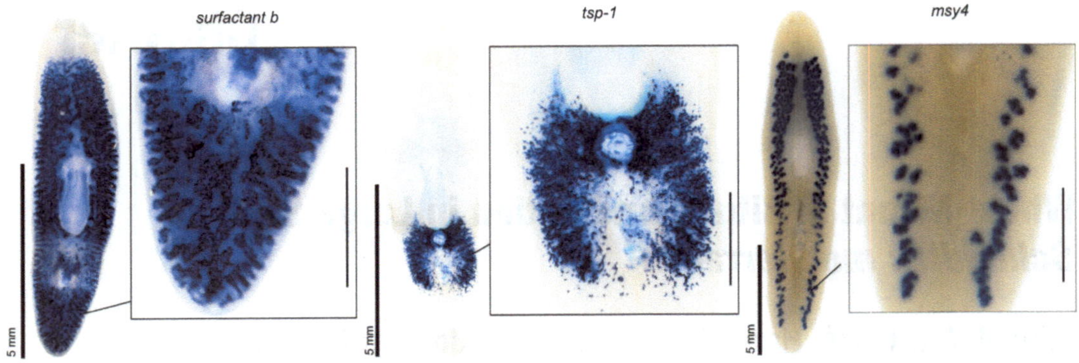

Fig. 1 Expression of marker genes for the reproductive system in large-sized *Smes* by WISH. Riboprobes for the genes *surfactant b*, *tsp-1*, and *msy4* labelling yolk glands, eggshell glands, and testes, respectively [12, 14, 15]. Inset: detail of selected posterior areas. Scale bar insets: 1 mm. Specimens were larger than 1.2 cm prior to fixation; riboprobe names are indicated on top of the inset

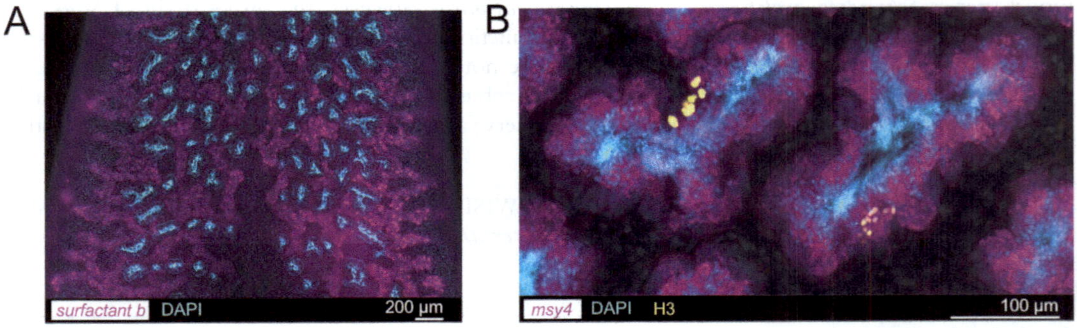

Fig. 2 Expression of marker genes for the reproductive system in large-sized *Smes* by FISH. (**a**) Yolk gland labelling with a *surfactant b* riboprobe (magenta) and nuclei stained with DAPI (cyan). The shown region is prepharyngeal. (**b**) Testis labelling with *msy4* riboprobe (magenta), phospho-H3(Ser10) (yellow), and nuclei stained with DAPI (cyan). Specimens were larger than 1.2 cm prior to fixation; riboprobe names are indicated at the bottom of the image

laboratory strains (*Smed*), *Smes* sexually reproduces by laying egg capsules. The ability of sexual *Smes*, and other planarians, to regenerate the germline de novo [8, 9] makes them a powerful model system to study germline development and evolution of reproductive strategies [10–13]. However, because *Smes* reaches sexual maturity only at around 10 mm in length, the existing WISH protocol optimized for smaller specimens is not ideal. Here, we describe a set of modifications and further optimizations of the *Smed* protocol [3] that collectively improve WISH and FISH in sexually mature *Smes* (Figs. 1 and 2). The sample preparation, which is more aggressive than for smaller animals, is followed by the hybridization steps that include long and stringent post-hybridization washes to reduce the background signal. Finally, detailed subsections are provided for colorimetric and fluorescent development. The result is a long but robust protocol that might offer a helpful starting point for adapting WISH and FISH to other large or hard-to-permeabilize planarian species.

2 Materials

Planarians: The inbred strain S2 of *Smes* was maintained in "Montjuïc planarian water" supplemented with gentamycin sulfate at 50 μg/ml, at 20 °C and in the dark. The animals were fed liver paste once a week and starved for 2 weeks before fixation (*see* **Note 1**). Selected animals were 12–16 mm long when selected for starvation.

Please refer to Chap. 8 for riboprobe synthesis. We routinely use autoclaved Milli Q water (here referred to as H_2O) for preparing any of the solutions and buffers listed below.

2.1 Sample Preparation

1. Chemical hood.
2. Water bath.
3. Lightbox.
4. "Taumel-rollenmischer" rolling nutator.
5. Tungsten wire or forceps.
6. Plastic transfer pipette with the tip cut to enlarge the tip opening's diameter.
7. 10 cm Petri dishes.
8. 50 ml conical tubes.
9. Racks for 50 ml conical tubes.
10. Aluminum foil.
11. 1× Montjuïc planarian water (PW): 1.6 mM NaCl, 1.0 mM $CaCl_2*2H_2O$, 1.0 mM $MgSO_4*7H_2O$, 0.1 mM $MgCl_2*6H_2O$, 0.1 mM KCl, 1.2 mM $NaHCO_3$ in H_2O. Adjust pH to ~7.5 with 1 M HCl.
12. 10× phosphate-buffered saline (10× PBS): 1.37 M NaCl, 27 mM KCl, 100 mM Na_2HPO_4, 18 mM KH_2HPO_4. Dissolve 80 g of NaCl, 2 g of KCl, 2.4 g of KH_2PO_4, 14.4 g of Na_2HPO_4 and fill to 1 l with H_2O. The pH of the 10× stock will be approximately 6.8.
13. 1× PBSTx0.5 (PBSTx0.5). Dissolve 50 ml of Triton X-100 in 1 l of 10× PBS. Dilute 100 ml of 10× PBSTx0.5 with 900 ml of H_2O for a final volume of 1 l of 1× PBSTx0.5. Triton X-100 dissolves faster in 10× PBS than in 1× PBS. After dilution, the 1× PBSTx0.5 pH should change to ~7.4. Adjust pH to 7.4 using HCl and sterile filter (0.22 μm) as necessary.

14. Neutralized N-acetyl-L-cysteine (NAC) solution (neutralized NAC solution): 0.5% NAC (w/v), pH 7.25. Dissolve 50 mg NAC in 8 ml of PW and add 200 μl 1 M HEPES-NaOH pH 7.25 and 5 μl phenol-red solution. Mix with the help of a stirring bar. Adjust pH to ~7 by adding 1 M NaOH drop by drop till the solution turns pink (yellow originally). Add PW to a final volume of 10 ml. Make fresh.
15. NAC solution: 7.5% NAC (w/v) in 1× PBST. Make fresh.
16. Fixative solution: 4% formaldehyde, PBSTx0.5. Dilute 25 ml of 16% EM-grade formaldehyde with 50 ml of 1× PBSTx0.5 and add H_2O to a final volume of 100 ml. Make fresh (*see* **Note 2**).
17. 1 M DTT: 1 M 1,4-dithiothreitol in H_2O. Dissolve 1.54 g of DTT in 8 ml of H_2O. Adjust the total volume to 10 ml. 0.22 μm filter. Prepare 1 ml aliquots and store them at −20 °C.
18. Reduction solution: 1% NP-40, 0.5% SDS, 50 mM DTT, 1× PBS. Make fresh and preheat to 37 °C in a water bath.
19. Formamide-bleaching solution: 5% nondeionized formamide, 1.2% hydrogen peroxide (H_2O_2), 0.5× SSC. Make fresh.
20. 100% methanol (100% MeOH).
21. 50% methanol (50% MeOH): 50% (v/v) methanol in PBSTx0.5.

2.2 In Situ Hybridization

1. Laboratory vacuum system connected with a plastic tube to a 1000 ul pipet tip, inserted into a 200 μl tip at its turn.
2. 4 °C room or incubator.
3. Hybridization oven.
4. Rocker.
5. Well plate stand to hold the 12 well plate.
6. Plastic transfer pipette.
7. 1000 μl pipette.
8. 1000 μl filter tips.
9. 12 well plates.
10. PCR sealing foil.
11. Proteinase K solution: 0.1% SDS, 2 μg/ml Proteinase K, 1× PBS. Make fresh.
12. 10% Tween-20: Dissolve 100 ml of 100% Tween-20 in a final volume of 1 l of H_2O (v/v).
13. 0.05×, 0.2×, and 2× SSC-Tw: 20× SSC stock diluted to 0.05×, 0.2×, and 2×, respectively, in H_2O. Add 10% Tween-20 to final concentration of 0.1% (v/v).
14. PreHyb: 50% (v/v) formamide, 1% (v/v) Tween-20, 5× SSC, 1× Denhardt's solution, 100 μg/ml heparin, 1 mg/ml yeast

RNA, 50 mM DTT. Use a whole fresh 500 ml bottle of formamide, 100 ml of 10% (v/v) Tween-20, 250 ml of 20× SSC (pH 6.0 or 7.0), 20 ml of 50× Denhardt's solution, 10 ml of 10 mg/ml heparin, 1 g of yeast RNA, 7.7 g of DTT, and add H_2O to a final volume of 1 l. Add DTT last as it will oxidize over time. Store at −20 °C.

15. Hyb: 50% (v/v) formamide, 1% (v/v) Tween-20, 5× SSC, 1× Denhardt's solution, 100 μg/ml heparin, 0.25 mg/ml yeast RNA, 50 mM DTT, 5% (w/v) dextran sulfate sodium salt from *Leuconostoc* spp. Use a whole fresh 500 ml bottle of formamide, 100 ml of 10% (v/v) Tween-20, 250 ml of 20× SSC (pH 6.0 or 7.0), 20 ml of 50× Denhardt's solution, 10 ml of 10 mg/ml heparin, 250 mg of yeast RNA, 7.7 g of DTT, 50 g of dextran sulfate, and add H_2O to a final volume of 1 l. Add DTT last, as it will oxidize over time. Make sure the dextran goes completely into the solution. Store aliquots of 50 ml at −20 ° C.
16. Riboprobe mix. Dilute the riboprobe at the desired concentration in Hyb (for riboprobe concentrations, *see* **Note 3**).
17. WashHyb: 50% (v/v) formamide, 0.5% (v/v) Tween-20, 5× SSC, 1× Denhardt's solution. Use a complete fresh 500 ml bottle of formamide, 50 ml of 10% (v/v) Tween-20, 250 ml of 20× SSC (pH 6.0 or 7.0), 20 ml of 50× Denhardt's solution, and add H_2O to a final volume of 1 l. Store at −20 °C.
18. 0.3× TNT: 100 mM Tris–HCl pH 7.5, 150 mM NaCl, 0.3% (v/v) Triton X-100. Dissolve 8.766 g of NaCl, 100 ml of 1 M Tris–HCl pH 7.5 and add 3 ml of Triton X-100 and add H_2O to a final volume of 1 l.

2.2.1 Colorimetric Development

1. Blocking solution (BS): 5% sterile horse serum, 0.5% Roche western blocking reagent (RWBR) in 0.3× TNT. Roche western blocking reagent is very viscous and cannot be filtered. Therefore, first, prepare the horse serum-0.3× TNT mixture and filter. Then, add Roche western blocking reagent.
2. Antibody solution: Anti-DIG-alkaline phosphatase (1:3000) in BS.
3. 10% polyvinyl alcohol (average molecular weight 30,000–70,000) (10% PVA) in H_2O (w/v).
4. Alkaline phosphatase buffer (AP buffer): 0.1 M Tris–HCl pH 9.5, 0.1 M NaCl, 0.05 M $MgCl_2$, 0.1% Tween-20 in H_2O. Make fresh.
5. Developmental buffer (DEV buffer): 0.1 M Tris–HCl pH 9.5, 0.1 M NaCl, 0.05 M $MgCl_2$, 0.1% Tween-20 in H_2O, with NBT/BCIP (5.3 μl BCIP and 2.7 μl NBT in 1 ml AP buffer). Make fresh.

6. Developmental buffer with PVA (DEV–PVA buffer): 0.1 M Tris–HCl pH 9.5, 0.1 M NaCl, 0.05 M $MgCl_2$, 0.1% Tween-20 in 7.8% PVA, with NBT/BCIP (5.3 μl BCIP and 2.7 μl NBT in 1ul AP buffer). Make fresh.
7. 100% ethanol (100% EtOH).
8. 50% ethanol (50% EtOH).
9. Scale A2: 2 M urea, 75% glycerol. Store at room temperature.

2.2.2 Fluorescent Development

1. Fluorescent blocking solution (FISH-BS): 5% sterile horse serum, 1% RWBR in 0.3× TNT. Mode of preparation as for BS adjusting RWBR concentration.
2. FISH antibody solution: antibody in FISH-BS. Antibody dilutions: anti-DIG-POD (1:2000), anti-DNP-HRP (1:200), anti-FITC-POD (1:2000).
3. TSA buffer: 2 M NaCl, 0.1 M boric acid, pH 8.5. Dissolve 1.24 g of boric acid and 23.38 g of NaCl in H_2O. Adjust pH to 8.5 using 1 M NaOH and add H_2O to a final volume of 200 ml. Store at 4 °C.
4. 1000× 4-iodophenylboronic acid stock (4-IPBA): 20 mg/ml 4-IPBA diluted in dimethylformamide.
5. Tyramide solution: fluorescent-conjugated tyramide (for tyramide concentrations, *see* **Note 4**), 0.006% H_2O_2 (1:5000 dilution from 30% stock solution), 20 μg/ml 4-IPBA (1:1000 dilution from 4-IPBA stock solution) in TSA buffer.
6. 1 M sodium azide. Dissolve 6.5 g sodium azide in 100 ml of H_2O. Store at 4 °C. Sodium azide is highly toxic. Prevent contact with skin.
7. 200 mM sodium azide. Dilute 20 ml of 1 M sodium azide in 0.3× TNT to a final volume of 100 ml.
8. 10% Triton X in H_2O.
9. Scale S4: 10% glycerol, 15%, dimethyl sulfoxide (DMSO), 40% sorbitol, 4 M urea, 0.1% Triton X-100, 2.5% DABCO. 50 ml of 100% glycerol, 75 ml of DMSO, 200 g of sorbitol, 120 g of urea, 5 ml of 10% Triton X-100, 12.5 g of 100% DABCO (antifade reagent), H_2O to a final volume of 500 ml. Store at 4 °C. Seal slides with nail polish to avoid evaporation.

3 Methods

All the steps are performed at room temperature unless otherwise stated. Every fixation step in the protocol is conducted under a chemical hood. If specimens stick together, use tungsten wire or fine forceps to gently pry them apart before proceeding.

3.1 Sample Preparation

1. Transfer 25 planarians to a fresh petri dish with the help of a pre-cut plastic transfer pipette and exchange the PW at least once to remove floating debris that can attach to specimens in the subsequent steps.
2. Transfer rinsed worms to a 50 ml conical tube (*see* **Note 5**).
3. Remove as much PW as possible, add 45 ml of neutralized NAC solution and gently rock the tube by hand for 10 min. This step removes mucus through NAC activity and liquid displacement (*see* **Note 6**).
4. Remove as much neutralized NAC solution as possible and add 45 ml of fresh NAC solution. Gently rock on a rolling nutator for 10 min at an appropriate speed, allowing specimen agitation (i.e., gentle movement of the worms without clumping at one end of the tube due to inadequate speed or breaking/shredding due to excessive speed). This step kills planarians and further removes mucus (*see* **Notes 7 and 8**).
5. Carefully remove as much NAC solution as possible without exposing the specimens to air, and add 25 ml of the fresh fixative solution to dilute out the residual NAC solution. Without agitating the fragile worms at the bottom of the tube, carefully rinse the sides and the lid with the NAC–fixative solution mixture (*see* **Note 9**).
6. Remove as much NAC–fixative solution as possible, add 40 ml of fresh fixative solution, gently agitate the tube by hand until specimens are well separated, and lay the tube on its side for 1 h. To prevent specimens from sticking to the tube wall during fixation, gently agitate every 15 min.
7. Wash in PBSTx0.5 two times for 5 min each time.
8. Replace PBSTx0.5 with 15 ml of reduction solution preheated to 37 °C and submerge the tube for 15 min in a water bath heated to 37 °C (*see* **Note 10**). Importantly, every 5 min, briefly and very gently shake the samples in the tube with a flick of the wrist while holding the tube by the cap to ensure sufficient exposure of the samples to the solution.
9. Remove reduction solution and wash in PBSTx0.5 twice for 5 min each time (*see* **Note 11**).
10. Remove PBSTx0.5 and rinse worms in 1× SSC for 10 min.
11. Remove 1× SSC, add 50 ml of formamide-bleaching solution, and expose samples to direct light (e.g., with a lightbox) for 2–2.5 h at room temperature to remove pigmentation. Refresh the formamide-bleaching solution every hour (*see* **Note 12**).
12. Remove the formamide-bleaching solution and rinse in 1× SSC for 10 min.

13. Remove 1× SCC and wash in PBSTx0.5 two times 5 min each time.
14. Remove PBSTx0.5 and rinse in 50% MeOH for 20 min.
15. Remove 50% MeOH and rinse in 100% MeOH for 5 min.
16. Replace the 100% MeOH with new 100% MeOH and store the samples at −20 °C at least overnight (*see* **Note 13**).

3.2 In Situ Hybridization

Samples are maintained in wells of 12 well plates with gentle rocking for the rest of the procedure unless otherwise stated. Each well comfortably accommodates 5–7 individuals in a minimum of 1.2 ml solution. To facilitate the many solution exchanges during the process, we recommend the use of a laboratory vacuum suction system. When aspirating, always leave ~15% of the current solution in the well, as aspirating more can cause the pipette tip to contact and damage specimens. Adding new solutions to the wells is done with a plastic transfer pipette, using a well plate stand to hold the 12 well plates.

1. Move the samples in 100% MeOH from −20 °C to room temperature for 5 min.
2. Transfer samples to wells in a 12 well plate. Ensure that they remain submerged in MeOH and do not dry out.
3. Replace 100% MeOH with 50% MeOH. Incubate for 10 min.
4. Replace 50% MeOH and wash in PBSTx0.5 two times 5 min each time.
5. Replace PBSTx0.5 with Proteinase K solution. Incubate for 25 min.
6. Replace Proteinase K solution with the fixative solution. Incubate for 10 min.
7. Remove fixative and wash in PBSTx0.5 two times 5 min each time.
8. Replace PBSTx0.5 and incubate in a solution 1:1 PBSTx0.5: PreHyb for 10 min.
9. Replace the mix with preheated PreHyb and incubate for 2 h at 56 °C in a hybridization oven (*see* **Note 14**).
10. Denature riboprobe in the riboprobe mix at 70 °C for 5 min and let it cool down before applying it to the samples.
11. Remove PreHyb and hybridize in riboprobe mix for ~16 h at 56 °C (*see* **Note 15**).
12. Remove as much riboprobe mix as possible and add WashHyb at 56 °C. Incubate for 15 min (*see* **Note 16**).
13. Remove the solution and incubate at 56 °C in WashHyb two times for 30 min each time.

14. Remove the WashHyb and incubate at 56 °C in 1:1 WashHyb:2× SSC-Tw twice for 30 min each time.
15. Remove the solution and incubate at 56 °C in 2× SSC-Tw two times for 30 min each time.
16. Remove the solution and incubate at 56 °C in 0.2× SSC-Tw two times for 30 min each time.
17. Remove the solution and incubate at 56 °C in 0.05× SSC-Tw two times for 30 min each time (*see* **Note 17**).
18. Leave the plate on the bench for 10 min to equilibrate the specimens to room temperature.
19. Remove the solution and wash in 0.3× TNT three times for 10 min each time.

For colorimetric development, continue to **step 1** in Subheading 3.2. For fluorescent development, skip **step 1** in Subheading 3.2 and proceed to **step 2** in Subheading 3.2.

3.2.1 Colorimetric Development

1. Replace 0.3× TNT with BS. Incubate for 1 h.
2. Replace BS with antibody solution and incubate a minimum of 12 h at 4 °C (*see* **Note 18**).
3. Remove antibody solution and wash in 0.3× TNT twice for 30 min each time, followed by three times for 1 h each time.
4. Incubate in AP buffer for 10 min.
5. Incubate in DEV buffer for 30 min. From now on, keep samples protected from light.
6. Incubate in DEV–PVA buffer until worms are finished developing (*see* **Note 19**).
7. Stop the reaction by removing as much DEV–PVA as possible and wash three times in PBSTx0.5 for 5 min each time.
8. Wash in PBSTx0.5 overnight at 4 °C. No need for rocking.
9. Postfix the sample in fixative solution for 45 min at room temperature.
10. Wash in PBSTx0.5 two times for 5 min each time.
11. Remove the PBSTx0.5 and clear in 100% EtOH for 30–40 min. Change the 100% EtOH if the solution turns bluish.
12. Wash in 50% EtOH in PBSTx0.5 for 5 min.
13. Wash in PBSTx0.5 two times for 5 min each time.
14. Transfer to Scale A2 and leave overnight at 4 °C. No need for rocking.
15. Mount (*see* **Note 20**).

3.2.2 Fluorescent Development

1. Replace 0.3× TNT with FISH-BS and incubate for 1 h.
2. Replace FISH BS with FISH antibody solution and incubate overnight at 4 °C (*see* **Note 21**).
3. Remove FISH antibody solution and wash in 0.3× TNT six or more times for 20 min each time.
4. Incubate in tyramide solution for 30–45 min.
5. Wash in 0.3× TNT six times for 10 min each time.
6. Kill peroxidase activity by incubating in 200 mM sodium azide for at least 1 h (*see* **Note 22**).
7. Wash in 0.3× TNT two times for 15–20 min each time.
8. Return to **step 2** for second riboprobe development if doing double or triple fluorescent in situ.
9. Transfer to Scale S4 and leave overnight at 4 °C. No need for rocking.
10. Mount (*see* **Note 23**).

4 Notes

1. In our hands, animals starved for only 1 week tend to show extensive nonspecific background in their intestines after colorimetric development. Therefore, we extended the starvation time to 2 weeks and thoroughly cleaned the boxes housing the animals at 1, 2, 7, and 12 days post-feeding to promote clearance of residual food from the intestine.
2. For the initial fixation of the animals, it is essential to use a freshly prepared fixative solution. Nevertheless, for subsequent fixation steps (after Proteinase K treatment and before transferring animals to scale), you can use a fixative solution prepared up to 1 month before and stored at 4 °C.
3. While riboprobe concentration should be empirically optimized for each target [3], we routinely achieve good results with a dilution factor of 1:3000. Alternatively, one can adjust riboprobe concentration to 0.05–0.1 ng/μl.
4. Tyramide dilutions: FAM–Tyr (1:2000), Cy3–Tyr (1:1000), Cy5–Tyr (1:1000), rhodamine–Tyr (1:2000).
5. The maximum number of worms we use per tube is 25–30. In our hands, higher numbers of individuals per tube tend to result in the occasional destruction of the animals when applying NAC solution.
6. It is essential to leave an air gap of at least 5 ml in the tube to achieve sufficient agitation for mucus removal.

7. We use a rolling nutator, but nutators with a different movement can be utilized after initial optimization.
8. This treatment results in most individuals being perfectly stretched. Even though the epidermis is frequently damaged, this seems to have little to no adverse effect on the in situ pattern for genes other than epidermal markers. However, further optimization is advised for worms close to 2 cm in length or bigger, as they frequently fix in a curved body posture rather than the desired stretched-out body shape.
9. The NAC solution is very acidic and can interfere with fixation if not sufficiently removed. Moreover, big planarians can leak some NAC solution during the fixation process. We found that exchanging fixative solutions once to dilute residual NAC helps maintain a suitable pH of ~6.5–7 throughout fixation.
10. We maintain the tubes submerged in a water bath in plastic racks. Keep the water level sufficiently low in order to prevent the rack from floating.
11. The reduction solution can turn slightly brownish due to the release of pigment from the samples, depending on the size and number of individuals in the tube.
12. We bleach the falcons lying horizontally on top of a lightbox that projects light from below, covered on top with aluminum foil. The sample temperature will increase during that process due to the heat transmitted by light, possibly affecting the bleaching or the characteristics of the sample. We have not personally experimented with this, but formamide-bleaching can likely be conducted right before the hybridization steps, as is regularly done in other WISH protocols. Note that as the 100% MeOH incubation at −20 °C will change the characteristics of the tissue, it may be necessary to adjust the bleaching conditions.
13. We recommend keeping samples at −20 °C in 100% MeOH for a maximum of 6 months. Nevertheless, we have successfully carried out in situ hybridizations for strongly expressed genes in wild-type animals that were stored for 2 years at −20 °C.
14. To avoid solution evaporation, we add 1 ml of H_2O to empty wells and 2 ml to the space between the wells, seal the plate with a PCR sealing foil, and finally wrap the plate in aluminum foil.
15. As the dextran sulfate in the hybridization buffer can form aggregates, we recommend equilibrating the hybridization buffer tube to room temperature over ~20 min and subsequent mixing to ensure that the dextran sulfate is in solution. To transfer the riboprobe mix to the wells, use 1000 ul filter pipettes.

16. At this point, the worms are floating in the solution. This step allows the worm to sink into the bottom of the well, making it easier to remove the solution without damaging the worms.
17. While longer times in 2× or 0.2x SSC-Tw do not harm the specimens, do not extend the time in 0.05× SSC since washes in low salt concentration can remove the hybridized probe and "erase" the signal.
18. We protect the plate from liquid evaporation as described in **Note 13**.
19. Developing times vary from riboprobe to riboprobe and should be empirically tested. We recommend overdeveloping a riboprobe the first time it is used to capture as much of its expression domain as possible. The development time can be shortened in subsequent experiments to maximize signal to noise for specific subdomains of the expression pattern.
20. To mount the planarians, we use double-sided adhesive tape. We first tape several layers together, the number of them depending on the specimens' size. We then punch a hole of the appropriate size and shape for the number and size of the specimens to be mounted (rectangular or circular) inside each piece. For this, we use punchers commercially available in hobby shops. Finally, the resulting piece is taped on a microscope glass slide, the specimens deposited in the hole with the mounting media and everything covered with a coverslip.
21. We protect the plate from liquid evaporation as described in **Note 14**.
22. This step is only required for double and triple fluorescent in situ.
23. We mount the planarians as described in **Note 20**.

Acknowledgements

We thank Dr. James Cleland for his scientific input and proofreading of the manuscript and Dr. Tobias Boothe for his support in imaging.

References

1. Ross KG, Omuro KC, Taylor MR, Munday RK, Hubert A, King RS, Zayas RM (2015) Novel monoclonal antibodies to study tissue regeneration in planarians. BMC Dev Biol 15: 2. https://doi.org/10.1186/s12861-014-0050-9
2. King RS, Newmark PA (2013) In situ hybridization protocol for enhanced detection of gene expression in the planarian *Schmidtea mediterranea*. BMC Dev Biol 13(8). https://doi.org/10.1186/1471-213X-13-8
3. King RS, Newmark PA (2018) Whole-mount in situ hybridization of planarians. Methods Mol Biol 1774:379–392. https://doi.org/10.1007/978-1-4939-7802-1_12

4. Benazzi M, Baguñà J, Ballester R (1970) First report on an asexual form of the planarian *Dugesia lugubris* s.l. Rend Arc Naz Lincei 48(2):282–284
5. Benazzi M, Baguñà J, Ballester R, Puccinelli I, Papa RD (1975) Further contribution to the taxonomy of the «*Dugesia lugubris-polychroa* group» with description of *Dugesia mediterranea* N.SP. (Tricladida, Paludicola). Boll Zool 42(1):81–89. https://doi.org/10.1080/11250007509430132
6. Newmark PA, Sánchez Alvarado A (2002) Not your father's planarian: a classic model enters the era of functional genomics. Nat Rev Genet 3(3):210–219. https://doi.org/10.1038/nrg759
7. Lázaro EM, Harrath AH, Stocchino GA, Pala M, Baguñà J, Riutort M (2011) *Schmidtea mediterranea* phylogeography: an old species surviving on a few Mediterranean islands? BMC Evol Biol 11:274. https://doi.org/10.1186/1471-2148-11-274
8. Morgan TH (1901) Growth and regeneration in *Planaria lugubris*. Arch Entwmech 13:179–212. https://doi.org/10.1007/BF02161982
9. Wang Y, Zayas RM, Guo T, Newmark PA (2007) *Nanos* function is essential for development and regeneration of planarian germ cells. PNAS 104(14):5901–5906. https://doi.org/10.1073/pnas.0609708104
10. Iyer H, Collins JJ 3rd, Newmark PA (2016) NF-YB regulates spermatogonial stem cell self-renewal and proliferation in the planarian *Schmidtea mediterranea*. PLoS Genet 12(6): e1006109. https://doi.org/10.1371/journal.pgen.1006109
11. Issigonis M, Newmark PA (2019) From worm to germ: germ cell development and regeneration in planarians. Curr Top Dev Biol 135: 127–153. https://doi.org/10.1016/bs.ctdb.2019.04.001
12. Rouhana L, Tasaki J, Saberi A, Newmark PA (2017) Genetic dissection of the planarian reproductive system through characterization of *Schmidtea mediterranea* CPEB homologs. Dev Biol 426(1):43–55. https://doi.org/10.1016/j.ydbio.2017.04.008
13. Issigonis M, Redkar AB, Rozario T, Khan UW, Mejia-Sanchez R, Lapan SW, Reddien PW, Newmark PA (2022) A Kruppel-like factor is required for development and regeneration of germline and yolk cells from somatic stem cells in planarians. PLoS Biol 20(7):e3001472. https://doi.org/10.1371/journal.pbio.3001472
14. Chong T, Stary JM, Wang Y, Newmark PA (2011) Molecular markers to characterize the hermaphroditic reproductive system of the planarian Schmidtea mediterranea. BMC Dev Biol 11:69. https://doi.org/10.1186/1471-213X-11-69
15. Steiner JK, Tasaki J, Rouhana L (2016) Germline defects caused by Smed-boule RNA-interference reveal that egg capsule deposition occurs independently of fertilization, ovulation, mating, or the presence of gametes in planarian flatworms. PLoS Genet 12(5): e1006030. https://doi.org/10.1371/journal.pgen.1006030

Chapter 8

Preparing Planarian Cells for High-Content Fluorescence Microscopy Using RNA in Situ Hybridization and Immunocytochemistry

Markus A. Grohme, Olga Frank, and Jochen C. Rink

Abstract

High-content fluorescence microscopy combines the efficiency of high-throughput techniques with the ability to extract quantitative information from biological systems. Here we describe a modular collection of assays adapted for fixed planarian cells that enable multiplexed measurements of biomarkers in microwell plates. These include protocols for RNA fluorescent in situ hybridization (RNA FISH) as well as immunocytochemical protocols for quantifying proliferating cells targeting phosphorylated histone H3 as well as 5-bromo-2′-deoxyuridine (BrdU) incorporated into the nuclear DNA. The assays are compatible with planarians of virtually any size, as the tissue is disaggregated into a single-cell suspension before fixation and staining. By sharing many reagents with established planarian whole-mount staining protocols, preparation of samples for high-content microscopy adoption requires little additional investment.

Key words Planarian, Maceration, High-content fluorescence microscopy, Formaldehyde fixation, RNA FISH, Immunocytochemistry, BrdU, Phospho-histone 3, Tyramide signal amplification

1 Introduction

Major technical advances in automated microscopy and image analysis have enabled screens for a wide range of visual phenotypes in cells and organisms using small-molecule or RNAi-based perturbation approaches [1]. High-content cell-based assays allow the parallel monitoring of multiple cellular phenotypes using dyes, antibodies, or RNA/DNA probes at the (sub)cellular level. Thus, they provide insight into the complexity of biological processes both in individual cells and on a population scale encompassing thousands of cells [2].

The planarian community has developed a wide range of protocols and assays for planarian whole mounts. Colorimetric and fluorescent whole-mount in situ hybridization are popular techniques for localizing RNA expression patterns within planarian tissue.

Luca Gentile (ed.), *Schmidtea Mediterranea: Methods and Protocols*, Methods in Molecular Biology, vol. 2680,
https://doi.org/10.1007/978-1-0716-3275-8_8,
© The Author(s), under exclusive license to Springer Science+Business Media, LLC, part of Springer Nature 2023

A general drawback of staining macroscopic organisms is the limited diffusion of reagents into the tissue, leading to experimental variability and misleading staining patterns. To overcome these limitations, staining can be performed on tissue sections [3, 4], although the cutting plane of a given cell might vary. Therefore, single-cell assays allow the unequivocal assignment of staining patterns to distinct cells at (sub)cellular resolution. In planarians, RNA fluorescence in situ hybridization (RNA FISH) or immunocytochemistry (ICC) on single cells have been described previously [5, 6]. The available protocols involve isolation of cells by fluorescence-activated cell sorting and/or fixation on coverslips, which is not practicable for high-throughput analyses. In an attempt to streamline the staining and imaging of single planarian cells, we have combined chemical dissociation of planarian tissue with subsequent formaldehyde fixation in multi-well plates. Cell suspensions are nonenzymatically generated using a "maceration solution" while retaining the composition of the original tissue [5, 7]. In its simplest form, the maceration solution contains a mixture of acetic acid, glycerol, and water, although several modified compositions exist [8, 9]. Its utility has been recently rediscovered for single-cell RNA sequencing applications [10]. All subsequent stainings are thereby rendered agnostic to the dimensions of the original tissue, which abolishes many problems associated with riboprobe or antibody penetration. An early version of this method has been used to characterize components of the mitotic spindle checkpoint in the planarian *Schmidtea mediterranea* [11]. Building on the excellent protocols developed for RNA FISH in planarians and other organisms [12–15], we have improved many aspects of sample preparation, storage, and staining conditions for multi-well plates. Using the popular peroxidase-based tyramide signal amplification, transcripts over a wide range of expression levels can be detected, generally on par with established whole-mount protocols. The full procedure takes 2–6 days depending on the degree of target multiplexing and chosen incubation times. A typical result for a dual RNA FISH staining combined with phospho-histone 3 ICC staining is presented in Fig. 1. While this method has been established using the planarian *S. mediterranea*, it should be transferable to other planarian species or soft organisms that can be easily disaggregated, including Hydra, for which the maceration solution was initially described [7]. We envision this method to be widely applicable for quantifying cell types and other cellular parameters. The flexibility and modularity of the plate format is ideal for high-throughput screens, including the quantification of proliferation markers, counting cell numbers, screening hybridoma supernatants, or the optimization of antigen retrieval or blocking conditions. Potentially, it could serve as a complementary assay for the validation of single-cell RNA sequencing experiments.

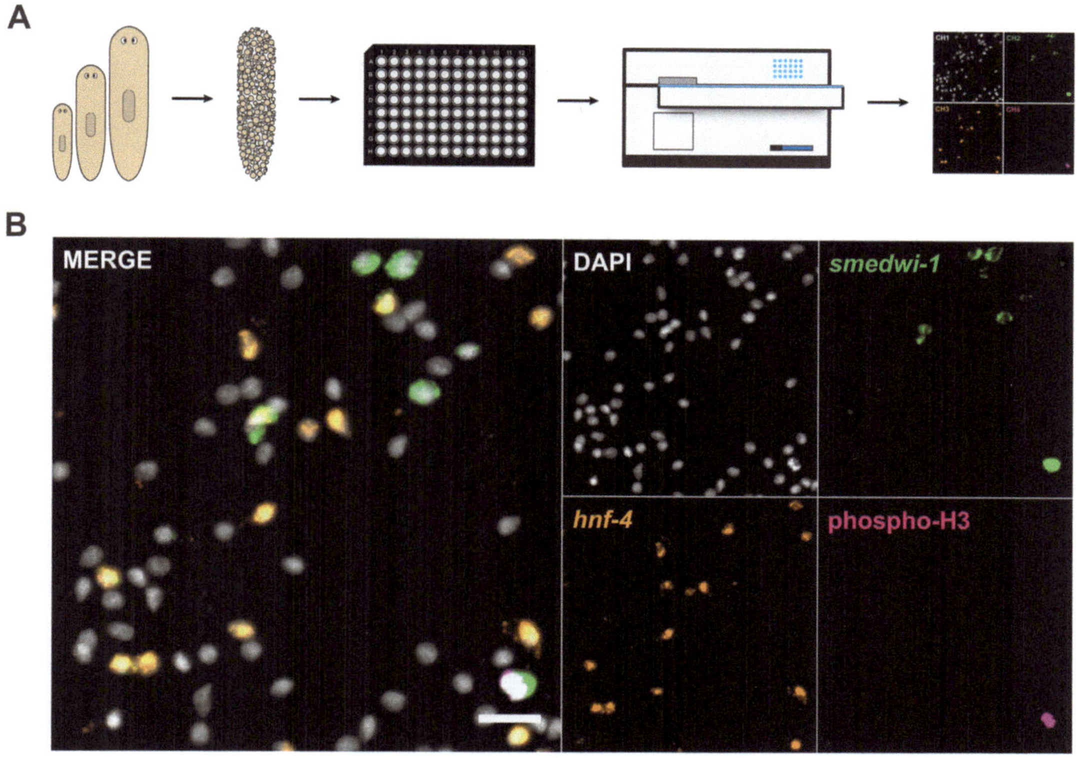

Fig. 1 Preparing planarian cells for high-content fluorescence microscopy. (**a**) General workflow for preparing planarians for high-content imaging. Planarians of arbitrary size and experimental background are dissociated and transferred to microwell plates, fixed and stained using RNA FISH or ICC. The plates are compatible with high-content imaging systems yielding multicolor fluorescence images. (**b**) Multicolor epifluorescence images of macerated planarians with two-color FISH combined with ICC for phospho-H3 (S10/T11). DIG-labeled riboprobe against Smed *hnf-4* and fluorescein-labeled riboprobe against *smedwi-1* were developed using AF568–tyramide (orange) and fluorescein–tyramide (green), respectively. Cells in M-phase were detected using rabbit anti-phospho-H3 (S10/T11) and goat anti-rabbit Alexa 647-labeled secondary antibody (magenta). Nuclei were stained using DAPI (gray). Scale bar: 20 μM

1.1 Experimental Considerations and Experimental Controls

The various assays in this chapter are organized into modules and can be combined into single, dual, or triple RNA FISH or a combination of FISH and immunocytochemistry (ICC) or ICC alone, as summarized in Fig. 2. In combined FISH and ICC experiments, the riboprobes should be developed before staining for other epitopes, due to the comparatively lower stability of RNA transcripts. A successful staining experiment is cumulatively affected by the purity of riboprobes, proper sample blocking, and specificity of antibodies. The starting point for riboprobe synthesis is the PCR products generated directly from gel purified amplicons or from cloned cDNA in a plasmid using a range of available haptens. This method chapter also includes protocols on how to store cell suspensions or cells fixed on plates for later analysis. Plates can be stored in an antifade medium and reimaged at later time-

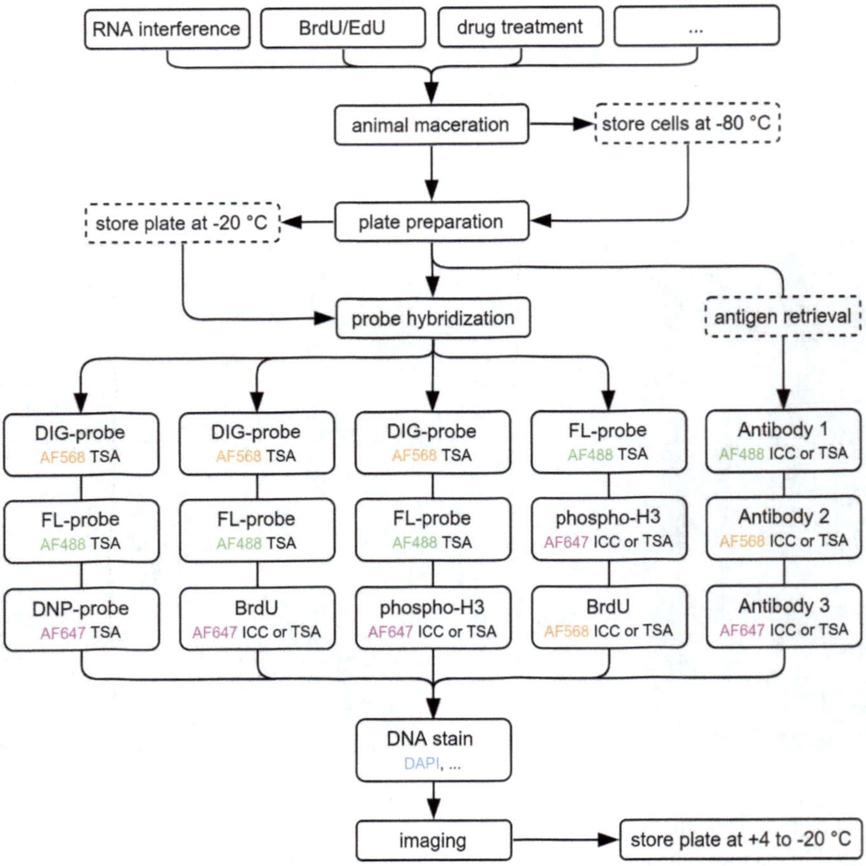

Fig. 2 Flowchart of potential applications and orders of staining for RNA FISH and ICC on planarian cells. Input samples may originate from a wide range of experiments as long as they do not disrupt cellular integrity. Two major extended pause points are storage of dissociated animals as cell suspension in glycerol at −80 °C, or fixed and permeabilized cells in formamide-containing pre-hybridization buffer in multi-well plates at −20 °C. Examples of typical experimental workflows of FISH and FISH+ICC combinations are presented. Optional steps are shown with a dashed outline. As a final step, nuclei are stained using a DNA-specific dye to quantify total cell numbers. Cells can be imaged or stored in an antifade medium. The figure was generated using draw.io-desktop (v14.6.13)

points with no apparent loss of fluorescent signal. Supplementary protocols for the synthesis of hapten-conjugated riboprobes as well as for fluorophore-conjugated tyramides are also included. Image acquisition, feature extraction, and data analysis strategies have been described elsewhere in detail [16–18].

Good controls are important for proper interpretation of any experiment. The simplest and most established approach in planarians is the depletion of transcripts or cell types using RNA interference (RNAi) [19–22]. RNAi-treated samples can serve as negative biological controls in conjunction with sense riboprobes to control for signal specificity (see Subheading 3.1 *Riboprobe preparation*). For establishing or troubleshooting RNA FISH, strong and/or reasonably abundant positive control transcripts are, e.g., *Smed-agat1*, *Smed-gata4/5/6*, or *Smedwi-1* [23]. For controlling staining consistency across plates, a large batch of a defined sample (starved, irradiated, or RNAi) can be prepared in advance and frozen in small aliquots to be included alongside experimental samples (see Subheading 3.3.1).

2 Materials

Equipment should be clean to prevent contamination by microorganisms or exogenous enzymes, especially RNAses (e.g., from plasmid isolation buffers). Stocks of buffers and solutions should be prepared from reagent grade or molecular biology grade reagents using RNAse-free H_2O and disposable filter pipette tips. In our hands, autoclaved 18.2 MOhm ultrapure water (referred to in the text as "ultrapure H_2O") is sufficiently pure, but if problems persist, treatment of H_2O with 0.1% (v/v) diethylpyrocarbonate might be required [24].

2.1 Riboprobe Preparation

Equipment, Consumables, and Reagents

1. Thermal cycler, water bath, or hybridization oven set at 37 °C.
2. Centrifuge with fixed-angle rotor for 1.5 mL tubes, coolable.
3. Horizontal agarose gel electrophoresis apparatus and power supply.
4. 1.5 mL tubes (recommended: DNA low-binding tubes).
5. Deionized formamide.
6. 6× gel loading buffer.
7. Absolute ethanol.
8. 75% (v/v) ethanol in ultrapure H_2O.
9. 7.5 M ammonium acetate.
10. RNase-free DNase I.
11. Pyrophosphatase, inorganic.
12. RNase inhibitor.
13. T7 RNA polymerase.
14. T3 RNA polymerase.
15. RNase-free agarose.
16. 10× DIG labeling mix (Roche, cat. no.: 11277073910) or 10× fluorescein labeling mix (Roche, cat. no.: 11685619910) or 10× DNP labeling mix: 3.5 mM DNP-11-UTP, 6.5 mM UTP, 10 mM of ATP, GTP, CTP each.

Table 1
Exemplary fluorescent tyramide conjugation reactions spanning the typical imaging spectrum

Dye ester	MW	mg	DMF (μL)	1.1 molar excess tyramine HCl in DMF–TEA (μL)	EtOH to 1 mg/mL (μL)
CF-dye 405S	1169	1.169[a]	100	19.14	1049.86
AZDye 488	632	1	100	25.03	874.97[b]
Sulfo-Cy3	736	1	100	21.49	878.51
AZDye 568	792	1	100	24.17	875.83
AZDye 594	804	1	100	23.81	876.19
AZDye 647	1250	1	100	12.65	887.35

[a]Commercially available as 1 μmol
[b]Will precipitate at −20 °C in EtOH, bring to RT before use or alternatively dilute in DMF

Buffers and Solutions

1. 1× TAE electrophoresis buffer: 40 mM Tris–HCl, pH 7.6, 20 mM acetate, 1 mM Na_2EDTA. Adjust pH to 7.6 with 1 M HCl and make up to the required volume with ultrapure H_2O.
2. 10× transcription buffer: 400 mM Tris–HCl pH 8.0, 100 mM $MgCl_2$, 100 mM DTT, 20 mM spermidine. Store at −20 °C.

2.2 Fluorophore-Conjugated Tyramide Synthesis

Equipment, Consumables, and Reagents

1. Tabletop centrifuge or minispin.
2. 1.5 mL tubes (recommended: DNA low-binding tubes).
3. NHS ester of the fluorescent dye of choice (*see* Table 1).
4. Absolute ethanol.

Buffers and Solutions

1. DMF–TEA coupling buffer: add 100 μL TEA to 9.90 mL DMF. Mix well and prepare fresh just prior to use.
2. Tyramine stock solution (10 mg/mL): Add 100 mg tyramine–HCl to 10 mL of DMF–TEA coupling buffer. Mix well until all the tyramine–HCl is dissolved.
3. DMF-dissolved NHS ester of the chosen dye (*see* Table 2): Dissolve the dry fluorophore NHS ester completely in DMF at a concentration of 10 mg/mL by vortexing and collecting drops by brief centrifugation in a tabletop centrifuge (*see* **Note 3**).

2.4.4 *Immunocytochemistry*

Equipment, Consumables, and Reagents

1. 200 μL multichannel pipette, 8 or 12 channels.
2. Multipette M4 repeating pipette or similar.
3. Optional: vacuum pump and collection bottle.
4. Rocking shaker.
5. Optional: well plate stand.

Staining of Phospho-Histone H3 in Metaphase Nuclei

A broad range of anti-phospho-histone H3 (Ser10) antibodies, which largely cross-react in *S. mediterranea*, are commercially available. The appropriate secondary antibody (and the blocking serum) should be chosen according to the cross-reactivity with the primary antibody and the desired excitation wavelength.

Buffers and Solutions

1. 10× PBS: 1.37 M NaCl, 27 mM KCl, 100 mM Na_2HPO_4, 18 mM KH_2HPO_4 (1 L). The pH of the 10× stock will be approximately 6.8; once diluted to 1× PBS, the pH should change to 7.4. In case, adjust pH to 7.4 using HCl and sterile filter (0.22 μM).
2. 1× PBSTx 0.1: Dissolve 10 mL of Triton X-100 in 100 mL of 10× PBS, and then add 900 mL ultrapure H_2O for 1 L of PBSTx 0.1. Tip: Triton X-100 dissolves faster in 10× PBS than in 1× PBS.
3. Optional: 1× TBSTx 0.1: 50 mM Tris–HCl, pH 7.5, 150 mM NaCl, 0.1% (v/v) Triton X-100. Adjust pH to 7.5 with 1 M HCl and fill up the required volume of ultrapure H_2O.
4. Blocking solution: 5% (v/v) goat serum, 0.5% (v/v) RWBR in PBSTx 0.1 or TBSTx 0.1.
5. Anti-phospho H3 solution: primary rabbit anti-phospho H3 antibody 1:1000, 1% (v/v) goat serum, 0.1% (v/v) RWBR in PBSTx 0.1 or TBSTx 0.1.
6. Secondary antibody solution: goat anti-rabbit Alexa Flour 647 secondary antibody 1:1000, 1% (v/v) goat serum, 0.1% (v/v) RWBR in PBSTx 0.1.

Staining of Phospho-Histone H3 in Metaphase Nuclei Combined with TSA

Buffers and Solutions

1. 10× phosphate-buffered saline (10× PBS): 1.37 M NaCl, 27 mM KCl, 100 mM Na_2HPO_4, 18 mM KH_2HPO_4 (1 L). The pH of the 10× stock will be approximately 6.8; once diluted to 1× PBS, the pH should change to 7.4. In case, adjust pH to 7.4 using HCl and sterile filter (0.22 μM).
2. 1× PBSTx 0.1: Dissolve 10 mL of Triton X-100 in 100 mL of 10× PBS, and then add 900 mL ultrapure H_2O for 1 L of PBSTx 0.1. Tip: Triton X-100 dissolves faster in 10× PBS than in 1× PBS.

3. 1× PBSTx 0.1SA: Dissolve 162.5 mg sodium azide in 100 mL of 1× PBSTx 0.1.
4. Blocking solution: 5% (v/v) goat serum, 0.5% (v/v) RWBR in PBSTx 0.1 or TBSTx 0.1.
5. Donkey anti-rabbit-POD solution: 1:1000–1:2000 donkey anti-rabbit-POD, 1% (v/v) horse serum, 0.1% (v/v) RWBR in PBSTx 0.1.
6. Anti-phospho H3 solution: primary rabbit anti-phospho H3 antibody 1:1000, 1% (v/v) goat serum, 0.1% (v/v) RWBR in PBSTx 0.1 or TBSTx 0.1.
7. Secondary antibody solution: goat anti-rabbit Alexa Flour 647 secondary antibody 1:1000, 1% (v/v) goat serum, 0.1% (v/v) RWBR in PBSTx 0.1.
8. 4-Iodophenylboronic acid. Dissolve at 20 mg/mL in N-N-dimethylformamide. Store in aliquots at −20 °C. Use at 1: 1000.
9. Tyramide developing TSA buffer: 0.5–1 μg/mL fluorophore-conjugated tyramide, 0.006% H_2O_2, 20 μg/mL of 4-iodophenylboronic acid, 2 M NaCl, 0.1 M boric acid. Adjust pH to 8.5 using 1 M NaOH. Store at 4 °C.
10. Tyramide developing TSA+ buffer: 0.5–1 μg/mL fluorophore-conjugated tyramide, 0.006% H_2O_2, 20 μg/mL of 4-iodophenylboronic acid, 2 M NaCl, 0.1 M boric acid, 2% (w/v) dextran sulfate, 0.1% (v/v) Tween-20. Slowly add the dextran sulfate while stirring, as it takes time to dissolve. Adjust pH to 8.5 using 1 M NaOH. Store at 4 °C.
11. Optional: 1× TBSTx 0.1: 50 mM Tris–HCl, pH 7.5, 150 mM NaCl, 0.1% (v/v) Triton X-100. Adjust pH to 7.5 with 1 M HCl.
12. SB (sodium borate) buffer: 0.1 M sodium borate in ultrapure H_2O. Adjust the pH to 8.5.

Detection of Incorporated BrdU During DNA Synthesis

A broad range of anti-5-bromo-2'-deoxyuridine (BrdU) antibodies are commercially available. The appropriate secondary antibody (and the blocking serum) should be chosen according to the cross-reactivity with the primary antibody and the desired excitation wavelength.

Buffers and Solutions

1. 2 N HCl (12 mL): Add 2 mL of HCl to 10 mL of ultrapure H_2O under a fume hood and wear protective glasses. The mixture will heat up and has to be cooled to RT before use. Prepare fresh.
2. 0.1 M sodium borate buffer. Dissolve the sodium borate in ultrapure H_2O. Adjust pH to 8.5.

3. Blocking solution: 5% (v/v) goat serum, 0.5% (v/v) RWBR in PBSTx 0.1 or TBSTx 0.1.
4. Primary antibody solution: primary antibody (e.g., mouse clone Bu20a, 2 μg/mL) 1:1000, 1% (v/v) horse serum, 0.1% (v/v) RWBR in PBSTx 0.1.
5. Secondary antibody solution: secondary antibody (e.g., donkey anti-mouse POD) 1:1000; 1% (v/v) horse serum, 0.1% (v/v) RWBR in PBSTx 0.1.
6. 4-Iodophenylboronic acid. Dissolve at 20 mg/mL in N-N-dimethylformamide. Store in aliquots at −20 °C. Use at 1: 1000.
7. Tyramide developing TSA buffer: 0.5–1 μg/mL fluorophore-conjugated tyramide, 0.006% H_2O_2, 20 μg/mL of 4-iodophenylboronic acid, 2 M NaCl, 0.1 M boric acid. Adjust pH to 8.5 using 1 M NaOH. Store at 4 °C.
8. Tyramide developing TSA+ buffer: 0.5–1 μg/mL fluorophore-conjugated tyramide, 0.006% H_2O_2, 20 μg/mL of 4-iodophenylboronic acid, 2 M NaCl, 0.1 M boric acid, 2% (w/v) dextran sulfate, 0.1% (v/v) Tween-20. Slowly add the dextran sulfate while stirring, as it takes time to dissolve. Adjust pH to 8.5 using 1 M NaOH. Store at 4 °C.

2.5 Nuclear DNA Staining

Dyes could be dissolved either in ultrapure H_2O, in DMSO, or in a mixture of equal parts of DMSO and ultrapure H_2O. The presence of DMSO allows storage at −20 °C in liquid form, as this eutectic mixture displays a strong freezing point depression [25] and works better for, e.g., Hoechst 33342 hydrochloride.

Equipment, Consumables, and Reagents

1. Aluminum foil tape for sealing plates for storage.

Buffers and Solutions

1. 10× phosphate-buffered saline (10× PBS): 1.37 M NaCl, 27 mM KCl, 100 mM Na_2HPO_4, 18 mM KH_2HPO_4 (1 L). The pH of the 10× stock will be approximately 6.8; once diluted to 1× PBS, the pH should change to 7.4. In case, adjust pH to 7.4 using HCl and sterile filter (0.22 μM).
2. 1× PBS: Dilute 100 mL of 10× PBS with 900 ml ultrapure H_2O for 1 L of PBS.
3. 1× PBSTx 0.1: Dissolve 10 mL of Triton X-100 in 100 mL of 10× PBS, and then add 900 mL ultrapure H_2O for 1 L of PBSTx 0.1. Tip: Triton X-100 dissolves faster in 10× PBS than in 1× PBS.
4. DAPI (4′,6-diamidino-2-phenylindole) (Ex_{max} 358 nM, Em_{max} 461 nM): 1 mg/mL in ultrapure H_2O.

5. Optional: Hoechst 33342 hydrochloride (Ex_{max} 350 nM, Em_{max} 461 nM): 1 mg/mL in a mixture of equal parts of DMSO and ultrapure H_2O.
6. Optional: Nuclear Violet LCS1 (Ex_{max} 401 nM, Em_{max} 460 nM): 5 mM in DMSO.
7. Optional: DRAQ5 (Ex_{max} 646 nM, Em_{max} 697 nM, intercalated into dsDNA)—5 mM in DMSO. Caution: sodium azide interferes with DRAQ5 staining.
8. Buffered glycerol storage medium (100 mL): Combine 10 mL of 10× PBS with 50–80 mL of glycerol and fill up to 100 mL with ultrapure H_2O.
9. DABCO antifade storage medium (100 mL): Combine 50–80 mL of glycerol, 20 mM Tris–HCl, pH 8.5, 50 mg sodium azide, and 2.5 g of DABCO (1,4-diazabicyclo[2.2.2] octane) and fill up to 100 mL with ultrapure H_2O. Store at −20 °C and use working solutions at 4 °C protected from light. Sodium azide is highly toxic! Prevent contact with skin!

3 Methods

3.1 Riboprobe Preparation

This protocol describes the steps for hapten-conjugated riboprobe synthesis. Targeting ~1 kb for riboprobe length is a good starting point, but even riboprobes of ~2 kb may work equally well, depending on the target mRNA. Since diffusion is not a problem with dissociated cells, riboprobe fragmentation is generally not required. Digoxigenin-labeled riboprobes give clean and robust signal and are optimal for single RNA FISH experiments. For dual RNA FISH, both digoxigenin- and fluorescein-labeled riboprobes can be used. For triple FISH, an additional DNP-labeled riboprobe can be included, but its corresponding detection antibody is generally more expensive. Biotin-labeled riboprobes have been used in planarian WISH [12] but are not popular due to endogenous biotin being present in some planarian tissues and we have not tested them. The integrity of the synthesized riboprobes can be visualized using denaturing agarose gel electrophoresis such as standard MOPS/formaldehyde [26], Bis–Tris/PIPES/glyoxal [27], or TAE/formamide gel electrophoresis [28], of which the latter is the most convenient and least toxic approach. Depending on the RNA polymerase, anti-sense (T7) or sense/control riboprobes (T3) can be synthesized from the same DNA template. For synthesizing riboprobes, prepare synthesis reactions according to Table 2.

1. In a 1.5 mL tube, assemble the in vitro transcription reaction at RT to prevent spermidine precipitation. Mix the reaction carefully by pipetting up and down.
2. Incubate the reaction 4 h or overnight at 37 °C.

3. To remove the template DNA at the end of riboprobe synthesis, add 1–2 units of DNase I to the reaction and mix well by pipetting.
4. Incubate for 45 min at 37 °C.
5. To precipitate the riboprobe, add 0.5 volumes of 7.5 M ammonium acetate, followed by 2.5 volumes of ice-cold absolute ethanol to the in vitro transcription reaction (*see* **Note 1**). Mix by vortexing and put at −80 °C for at least 30 min.
6. Spin at max speed for 30 min at 4 °C. Completely aspirate the supernatant.
7. Wash the pellet extensively with 1 mL of 75% ethanol. Vortex to dislodge the pellet.
8. Centrifuge at 12,000–16,000 × g for 5 min at 4 °C and completely aspirate the supernatant.
9. Repeat the 75% ethanol wash and centrifugation step.
10. Resuspend the riboprobe in 100 μL of formamide (*see* **Note 2**).
11. To analyze the integrity of the riboprobe, prepare a gel boiling 0.8% (w/v) agarose in 1× TAE buffer.
12. Combine 4 μL of formamide, 1 μL of 6× gel loading buffer, and 1 μL of purified riboprobe. Mix well by vortexing or pipetting.
13. Load and electrophorese the sample(s) with an appropriate DNA marker at 5–10 V/cm in 1× TAE electrophoresis buffer. The single-stranded riboprobe runs faster than the DNA template (appears ~1/2 the size). Optionally, measure riboprobe concentration spectrophotometrically (e.g., NanoDrop).
14. Store riboprobes at −80 °C for best long-term storage. Use riboprobes at dilutions of 1:2000–1:4000.

3.2 Fluorophore-Conjugated Tyramide Synthesis

Tyramide signal amplification, also known as catalyzed reporter deposition (CARD), is a highly sensitive method enabling the detection of low-abundance targets in fluorescent applications (e.g., FISH) [29]. In the presence of H_2O_2, it involves the peroxidase-catalyzed covalent deposition of highly reactive fluorescently labeled tyramide radicals onto tyrosines surrounding the enzyme. By sequential development of tyramides coupled to spectrally different fluorophores and intermittent peroxidase inactivation, multicolor staining can be achieved. In addition to the classical fluorophores, such as AMCA (aminomethylcoumarin acetate), fluorescein, rhodamine, and cyanine dyes (Cy2, Cy3, and Cy5), a range of superior fluorophores have become affordable. These include, e.g., the palette of sulfonated Alexa Fluor dyes [30, 31]. Alexa Fluor dyes are largely pH insensitive, brighter, and more photostable derivatives of classical dye backbones. They show reduced "dye stacking" which can lead to self-quenching and in turn decreases signal at higher labeling density. Sulfonation adds negative charges and makes the tyramides also highly water soluble and prevents trapping of the fluorophores in lipid-rich structures [32]. In our hands, the classical fluorophores can be used interchangeably with their respective dye backbone derivatives (e.g., fluorescein and Alexa Fluor 488). Depending on the final fluorophore-tyramide concentration, 1 mg of dye succinimidyl ester (or NHS ester) is sufficient for at least 1–2 L of tyramide development buffer. There are a range of vendors advertising their own proprietary dyes claiming improved properties (brightness, photostability, water solubility), including but not limited to iFluor (AAT Bioquest), CF Dyes (Biotium), DyLight (Thermo Fisher Scientific), ATTO dyes (ATTO-TEC), and AZDye/MB dyes (Fluoroprobes). In our hands, tyramides prepared from NHS esters of Alexa Fluor (or other sulfonated analogues) are compatible with planarian cells and tissues. When in doubt, the compatibility and retention with planarian cells should be tested for new types of fluorophores. The history and principles of fluorescence, as well as the chemistry of various fluorescent dyes used in biology, have been recently reviewed [33]. Iterative TSA amplification, e.g., depositing DNP-conjugated tyramide, which can then be detected with peroxidase-conjugated anti-DNP antibody and developed with a fluorophore-conjugated tyramide, has been described elsewhere [14]. For spectral properties of the fluorophores, we refer to the respective manufacturers' datasheet.

1. Add a 1.1 molar excess of the tyramine–HCl stock in DMF–TEA to the respective fluorophore NHS ester dissolved in DMF. Mix well by vortexing and collect by brief centrifugation (*see* Table 2 for exemplary conjugations reactions).
2. Incubate the conjugation reaction protected from light at RT for at least 2 h. Extended incubations in the dark are also not problematic and will ensure the coupling reaction goes to completion and the hydrolysis of any unreacted NHS ester.
3. Dilute the fluorophore-conjugated tyramide by adding absolute ethanol for a final concentration of 1 mg/mL and store in the dark at −20 °C. (see **Note 4**).

3.3 Preparing Macerated *Schmidtea mediterranea*

Although maceration, fixation, and staining can be performed on successive days, the protocols offer some flexibility and contain two major potential pause points. *Schmidtea mediterranea* can be macerated in advance with excess glycerol, frozen in liquid nitrogen, and stored at −80 °C. This is helpful when many time-points or RNAi experiments are to be analyzed on a single plate, but parallel sample collection is logistically challenging or impossible. Furthermore, after formaldehyde fixation, the plates containing the cells can be stored at −20 °C in a formamide-containing pre-hybridization buffer (see Subheading 3.4.2 *Probe hybridization and washing*), which effectively protects cellular RNAs from degradation [34]. By this approach, many plate copies from the same sample can be prepared and analyzed using different riboprobes or antibodies.

In general, starved animals (≥7 d) show the cleanest signal and are a good background control to be included in an experiment. Feeding might increase background and unspecific binding of riboprobes and antibodies as a result of mucus production or other digestive processes. Furthermore, after excessive cell death of a given cell type (e.g., stem cell depletion), abundant transcripts (e.g., *Smedwi-1*) might still be detectable as speckles and cellular debris in the first few days. It is generally not recommended to include dyes for tracking RNAi food uptake [22, 35], as this will generate fluorescent background and/or stain unspecific structures such as lipid droplets.

3.3.1 Preparing Macerated *S. mediterranea* for Storage and Later Use

Use this option if the samples have to be stored frozen and processed at a later time point. For preparing *S. mediterranea* for storage, the animals are initially dissociated and frozen at −80 °C in a smaller volume containing a higher glycerol concentration as cryoprotectant. After thawing, the glycerol is diluted with a glycerol-free maceration dilution solution to yield the desired cell densities. The diluted cell suspension is equivalent to freshly macerated animals.

1. Transfer the animals into a 2 mL tube, remove all liquid, and add ice-cold maceration solution at 1/10th of the volume of final desired maceration solution volume (e.g., 100 μL for 1 mL final volume) (*see* **Notes 5, 6**, and 7).
2. Let the tubes incubate for 15 min on ice in the cold room.
3. Afterwards, place the tube(s) on a rotator in the cold room and incubate them for an additional 15–30 min.
4. Most of the tissue should be dissociated by then. Carefully pipette up and down using a p1000 to dissociate remaining tissue clumps or when there are dark "clouds" in the cell suspension. Make sure to mix the macerate to homogeneity before proceeding further. Pharynx tissue takes the longest to dissociate.
5. Briefly collect cells by centrifugation for 10 s/500 × g/4 °C in a fixed-angle tabletop centrifuge and then add 0.6 volumes of glycerol to the cells and put on a rotator in the cold room for an additional 15 min (e.g., 60 μL for 1 mL final volume) (*see* **Note 8**).
6. Briefly collect cells by centrifugation for 10 s/500 × g/4 °C in a fixed-angle tabletop centrifuge. Snap-freeze in liquid nitrogen and store at −80 °C until needed.

7. Before starting the experiment, place the frozen aliquots on ice and add fresh, glycerol-free maceration solution to the final volume (e.g., 840 μL for 1 mL final volume).
8. Place tube(s) on a rotator in the cold room for 10 min/4 °C.
9. Briefly collect the cells by centrifugation for 10 s/500 × g/4 °C in a fixed-angle tabletop centrifuge, put the cells on ice, and proceed with **step 1** of Subheading 3.4.1 (Preparing the plate(s) and fixation of cells).

3.3.2 Preparing Macerated S. mediterranea for Subsequent Use

1. Cool down the fixed-angle centrifuge to 4 °C.
2. Transfer the animals into a 2 mL tube, remove all liquid, and add ice-cold maceration solution to the animals. A good starting point is generating ~500,000 cells/ml of maceration which corresponds to a ~ 5 mM *S. mediterranea* (CIW4 strain) dissociated in 1 ml of maceration solution (*see* **Notes 5, 6**, and 7).
3. Let the tubes incubate for 15 min on ice in the cold room.
4. Place the tube(s) on a rotator in the cold room and incubate them for an additional 15–30 min.
5. Most of the tissue should be dissociated by now. Carefully pipette up and down using a p1000 to dissociate remaining tissue clumps or when there are dark "clouds" in the cell suspension. Make sure to mix the macerate to homogeneity before proceeding. Pharynx tissue takes the longest to dissociate.
6. If necessary, dilute the cell suspension with ice-cold maceration solution.
7. Briefly collect cells by spinning for 10 s/500 × g/4 °C in a fixed-angle centrifuge, put the cells on ice, and proceed with **step 1** of Subheading 3.4.1 (Preparing the plate(s) and fixation of cells).

3.4 Fluorescent In Situ Hybridization and Immunocytochemistry

Once hybridization of the riboprobes has been performed, pauses should be limited to a maximum of 1–2 days at 4 °C in the dark until all riboprobes have been developed to limit potential RNA degradation by traces of RNAses. In contrast, immunocytochemistry is less critical and phospho-histone 3 ICC can be developed after some days of storage or extended primary antibody incubation at 4 °C. The full staining protocol takes roughly 4–6 days when developing all three channels using FISH riboprobes or with ICC of phospho-H3 and/or BrdU. The time estimates are for manually staining 1–2 full 96-well plates, which can be conveniently handled without automated liquid handling. The protocols are organized in a modular fashion and require roughly a day per module and can be linked together by overnight incubation steps. Cells should be covered by liquid at all times and not be allowed to dry for more than a few seconds. Most of the steps use generous volumes to prevent evaporation. Costly antibody or tyramide development solutions can likely be scaled down further, but should be tested empirically.

3.4.1 Preparing the Plate (S) and Fixation of Cells

This protocol module contains all the steps required for fixing the cells onto the bottom of a 96-well plate. Due to spatial constraints, all steps in 384-well plates are performed in ~25% of the volume and require automated liquid handling. We refer to the respective manufacturer's documentation for setting up automated liquid handling solutions.

Initial Note About the Density of Cell Seeding The probability of identifying a cell type/state of interest largely depends on its abundance within the animal. While in starved *S. mediterranea* fewer than 1% of the cells are in M-phase, some abundant cells of interest can make up 10–20% of the animal (e.g., gut cells). A single 5–7 mM *S. mediterranea* (CIW4 strain) contains ~640,000–1,100,000 cells [36] and 1 mL of maceration solution will typically lead to an acceptable cell density. Plating 120 μL of macerate per well in a 96-well plate is a good starting point for further optimizations. Generally, a density of ~30 cells per 100 μM^2 in the well is sufficient for most analyses. Overloading wells should be avoided, as cell clumps will lead to riboprobe and antibody trapping and complicates downstream automated image segmentation. Imaging of ~2000 cells at independent positions in the well is sufficient for detecting most cell types, although capturing rare cells/events might require more cells.

Initial Note About Fixation and Permeabilization For fixation we have not explored glutaraldehyde due to its well-known autofluorescence generation [37]. It has been shown that short RNAs such as miRNAs can be efficiently retained in tissue or in blotting applications after postfixation with EDC (1-ethyl-3-(3-dimethylaminopropyl) carbodiimide) [38, 39]. In our hands, EDC fixation completely abolished staining using tyramide signal amplification under our conditions. Staining strategies using EDC fixation might require different permeabilization steps or antigen retrieval, but these were not further explored. Compared to earlier versions of this protocol [11], we have found that omission of Proteinase K treatment, which is a common source of batch-to-batch variability, does not negatively affect staining. This shortens the protocol and makes cell treatment less time critical and variable. Permeabilization is ensured by the inclusion of detergents in washing buffers and heating during riboprobe hybridization.

1. *Distribute 120 μL of cell suspension from macerated animals* (see *Subheading 3.1) into each of the wells of a 96-well plate.*
2. *Gently pellet cells for 5 min/2000 × g/4 °C in a swing-out plate centrifuge. For a single plate experiment, fill up a counterbalance plate with* H_2O (see ***Note 9***).
3. *Under a fume hood, gently add 45 μL of 15% PFA per well (final concentration: 4%) and incubate the cells for 15–20 min at RT* (see ***Note 10***).
4. *Carefully remove 4% PFA/maceration solution and slowly add 200 μL of 1× PBS.*
5. *Repeat the 1× PBS wash by carefully removing 1× PBS and slowly adding fresh 200 μL of 1× PBS* (see ***Note 11***).
6. *Remove 1× PBS and postfix the cells by adding 100 μL of 4% PFA in 1× PBS for 15–20 min at RT.*
7. *Remove fixative and incubate with 200 μL of quenching buffer for 15 min at RT* (see ***Notes 12 and 13***).
8. *Remove quenching buffer and rinse with 200 μL PBSTx 0.1 for 5 min.*
9. *Repeat **step 8**.*

3.4.2 Probe Hybridization and Washing

This protocol module contains all the steps required for preparing the plates for riboprobe hybridization and subsequent washing steps. Optionally, the experiment can be paused by storing the plates at −20 °C in PreHyb before the addition of the riboprobes (*see* **Note 14**).

1. Bring sufficient WashHyb (25 μL/well) and PreHyb (100 μL/well) from −20 °C to RT.
2. Remove PBSTx 0.1 and incubate the cells with 50 μL of equal parts PBSTx 0.1/WashHyb for 5 min at RT.
3. Incubate with 100 μL of PreHyb for 45–60 min at 56–58 °C in an incubation oven.
4. In the meantime, bring Hyb (or Hyb+) to 56–58 °C and start preparing the riboprobes.
5. Briefly denature the riboprobes 5 min at 70 °C and immediately dilute the riboprobes of choice in Hyb or Hyb + at 1:2000 to 1:4000, depending on the riboprobe (*see* **Note 15**).
6. Remove PreHyb and incubate with 150–200 μL Hyb (or Hyb +), including riboprobes for 12–16 h at 56–58 °C in an incubation oven, preferable with shaking. To prevent evaporation, fill empty wells with water and seal the plate using adhesive PCR plate seals.
7. On the next day, before you begin, pre-warm the following solutions at 56–58 °C:

   (a) WashHyb (100 μL/well).
   (b) 2× SSCTx 0.1 (400 μL/well).
   (c) 0.2× SSCTx0.1 (400 μL/well).
   (d) 1:1 2× SSCTx 0.1/WashHyb (100 μL/well).

8. Remove Hyb (or Hyb+) and wash 1× with 100 μL of pre-warmed WashHyb for 15 min at 56–58 °C (*see* **Note 16**).
9. Remove the solution and wash 1× with 100 μL of pre-warmed 1:1 2× SSC/WashHyb for 15 min at 56–58 °C.
10. Remove the solution and wash 2× with 200 μL of pre-warmed 2× SSCTx 0.1 for 10 min at 56–58 °C.
11. Remove the solution and wash 2× with 200 μL of pre-warmed 0.2× SSCTx 0.1 for 10 min at 56–58 °C.
12. Bring the plate(s) to RT. Remove the solution and rinse once with 200 μL PBSTx 0.1 followed by 1× wash with 200 μL PBSTx 0.1 for 10 min.

3.4.3 Riboprobe Development Using Tyramide Signal Amplification

Initial Note About Probe Development, Order of Staining, and Signal Specificity Recipes for two alternative hybridization buffers (Hyb and Hyb+) and two tyramide signal amplification buffers (TSA and TSA+) are provided. Hyb + is a slightly acidified (pH 6.0) variation of Hyb [40], as formamide becomes basic upon heating, which is detrimental to RNA stability. The pH of Hyb (pH 7.0) or Hyb + (pH 6.0) can be controlled through the pH of the SSC buffer or alternatively by acidification of already prepared Hyb (pH 7.0) through addition of 460 μL of 1 M citric acid solution for each 50 mL of Hyb. The standard TSA buffer gives good staining with a wide range of riboprobes without generating excessive background. TSA+ buffer generates a stronger signal from lower tyramide concentrations and might be a good choice when maximum sensitivity is required. Extensive tyramide signal development is a double-edged sword and might lead to unacceptable background for weak riboprobes; therefore, proper antibody blocking and clean riboprobes are required. TSA+ seems to be generally safe for developing DIG-labeled riboprobes. Riboprobes containing different haptens (DIG, fluorescein, DNP) might behave differently and are not readily interchangeable as riboprobe and respective antibody affinity might differ. Figure 2 provides an overview and recommendations regarding the order of stainings or choice of fluorophores. In dual or triple FISH, weakly expressed targets should be developed first. It is generally advisable to develop the BrdU signal last, as the harsh antigen retrieval using HCl would likely affect most epitopes. For experiments involving BrdU detection, it is recommended to perform ICC in conjunction with tyramide deposition instead of fluorophore-conjugated secondary antibodies as antibody complexes might dissociate during BrdU antigen retrieval. Some antibodies might no longer work after maceration or in situ hybridization. Therefore, conditions for new antibodies should be tested individually. If the maceration solution destroys the target epitope, the cell suspensions have to be prepared in a different way, e.g., mechanically [41] or enzymatically [23, 42–44], and optionally flow sorted. The protocol should then start at **step 4** of Subheading 3.4.1.

Digoxigenin Riboprobe Development (DIG-Probe)

1. Remove PBSTx 0.1 and incubate in blocking solution for 30–60 min at RT.
2. Remove the blocking solution, add 100 μL of anti-DIG-POD solution, and incubate for 3–4 h at RT (*see* **Note 17**).
3. Remove solution and rinse with 200 μL of PBSTx 0.1.
4. Wash with 200 μL of PBSTx 0.1 for 20 min.
5. Repeat **step 4** twice.
6. Remove PBSTx 0.1 and incubate with 100 μL of tyramide developing TSA (or TSA+) buffer for 15–30 min at RT (*see* **Notes 18** and **19**).
7. Remove solution and rinse with 200 μL of PBSTx 0.1.
8. Wash with 200 μL of PBSTx 0.1 for 10 min.
9. Incubate with 100 μL of PBSTx 0.1SA for 15 min at RT or overnight at 4 °C (*see* **Note 20**).
10. Remove the solution and rinse twice with 200 μL of PBSTx 0.1.
11. Wash 10 min with 200 μL of PBSTx 0.1 for 10 min.
12. Repeat **step 11** twice.

Fluorescein Riboprobe Development (FL-Probe)

This protocol module is designed for the detection of fluorescein-conjugated riboprobes (FL-probes). As fluorescein is a fluorophore by itself, the riboprobe is generally developed using a spectrally identical fluorescein- or AF488-conjugated tyramide. Please note the slightly different blocking conditions for this assay (*see* **Note 21**).

1. Remove PBSTx 0.1 and incubate in blocking solution for 30–60 min at RT.
2. Remove the blocking solution, add 100 μL of anti-FL-POD solution, and incubate for 3–4 h at RT (*see* **Note 17**).
3. Remove solution and rinse with 200 μL of PBSTx 0.1.
4. Wash with 200 μL of PBSTx 0.1 for 20 min.
5. Repeat **step 4** twice.
6. Remove PBSTx 0.1 and incubate with 100 μL of tyramide developing TSA (or TSA+) buffer for 15–30 min at RT (*see* **Notes 18** and **19**).

7. Remove solution and rinse with 200 μL of PBSTx 0.1.
8. Wash with 200 μL of PBSTx 0.1 for 10 min.
9. Incubate with 100 μL of PBSTx 0.1SA for 15 min at RT or overnight at 4 °C (*see* **Note 20**).
10. Remove the solution and rinse twice with 200 μL of PBSTx 0.1.
11. Wash 10 min with 200 μL of PBSTx 0.1 for 10 min.
12. Repeat **step 11** twice.

Dinitrophenyl Riboprobe Development (DNP-Probe)

This protocol module is designed for the detection of dinitrophenyl-conjugated riboprobes (DNP-probes). While it is compatible with the plate format, the commonly used antibody (PerkinElmer, cat. no.: FP1128) makes this riboprobe/antibody combination more expensive in high-content assays (*see* **Note 22**).

1. Remove PBSTx 0.1 and incubate in blocking solution for 30–60 min at RT.
2. Remove the blocking solution, add 100 μL of anti-DNP-POD solution, and incubate for 3–4 h at RT (*see* **Note 17**).
3. Remove solution and rinse with 200 μL of PBSTx 0.1.
4. Wash with 200 μL of PBSTx 0.1 for 20 min.
5. Repeat **step 4** twice.
6. Remove PBSTx 0.1 and incubate with 100 μL of tyramide developing TSA (or TSA+) buffer for 15–30 min at RT (*see* **Notes 18** and **19**).
7. Remove solution and rinse with 200 μL of PBSTx 0.1.
8. Wash with 200 μL of PBSTx 0.1 for 10 min.
9. Incubate with 100 μL of PBSTx 0.1SA for 15 min at RT or overnight at 4 °C (*see* **Note 20**).
10. Remove the solution and rinse twice with 200 μL of PBSTx 0.1.
11. Wash 10 min with 200 μL of PBSTx 0.1 for 10 min.
12. Repeat **step 11** twice.

3.4.4 Immuno-cytochemistry

To prevent nonspecific binding of probes and antibodies, samples can be blocked using a variety of protein or nonprotein mixtures. Common blocking reagents are bovine serum album (BSA), gelatin (porcine or cold-water fish), serum (calf, sheep, goat, horse, donkey), skim milk, or purified casein. The type of blocking reagent should be optimized for each antibody or assay [45, 46]. For example, in the case of the commonly used anti-fluorescein antibody (Roche, cat. no.: 11426346910), inclusion of porcine gelatin suppressed nonspecific binding, while cold-water fish gelatin at the same concentration did not. Nonprotein blocking buffers often contain polyvinylpyrrolidone and detergents [47]. For some phosphoprotein antibodies, the Roche Western Blocking Reagent, skim milk (which contains the phosphoprotein casein), or phosphate buffers in general might interfere with staining. Here, Tris-based buffers are often recommended. Staining conditions have to be tested empirically, but fortunately, the plate format allows rapid identification of optimal conditions. Optimizing the composition of PreHyb and Hyb buffers for RNA FISH, which contain a number of different blocking reagents (Denhardt's Reagent [48], tRNA, heparin), is generally not necessary.

Initial Note About Fluorophores and Secondary Antibodies Optimal combinations of fluorophores and light sources and detection systems should be chosen to maximize sensitivity and prevent signal bleed-through. A common combination of DAPI (4′,6-diamidino-2-phenylindole), Alexa Fluor 488, Alexa Fluor 568, and Alexa Fluor 647 is spectrally well separated and compatible with most imaging setups. Although fluorophores emitting in blue/cyan are available (AMCA, Alexa Fluor 350, Alexa Fluor 405), this channel is generally reserved for nuclear DNA staining using DAPI. As spatial resolution is lost after dissociation, a nuclear DNA marker is useful for discriminating between bona fide cells and debris, as well as to quantify total cell numbers. Nuclei are readily stained with dsDNA-specific dyes in the blue/cyan spectrum (DAPI, Hoechst 33342, Nuclear Violet™ LCS1, or similar). Alternatively, DRAQ5 a far-red emitting DNA dye can be used [49], although this sacrifices the channel with generally lower autofluorescence. For ICC, primary antibodies should be chosen from different species and fluorophore-conjugated secondary antibodies should be highly cross-adsorbed to minimize nonspecific binding. If two or more murine monoclonal antibodies belong to different IgG subclasses, staining with IgG subclass-specific secondary antibodies is possible [50]. Mice express 4 of the 5 available IgG subclasses, which typically encode for IgG1, IgG2b, and IgG3 and depending on their strain will also express either IgG2a (BALB/c and Swiss Webster) or IgG2c (C57Bl/6, C57Bl/10, SJL, and NOD strain) [51, 52]. We have successfully used murine IgG1 and IgG2a primary antibodies with subclass-specific secondary antibodies and found their performance comparable to multispecies detection setups, e.g., murine vs rabbit primary and secondary antibodies. In the Appendix, some commercially available primary and secondary antibodies that are validated for whole-mount IHC of *S. mediterranea* are provided.

Initial Note About Antigen Retrieval for Immunocytochemistry Some antibodies might require antigen retrieval, and in pure ICC experiments (without hybridization), the mild conditions employed might leave antigens masked. There is a wide range of antigen retrieval techniques employed in the histology field. We have had success with chemical antigen retrieval strategies, including 0.1% (w/v) sodium dodecyl sulfate (SDS) [53], 6 M urea [54], or 6 M guanidine hydrochloride [55]. Treatment for 30 min at RT was generally sufficient to improve staining. Heat-induced antigen retrieval has also been described for planarians [14], which we have not explored. Some antibodies work better after post-in situ hybridization, such as the anti-BrdU antibody (clone Bu20a) [56]. If RNA FISH is not required, a shorter mock in situ hybridization without riboprobes can act as a mild antigen retrieval step in place of the full protocol.

Staining of Phospho-Histone H3 in Metaphase Nuclei

This protocol module uses primary antibodies against phospho-histone H3 as a marker for cells in metaphase (M-phase). Detection is accomplished using fluorophore-conjugated secondary antibodies, preferably using a far-red emitting fluorophore (e.g., Alexa Fluor 647), although other detection schemes are possible. This protocol also serves as a basis for ICC using a range of other antibodies.

1. Remove the PBSTx 0.1 and incubate in 50 μL blocking solution for 30 min at RT (*see* **Notes 23** and **24**).
2. Remove blocking solution and incubate with 50 μL of anti-phospho H3 solution at 4 °C overnight or 3–4 h at RT (*see* **Note 25**).
3. Remove solution and rinse once with 200 μL of PBSTx 0.1.
4. Wash with 200 μL of PBSTx 0.1 for 10 min at RT.
5. Repeat **step 4** twice.
6. Remove the PBSTx 0.1 and add 50 μL of secondary antibody solution; incubate for 3–4 h at RT (*see* **Note 26**).
7. Remove solution and rinse once with 200 μL of PBSTx 0.1.
8. Wash with 200 μl of PBSTx 0.1 for15 min.
9. Repeat **step 8** twice.

Staining of Phospho-Histone H3 in Metaphase Nuclei Combined with TSA

This protocol module is a more sensitive modification of the ICC protocol for the detection of phospho-histone H3 and is required in combination with BrdU detection in the same experiment (*see* Subheading 3.4.4.3). The phospho-histone H3-positive nuclei are covalently labeled using tyramide signal amplification as the BrdU antigen retrieval would otherwise destroy noncovalently bound antibody staining.

1. Remove the PBSTx 0.1 and incubate in 50 μL blocking solution for 30 min at RT (*see* **Notes 23** and **24**).
2. Remove blocking solution and incubate with 50 μL of anti-phospho H3 solution at 4 °C overnight or 3–4 h at RT (*see* **Note 25**).

3. Remove solution and rinse once with 200 μL of PBSTx 0.1.
4. Wash with 200 μL of PBSTx 0.1 for 10 min at RT.
5. Repeat **step 4** twice.
6. Replace PBSTx 0.1 with 50 μL of donkey anti-rabbit-POD solution; incubate for 3–4 h at RT (*see* **Note 25**).
7. Remove solution and rinse once with 200 μL of PBSTx 0.1.
8. Wash with 200 μL of PBSTx 0.1 for 20 min at RT.
9. Repeat **step 8** twice.
10. Remove PBSTx 0.1 and incubate with 100 μL of tyramide developing TSA (or TSA+) buffer for 15–30 min at RT.
11. Remove solution and rinse once with 200 μL PBSTx 0.1.
12. Wash with 200 μL PBSTx 0.1 for 10 min at RT.
13. Incubate with 100 μL PBSTx 0.1SA for 15 min at RT or overnight at 4 °C (*see* **Note 20**).
14. Remove solution and rinse twice with 200 μL of PBSTx 0.1.
15. Wash with 200 μL of PBSTx 0.1 for15 min.
16. Repeat **step 8** twice.

Detection of Incorporated BrdU during DNA Synthesis

This protocol module is designed for the detection of cells that have incorporated BrdU into their nuclear DNA. It includes antigen retrieval (DNA denaturation) and the detection of the exposed BrdU antigen using the murine Bu20a antibody and fluorophore-conjugated tyramide signal amplification. BrdU delivery strategies and BrdU pulse-chase durations are largely dependent on the biological question and have been covered in detail elsewhere [57]. Existing BrdU pulse chase protocols for whole mounts should be directly adaptable to the plate format without additional modifications.

1. Remove the PBSTx 0.1 and incubate in 50–100 μL of freshly prepared 2 N HCl for 20–30 min at RT to denature the DNA and expose the BrdU epitope (*see* **Note 26**).
2. Remove 2 N HCl and neutralize with 200 μL of SB buffer for 10 min at RT.
3. Remove the SB buffer and wash the cells with 200 μL of PBSTx 0.1 for 5 min.
4. Repeat **step 3**.
5. Remove the PBSTx 0.1 and incubate in 50 μL blocking solution for 30 min at RT (*see* **Notes 23** and **24**).
6. Remove blocking solution and incubate with 50 μL of primary antibody solution; incubate at 4 °C overnight (*see* **Note 25**).

7. Remove the primary antibody solution and rinse once with 200 μL PBSTx 0.1.
8. Wash with 200 μL of PBSTx 0.1 for 10 min at RT.
9. Repeat **step 8**.
10. Replace PBSTx 0.1 with 100 μL of secondary antibody solution; incubate for 3–4 h at RT.
11. Remove solution and rinse with 200 μL of PBSTx 0.1.
12. Wash with 200 μL of PBSTx 0.1 for15 min.
13. Repeat **step 12** twice.
14. Remove PBSTx 0.1 and incubate with 100 μL of tyramide developing TSA (or TSA+) buffer for 15–30 min at RT.
15. Remove solution and rinse with 200 μL of PBSTx 0.1.
16. Wash with 200 μL of PBSTx 0.1 for 15 min.
17. Repeat **step 16**.

3.5 Nuclear DNA Staining

This protocol module describes methods for fluorescently staining nuclear DNA using dsDNA-specific dyes. This is commonly done using DAPI, but spectrally similar dyes include Hoechst 33342 and Nuclear Violet™ LCS1. Alternatively, DRAQ5 can be used for far-red imaging. To achieve consistent labeling, nuclear DNA staining should be performed as a final step before imaging. A glycerol-based antifade storage medium containing DABCO prevents fluorophore bleaching. Both DABCO and sodium azide are singlet oxygen quenchers and used as antifade reagents [58–60], although the main function of the latter is to prevent bacterial growth. An alternative popular antifading reagent is n-propyl gallate, which we have not tested. Properly sealed plates stored at 4 °C, protected from light, can be reimaged after several months without obvious loss of signal.

1. Remove PBSTx 0.1 and add 50 μL of either 0.1–1 μg/mL DAPI, 1 μg/mL Hoechst 33342, 0.5–5 μM Nuclear Violet LCS1, or 5–20 μM DRAQ5 in PBSTx 0.1; incubate for 30 min at RT.
2. Remove the nuclear counterstaining solution and rinse once with 200 μL of PBSTx 0.1.
3. Remove PBSTx 0.1 and wash with 200 μL 1× PBS for 5 min.
4. Remove PBS and store cells in 100–200 μL of buffered glycerol (or DABCO antifade) storage medium at 4 °C (or −20 °C), protected from light (*see* **Note 27**). Seal the plate with an adhesive PCR plate seal to avoid spillage or evaporation.

4 Notes

1. Ammonium acetate is the salt of choice for ethanol precipitation of riboprobes as it tends to keep unincorporated hapten-NTPs in solution, which could lead to the deposition of fluorescent speckles or a bright "glow" of the whole well surface after tyramide signal amplification.
2. Once dissolved in formamide, riboprobes are extremely difficult to reprecipitate again, so the pellet should be washed thoroughly before resuspension [61].
3. NHS esters are prone to hydrolysis, and therefore, the entire content of the tube should be dissolved in anhydrous DMF and used immediately for the coupling reaction.
4. Some fluorophore-coupled tyramides will partially precipitate upon storage at −20 °C (e.g., Alexa Fluor 488–tyramide). Make sure to warm the tube to RT and fully resuspend the tyramide before use. Alternatively, dilute the coupling reaction in DMF instead of ethanol to the desired concentration.
5. As animal cell number does not scale linearly with animal length, we refer to the literature for cell numbers over a broad range of animal sizes [24]. For a given *S. mediterranea* (CIW4; 1–2 weeks starved), the number of cells (y) of animal length in mm (x) can be approximated using the equation: $y = 15981x^2 + 52{,}774x - 25{,}382$. Note that some RNAi phenotypes, especially of genes important for stem cell function, will prevent the animals from growing. Compared to control animals, these will have fewer cells; therefore, volumes or number of animals have to be adjusted accordingly.
6. The round bottom of 2 mL tubes helps the liquids to mix more thoroughly. Conical 1.5 mL tubes often lead to tissue being trapped at the bottom of the tube.
7. *Schmidtea mediterranea* can also be macerated using methanol-containing solutions (13:3:2:2 ratio of distilled water, methanol, glacial acetic acid, and glycerol [9]). While inclusion of methanol preserves cellular morphology slightly better, dissociation is much slower and might require cumbersome manual pipetting for each sample. Furthermore, the chemical compatibility of the polystyrene plates with methanol is limited; therefore, caution is advised. As we have not extensively tested other maceration solutions, we refer to the composition described by David [7]. To ensure more consistent results and limit endogenous enzyme activity, maceration at cold temperatures is recommended.

8. Glycerol is very viscous at cold temperatures. Using a repeater pipette with Combitips makes dispensing glycerol more efficient and consistent. Note that Combitips have fixed volume increments, so volume choices are not arbitrary.
9. Do not spin the plates above 2000 × g as they might crack. Changing this parameter might require adjusted centrifugation times.
10. To prevent cells from aggregating at the edges of the wells, do not move or shake the plates and keep them in a horizontal position during fixation.
11. Formaldehyde fixation is inefficient at low pH [62] and cells are pre-infiltrated with fixative at this step. Take care to pipette down the wall of the well and not to disturb the cells to prevent cell loss. It is fine to not completely remove all of the liquid during 1× PBS washes and keep a ring of liquid around the well edges. The 1× PBS washes neutralize the pH for efficient postfixation. Suboptimally fixed cells might lose RNA and protein epitopes during downstream processing. When the plate is propped up at an angle, underfixed cells slowly slide along the bottom of the well forming a gradient or are washed off completely.
12. A repeater pipette is safe to use from this point on and the plates can now be propped up at a 45° angle for washing in a well plate stand or as a cheap alternative in a tip box lid. When possible, the plate should be rocked gently on a nutator from now on throughout the process, but it is not critical. Liquid removal speed using a vacuum pump can be tuned by attaching different tips (p10 or p200).
13. Tris and glycine quench PFA-induced autofluorescence and the sodium azide inactivates endogenous peroxidases that might have survived and could generate unspecific background in TSA. Do not use any detergents up to this point.
14. Instead of proceeding with blocking and hybridization, the plates can be stored in PreHyb at −20 °C after addition of PreHyb. To continue staining after storage at −20 °C, bring the plate to RT and proceed with heating the plate up to hybridization temperature in an incubator oven.
15. For very complex riboprobe combinations and plate layouts, pre-aliquot the riboprobes in Hyb in a PCR plate and use a multichannel pipette for transfer.
16. The repeater pipette tips can be preheated together with the solutions in an incubation oven to prevent volume changes when taken up heated solutions with the RT tip.
17. Optionally filter the antibody using a syringe filter (PES, 0.22 μM). Then afterwards add Roche Western Blocking Reagent.

18. Optionally filter the TSA (or TSA+) buffer with the tyramide using a syringe filter (PES, 0.22 μM) to get rid of fluorescent precipitates if that is a problem.
19. Old bottles of H_2O_2 or aging POD-conjugated antibodies can lead to declining over time. Replace H_2O_2 or increase its concentration and/or use a fresh aliquot of POD-conjugated antibody.
20. This inactivates the peroxidase to allow subsequent tyramide development for additional targets. The omission of this step can lead to unspecific tyramide deposition in the next tyramide signal amplification reaction. If no further tyramide signal amplification is required in the experiment, azide treatment can be skipped. Make sure to not have any sodium azide present when working with peroxidase-conjugated antibodies as this will inactivate the enzyme.
21. In our hands, the anti-fluorescein-POD antibody (Roche, cat. no.:11426346910) generally requires more stringent blocking conditions and skipping the blocking step or diluting the blocking reagents often leads to a higher background. We have found that the inclusion of porcine skin gelatin in addition to RWBR improved staining. Pig-derived gelatin also improved specificity of some fluorophore-conjugated secondary antibodies. The gelatin becomes quite viscous at colder temperatures, so blocking and incubation should be done at RT to maximize diffusion. Potentially, the antibody concentration could be reduced further to improve specificity, but we have not systematically tested this.
22. For routine work we use a combination of DIG- and FL-probes as the commonly used anti-DNP-POD antibody (Perkin Elmer, cat. no.: FP1128) is required at much higher concentrations. Anti-DNP POD-conjugated antibodies are only available from a few commercial vendors. SYnabs S.A. (Gosselies, Belgium) has a large selection of commercial anti-DNP antibodies which can also be ordered as POD conjugates and might represent an alternative.
23. Buffer choice of PBSTx 0.1 vs TBSTx 0.1 depends on the primary antibody and should be tested, although we routinely use PBSTx 0.1 for detecting phospho-histone H3 (S10/T11). We prefer PBSTx 0.1 as a standard buffer because it is very cheap, but standard TBSTx 0.3 used in WISH performs equally well [14].
24. After developing the second riboprobe, blocking can be kept short. It is still advisable to reblock after each tyramide development reaction as the TSA/TSA+ buffer has high ionic strength, which likely washes off bound blocking proteins.

25. Elongated incubations at 4 °C, such as overnight or weekends for primary and/or secondary antibodies, do not negatively affect staining and, if at all, improve signal. If there are problems with precipitates, filter the solution (PES, 0.22 μM) before adding RWBR.
26. The BrdU antigen retrieval time should be carefully controlled. Too short treatment will not expose the antigen optimally, whereas extended incubations will increase background and eventually hydrolyze the DNA, abolishing staining with nuclear DNA dyes. Different anti-BrdU antibodies might slightly differ in their affinity for BrdU and their performance with different antigen retrieval strategies [63, 64].
27. Samples that went through the BrdU protocol are incompatible with DABCO-containing storage medium and generate irreversible and unacceptable high background signal when excited at 385/405 nM measured with DAPI emission filters (BP445/45). This is specific for UV/violet light excitation in combination with HCl antigen retrieval as channels excited at 488, 561, and 640 nM are not affected. Therefore, we recommend imaging and storing cells after BrdU staining in buffered glycerol without antifade or 1XPBS. We have not tested whether n-propyl gallate as an antifade might be substituted for DABCO in experiments involving BrdU-stained cells.

Acknowledgements

We thank the Rink lab for comments and specifically Albert Thommen for establishing the initial cell dissociation protocol and Hanh Thi-Kim Vu for technical advice on tyramide and riboprobe synthesis as well as helpful suggestions throughout the establishment of this method. Furthermore, we thank Marc Bickle, Martin Stöter, and the team of the Technology Development Studio (TDS) at MPI CBG for help with imaging and automation.

Appendix

Antibody	Supplier	Catalogue no.
Anti-DIG-POD	Roche	11,207,733,910
Anti-fluorescein-POD	Roche	11,426,346,910
Anti-DNP-POD	PerkinElmer	FP1128
Rabbit phospho-histone H3 (phospho S10 + T11) pAB	Abcam	ab32107

(continued)

Antibody	Supplier	Catalogue no.
Rabbit phospho-histone H3 (Ser10) (D7N8E) mAb	Cell signaling technology	53,348
Rabbit phospho-histone H3 (Ser10) (D2C8) mAb	Cell signaling technology	3377
Goat anti-mouse IgG (H + L), Alexa Fluor 488	Thermo fisher scientific	A-11001
Goat anti-rabbit IgG (H + L), Alexa Fluor 555	Thermo fisher scientific	A-21429
Goat anti-mouse IgG (H + L), Alexa Fluor plus 555	Thermo fisher scientific	A-A32727
Goat anti-mouse IgG (H + L), Alexa Fluor 568	Thermo fisher scientific	A-11004
Goat anti-rabbit IgG (H + L), Alexa Fluor 647	Thermo fisher scientific	A-21245
Goat anti-rabbit IgG (H + L), Alexa Fluor plus 647	Thermo fisher scientific	A32733
Goat anti-mouse IgG1, Alexa Fluor 568	Thermo fisher scientific	A-21124
Goat anti-mouse IgG2a, Alexa Fluor 647	Thermo fisher scientific	A-21241
Donkey anti-mouse IgG (H + L) F (ab′)$_2$ fragment, peroxidase	Jackson Immuno Research	715-036-151
Donkey anti-rabbit IgG (H + L) F (ab′)$_2$ fragment	Jackson Immuno Research	711-036-152
Donkey anti-rat IgG (H + L), peroxidase	Jackson Immuno Research	712-035-153
Anti-BrdU antibody, rat (BU1/75 [ICR1])	Abcam	ab6326
Anti-BrdU antibody, mouse IgG1 (Bu20a)	Multiple vendors	NA

References

1. Boutros M, Heigwer F, Laufer C (2015) Microscopy-based high-content screening. Cell 163:1314–1325. https://doi.org/10.1016/j.cell.2015.11.007
2. Liberali P, Snijder B, Pelkmans L (2015) Single-cell and multivariate approaches in genetic perturbation screens. Nat Rev Genet 16:18–32. https://doi.org/10.1038/nrg3768
3. Solana J (2018) RNA in situ hybridization on planarian paraffin sections. Methods Mol Biol 1774:393–404. https://doi.org/10.1007/978-1-4939-7802-1_13
4. Adell T, Barberán S, Sureda-Gómez M et al (2018) Immunohistochemistry on paraffin-embedded planarian tissue sections. Methods Mol Biol 1774:367–378. https://doi.org/10.1007/978-1-4939-7802-1_11

5. Newmark PA, Sánchez Alvarado A (2000) Bromodeoxyuridine specifically labels the regenerative stem cells of planarians. Dev Biol 220: 142–153. https://doi.org/10.1006/dbio.2000.9645
6. Scimone ML, Kravarik KM, Lapan SW, Reddien PW (2014) Neoblast specialization in regeneration of the planarian Schmidtea mediterranea. Stem Cell Rep 3:339–352. https://doi.org/10.1016/j.stemcr.2014.06.001
7. David CN (1973) A quantitative method for maceration of hydra tissue. Wilhelm Roux Arch Entwickl Mech Org 171:259–268. https://doi.org/10.1007/BF00577724
8. Schneider KC (1890) Histologie von Hydra fusca mit besonderer Berücksichtigung des Nervensystems der Hydropolypen. Archiv f mikrosk Anatomie 35:321–379. https://doi.org/10.1007/BF02955882
9. Baguñà J (1976) Mitosis in the intact and regenerating planarian Dugesia mediterranea n.sp. I. Mitotic studies during growth, feeding and starvation. J Exp Zool 195:53–64. https://doi.org/10.1002/jez.1401950106
10. García-Castro H, Kenny NJ, Iglesias M et al (2021) ACME dissociation: a versatile cell fixation-dissociation method for single-cell transcriptomics. Genome Biol 22:89. https://doi.org/10.1186/s13059-021-02302-5
11. Grohme MA, Schloissnig S, Rozanski A et al (2018) The genome of Schmidtea mediterranea and the evolution of core cellular mechanisms. Nature 554:56–61. https://doi.org/10.1038/nature25473
12. Pearson BJ, Eisenhoffer GT, Gurley KA et al (2009) Formaldehyde-based whole-mount in situ hybridization method for planarians. Dev Dyn 238:443–450. https://doi.org/10.1002/dvdy.21849
13. Lauter G, Söll I, Hauptmann G (2011) Multicolor fluorescent in situ hybridization to define abutting and overlapping gene expression in the embryonic zebrafish brain. Neural Dev 6: 10. https://doi.org/10.1186/1749-8104-6-10
14. King RS, Newmark PA (2013) In situ hybridization protocol for enhanced detection of gene expression in the planarian Schmidtea mediterranea. BMC Dev Biol 13:8. https://doi.org/10.1186/1471-213X-13-8
15. Brown DDR, Pearson BJ (2015) One FISH, dFISH, three FISH: sensitive methods of whole-mount fluorescent in situ hybridization in freshwater planarians. In: Hauptmann G (ed) In situ hybridization methods. Springer, New York, pp 127–150
16. Caicedo JC, Singh S, Carpenter AE (2016) Applications in image-based profiling of perturbations. Curr Opin Biotechnol 39:134–142. https://doi.org/10.1016/j.copbio.2016.04.003
17. Bougen-Zhukov N, Loh SY, Lee HK, Loo L-H (2017) Large-scale image-based screening and profiling of cellular phenotypes. Cytometry A 91:115–125. https://doi.org/10.1002/cyto.a.22909
18. Caicedo JC, Cooper S, Heigwer F et al (2017) Data-analysis strategies for image-based cell profiling. Nat Methods 14:849–863. https://doi.org/10.1038/nmeth.4397
19. Sánchez Alvarado A, Newmark PA (1999) Double-stranded RNA specifically disrupts gene expression during planarian regeneration. Proc Natl Acad Sci USA 96:5049–5054. https://doi.org/10.1073/pnas.96.9.5049
20. Newmark PA, Reddien PW, Cebrià F, Sánchez Alvarado A (2003) Ingestion of bacterially expressed double-stranded RNA inhibits gene expression in planarians. Proc Natl Acad Sci USA 100(Suppl 1):11861–11865. https://doi.org/10.1073/pnas.1834205100
21. Orii H, Mochii M, Watanabe K (2003) A simple "soaking method" for RNA interference in the planarian Dugesia japonica. Dev Genes Evol 213:138–141. https://doi.org/10.1007/s00427-003-0310-3
22. Rouhana L, Weiss JA, Forsthoefel DJ et al (2013) RNA interference by feeding in vitro-synthesized double-stranded RNA to planarians: methodology and dynamics. Dev Dyn 242:718–730. https://doi.org/10.1002/dvdy.23950
23. Wagner DE, Wang IE, Reddien PW (2011) Clonogenic neoblasts are pluripotent adult stem cells that underlie planarian regeneration. Science 332:811–816. https://doi.org/10.1126/science.1203983
24. Wolf B, Lesnaw JA, Reichmann ME (1970) A mechanism of the irreversible inactivation of bovine pancreatic ribonuclease by diethylpyrocarbonate. A general reaction of diethylpyrocarbonate. A general reaction of diethylpyrocarbonate with proteins. Eur J Biochem 13:519–525. https://doi.org/10.1111/j.1432-1033.1970.tb00955.x
25. Havemeyer RN (1966) Freezing point curve of dimethyl sulfoxide--water solutions. J Pharm Sci 55:851–853. https://doi.org/10.1002/jps.2600550822
26. Green MR, Sambrook J (2012) Molecular cloning: a laboratory manual (fourth edition), volume 1, 2 & 3, 4th edn. Cold Spring Harbor Laboratory Press, Cold Spring Harbor
27. Burnett WV (1997) Northern blotting of RNA denatured in glyoxal without buffer recirculation. BioTechniques 22:668–671

28. Masek T, Vopalensky V, Suchomelova P, Pospisek M (2005) Denaturing RNA electrophoresis in TAE agarose gels. Anal Biochem 336: 46–50. https://doi.org/10.1016/j.ab.2004.09.010
29. Hopman AH, Ramaekers FC, Speel EJ (1998) Rapid synthesis of biotin-, digoxigenin-, trinitrophenyl-, and fluorochrome-labeled tyramides and their application for in situ hybridization using CARD amplification. J Histochem Cytochem 46:771–777. https://doi.org/10.1177/002215549804600611
30. Panchuk-Voloshina N, Haugland RP, Bishop-Stewart J et al (1999) Alexa dyes, a series of new fluorescent dyes that yield exceptionally bright, photostable conjugates. J Histochem Cytochem 47:1179–1188. https://doi.org/10.1177/002215549904700910
31. Berlier JE, Rothe A, Buller G et al (2003) Quantitative comparison of long-wavelength Alexa Fluor dyes to cy dyes: fluorescence of the dyes and their bioconjugates. J Histochem Cytochem 51:1699–1712. https://doi.org/10.1177/002215540305101214
32. Hughes LD, Rawle RJ, Boxer SG (2014) Choose your label wisely: water-soluble fluorophores often interact with lipid bilayers. PLoS One 9:e87649. https://doi.org/10.1371/journal.pone.0087649
33. Grimm JB, Lavis LD (2022) Caveat fluorophore: an insiders' guide to small-molecule fluorescent labels. Nat Methods 19:149–158. https://doi.org/10.1038/s41592-021-01338-6
34. Chomczynski P (1992) Solubilization in formamide protects RNA from degradation. Nucleic Acids Res 20:3791–3792. https://doi.org/10.1093/nar/20.14.3791
35. Hattori M, Miyamoto M, Hosoda K, Umesono Y (2018) Usefulness of multiple chalk-based food colorings for inducing better gene silencing by feeding RNA interference in planarians. Develop Growth Differ 60:76–81. https://doi.org/10.1111/dgd.12413
36. Thommen A, Werner S, Frank O et al (2019) Body size-dependent energy storage causes Kleiber's law scaling of the metabolic rate in planarians. eLife 8. https://doi.org/10.7554/eLife.38187
37. Collins JS, Goldsmith TH (1981) Spectral properties of fluorescence induced by glutaraldehyde fixation. J Histochem Cytochem 29: 411–414
38. Pall GS, Hamilton AJ (2008) Improved northern blot method for enhanced detection of small RNA. Nat Protoc 3:1077–1084. https://doi.org/10.1038/nprot.2008.67
39. Pena JTG, Sohn-Lee C, Rouhanifard SH et al (2009) miRNA in situ hybridization in formaldehyde and EDC-fixed tissues. Nat Methods 6: 139–141. https://doi.org/10.1038/nmeth.1294
40. Thisse C, Thisse B (2008) High-resolution in situ hybridization to whole-mount zebrafish embryos. Nat Protoc 3:59–69. https://doi.org/10.1038/nprot.2007.514
41. Fernandéz-Taboada E, Moritz S, Zeuschner D et al (2010) Smed-SmB, a member of the LSm protein superfamily, is essential for chromatoid body organization and planarian stem cell proliferation. Development 137:1055–1065. https://doi.org/10.1242/dev.042564
42. Reddien PW, Oviedo NJ, Jennings JR et al (2005) SMEDWI-2 is a PIWI-like protein that regulates planarian stem cells. Science 310:1327–1330. https://doi.org/10.1126/science.1116110
43. Hayashi T, Asami M, Higuchi S et al (2006) Isolation of planarian X-ray-sensitive stem cells by fluorescence-activated cell sorting. Develop Growth Differ 48:371–380. https://doi.org/10.1111/j.1440-169X.2006.00876.x
44. Moritz S, Stöckle F, Ortmeier C et al (2012) Heterogeneity of planarian stem cells in the S/G2/M phase. Int J Dev Biol 56:117–125. https://doi.org/10.1387/ijdb.113440sm
45. Hauri HP, Bucher K (1986) Immunoblotting with monoclonal antibodies: importance of the blocking solution. Anal Biochem 159:386–389. https://doi.org/10.1016/0003-2697(86)90357-x
46. Vogt RF, Phillips DL, Henderson LO et al (1987) Quantitative differences among various proteins as blocking agents for ELISA microtiter plates. J Immunol Methods 101:43–50. https://doi.org/10.1016/0022-1759(87)90214-6
47. Haycock JW (1993) Polyvinylpyrrolidone as a blocking agent in immunochemical studies. Anal Biochem 208:397–399. https://doi.org/10.1006/abio.1993.1068
48. Denhardt DT (1966) A membrane-filter technique for the detection of complementary DNA. Biochem Biophys Res Commun 23: 641–646. https://doi.org/10.1016/0006-291x(66)90447-5
49. Smith PJ, Wiltshire M, Davies S et al (1999) A novel cell permeant and far red-fluorescing DNA probe, DRAQ5, for blood cell discrimination by flow cytometry. J Immunol Methods 229:131–139. https://doi.org/10.1016/S0022-1759(99)00116-7
50. Manning CF, Bundros AM, Trimmer JS (2012) Benefits and pitfalls of secondary

antibodies: why choosing the right secondary is of primary importance. PLoS One 7:e38313. https://doi.org/10.1371/journal.pone.0038313

51. Martin RM, Brady JL, Lew AM (1998) The need for IgG2c specific antiserum when isotyping antibodies from C57BL/6 and NOD mice. J Immunol Methods 212:187–192. https://doi.org/10.1016/s0022-1759(98)00015-5
52. Collins AM (2016) IgG subclass co-expression brings harmony to the quartet model of murine IgG function. Immunol Cell Biol 94: 949–954. https://doi.org/10.1038/icb.2016.65
53. Brown D, Lydon J, McLaughlin M et al (1996) Antigen retrieval in cryostat tissue sections and cultured cells by treatment with sodium dodecyl sulfate (SDS). Histochem Cell Biol 105: 261–267. https://doi.org/10.1007/BF01463929
54. Hausen P, Dreyer C (1982) Urea reactivates antigens in paraffin sections for immunofluorescent staining. Stain Technol 57:321–324. https://doi.org/10.3109/10520298209066731
55. Peränen J, Rikkonen M, Kääriäinen L (1993) A method for exposing hidden antigenic sites in paraformaldehyde-fixed cultured cells, applied to initially unreactive antibodies. J Histochem Cytochem 41:447–454. https://doi.org/10.1177/41.3.8429208
56. Magaud JP, Sargent I, Clarke PJ et al (1989) Double immunocytochemical labeling of cell and tissue samples with monoclonal anti-bromodeoxyuridine. J Histochem Cytochem 37:1517–1527. https://doi.org/10.1177/37.10.2476478
57. Cheng L-C, Alvarado AS (2018) Whole-mount BrdU staining with fluorescence in situ hybridization in planarians. Methods Mol Biol 1774:423–434. https://doi.org/10.1007/978-1-4939-7802-1_15
58. Catherine O, Therese W (1968) Quenching of singlet oxygen by tertiary aliphatic amines. Effect of DABCO (1,4-diazabicyclo[2.2.2] octane). J Am Chem Soc 90:6527–6528. https://doi.org/10.1021/ja01025a059
59. Johnson GD, Davidson RS, McNamee KC et al (1982) Fading of immunofluorescence during microscopy: a study of the phenomenon and its remedy. J Immunol Methods 55:231–242. https://doi.org/10.1016/0022-1759(82)90035-7
60. Langanger G, De Mey J, Adam H (1983) 1,4-Diazobicyclo-(2,2,2)-octane (DABCO) retards the fading of immunofluorescence preparations. Mikroskopie 40:237–241
61. Nadin-Davis S, Mezl VA (1982) Optimization of the ethanol precipitation of RNA from formamide containing solutions. Prep Biochem 12: 49–56. https://doi.org/10.1080/00327488208065549
62. Thavarajah R, Mudimbaimannar VK, Elizabeth J et al (2012) Chemical and physical basics of routine formaldehyde fixation. J Oral Maxillofac Pathol 16:400–405. https://doi.org/10.4103/0973-029X.102496
63. Ligasová A, Liboska R, Rosenberg I, Koberna K (2015) The fingerprint of anti-bromodeoxyuridine antibodies and its use for the assessment of their affinity to 5-Bromo-2'--Deoxyuridine in cellular DNA under various conditions. PLoS One 10:e0132393. https://doi.org/10.1371/journal.pone.0132393
64. Ligasová A, Konečný P, Frydrych I, Koberna K (2017) Looking for ugly ducklings: the role of the stability of BrdU-antibody complex and the improved method of the detection of DNA replication. PLoS One 12:e0174893. https://doi.org/10.1371/journal.pone.0174893

Chapter 9

An RNA/DNA-Based Flow Cytometry Approach for Isolating Slow-Cycling Stem Cells

Nicole Lindsay-Mosher, Alyssa M. Molinaro, and Bret J. Pearson

Abstract

Flow cytometry methods for sorting specific populations of cells based on fluorescence or physical properties have been a widely used technique for decades. Flow cytometry has been particularly vital to the study of planarians, which remain refractory to transgenic transformation, as it has provided a work-around solution for studying stem cell biology and lineage relationships in the context of regeneration. Many flow cytometry applications have been published in planarians, beginning with broad Hoechst-based strategies for isolating cycling stem cells and progressing to more function-based approaches involving vital dyes and surface antibodies. In this protocol, we look to build on the classic DNA-labeling Hoechst staining strategy by adding pyronin Y staining to label RNA. While Hoechst labeling alone allows for the isolation of stem cells in the S/G2/M phases of the cell cycle, heterogeneity within the population of stem cells with 2 C DNA content is not resolved. By considering RNA levels, this protocol can further divide this population of stem cells into two groups: G1 stem cells with relatively high RNA content and a slow-cycling population with low RNA content, which we call RNA^{low} stem cells. In addition, we provide instruction for combining this RNA/DNA flow cytometry protocol with EdU labeling experiments and describe an optional step for incorporating immunostaining prior to cell sorting (in this case with the pluripotency marker TSPAN-1). This protocol adds a new staining strategy and examples of combinatorial flow cytometry approaches to the repertoire of flow cytometry techniques for studying planarian stem cells.

Key words Flow cytometry, Planarians, Hoechst, Pyronin Y, CellMask, EdU, Stem cells

1 Introduction

The ability to automatically sort single cells based on their physical characteristics was developed almost 50 years ago [1]. The main principle is that cells in suspension can be flowed through capillaries until there is 1 cell per microdroplet. Laser light is directed on the droplets as they flow through, and the absorption, emission, and scattering of the laser light provide information about the size and granularity of each cell [2]. This was a key development that enabled the analysis and purification of many thousands of cells at once and was particularly useful prior to the advent of fluorescent

Luca Gentile (ed.), *Schmidtea Mediterranea: Methods and Protocols*, Methods in Molecular Biology, vol. 2680,
https://doi.org/10.1007/978-1-0716-3275-8_9,
© The Author(s), under exclusive license to Springer Science+Business Media, LLC, part of Springer Nature 2023

transgenes, with major impacts on the fields of stem cell and cancer cell biology [1]. Over the past decade, and with the increasing use of single-cell genomics, flow cytometry is as useful as ever. Cells can now be sorted by a vast array of fluorescent proteins and vital dyes and then subjected to a number of single-cell techniques [3]. In total, although still only used by a low percentage of laboratories, flow cytometry has become an essential technique to modern molecular biology.

One model system where flow cytometry is essential is the freshwater planarian (flatworm of the phylum Platyhelminthes), which has become a powerful model of stem cell biology and regeneration [4]. Despite being amenable to most molecular biological techniques, planarians have remained immune to transformation with exogenous transgenes. In order to get around this, flow cytometry methods have been developed and used to identify important cell populations in the animal. Over the past few years, single-cell genomics has dominated the planarian field, and the wide use of these techniques ultimately relies on robust protocols for flow cytometry [5–9].

In planarians, the first flow cytometry protocol to isolate stem cells was developed by Hayashi et al. in 2006, using the vital dyes Hoechst to measure DNA content and calcein to label metabolically active cells [10]. From this, the authors described three different cell populations that they could reliably gate based on their sensitivity or insensitivity to irradiation: actively cycling stem cells (called neoblasts) with >2 C DNA content (the "X1" gate), a classic side population of cells that efflux Hoechst dye (the "X2" gate), and fully differentiated, irradiation-insensitive cells (the "Xins" gate) [10]. With this protocol in hand, it was immediately combined with molecular markers to confirm that the X1 gate was in fact relatively homogenous for the neoblast marker *piwi-1* [11]. Additional dyes have since been applied to this technique for further analysis of stem and progenitor cell populations, including Mitotracker Green to measure mitochondrial mass and Annexin V to quantify cell death [12, 13]. The ability to isolate cell populations has led to many significant advances in the planarian field, including the discovery of the clonogenic neoblast [14], descriptions of lineage-committed stem cells [15, 16], and elucidation of the first cell type lineages [17, 18]. Recently, this technique was used to identify a cell-surface marker of pluripotent neoblasts, TSPAN-1, allowing the prospective isolation of these cells using fluorescence-activated cell sorting (FACS) [8]. RNA sequencing of X1 and X2 populations has played a key role in many of these discoveries; as sequencing technology becomes increasingly more powerful and widely available, techniques used for isolating cell populations of interest are more important than ever.

Here, we build on the classical Hoechst separation method by adding the RNA dye pyronin Y, allowing for simultaneous analysis of RNA and DNA content. This strategy can be used to isolate a population of slow-cycling stem cells with very small physical size and low RNA content, termed the "RNA^{low}" gate, allowing for further study of heterogeneity within the planarian stem cell compartment [19]. We show how this method can be combined with antibody staining for targets such as TSPAN-1 and describe how this method can be coupled with in vivo cell cycle analyses through administration of EdU, isolation of cells by FACS, and quantification of EdU$^+$ cells. Finally, we outline methods for visualization and measurement of cell size following isolation by FACS using the vital dye CellMask. Together, these protocols provide robust new tools for the analysis and isolation of planarian cell populations by flow cytometry.

2 Materials

Prepare all stock solutions and buffers ahead of time and store at room temperature, unless stated otherwise.

2.1 EdU Administration

1. High-salt medium: 5 g Instant Ocean Aquarium Salt in 1 L water.
2. DMSO (dimethyl sulfoxide).
3. EdU stock solution: 50 mg/mL 5-ethynyl-2′-deoxyuridine in DMSO. Aliquot and store at −20 °C.
4. Food coloring.
5. Automatic nanoliter injector.
6. Pulled glass micropipets.
7. Dissecting microscope.
8. Mineral oil.
9. Cold plate.
10. Filter paper.

2.2 Cell Dissociation and Fluorescence-Activated Cell Sorting

1. CMF (calcium/magnesium-free media): 200 mg NaH_2PO_4, 400 mg NaCl, 600 mg KCl, 400 mg $NaHCO_3$, 1.2 g glucose, 5 mg BSA, 15 mM HEPES. Add water to 500 mL. Mix and adjust pH to 7.3 with HCl. Store at 4 °C.
2. Plastic pestles.
3. Hoechst stock solution: 10 mg/mL Hoechst 342 in water. Store in the dark at 4 °C.
4. Pyronin Y stock solution: 10 mg Pyronin Y in 33 mL water. Store in the dark at 4 °C.
5. Anti-TSPAN-1 antibody.

6. Anti-rabbit secondary antibody conjugated to APC (allophycocyanin).
7. 40-μM cell strainers.

2.3 CellMask Staining and Cell Area Quantification

1. 96-well Cellview microplate.
2. CMF (calcium/magnesium-free media): 200 mg NaH_2PO_4, 400 mg NaCl, 600 mg KCl, 400 mg $NaHCO_3$, 1.2 g glucose, 5 mg BSA, 15 mM HEPES. Add water to 500 mL. Mix and adjust pH to 7.3 with HCl. Store at 4 °C.
3. Phosphate-buffered saline (PBS).
4. Fixation solution: 4% formaldehyde in PBS. Prepare fresh.
5. CellMask staining solution: 1:500 in PBS. Prepare fresh.
6. DAPI (4′,6-diamidino-2-phenylindole) staining solution: 1: 1000 in PBS.
7. Click reaction solution: 1 mM $CuSO_4$, 0.01 mM azide-fluor 488, 10 mM ascorbic acid in water. 200 μL of click reaction solution is required per sample. Prepare fresh.
8. Mounting medium: 100 μL EDTA, 500 μL Tris–HCl pH 7.5, 40 mL glycerol, 9.4 mL water.
9. ImageJ software.

3 Methods

Carry out all procedures at room temperature. Generally, 30 medium-sized planarians (6–8 mM in length), starved for at least 3 days, are used per sample (*see* **Note 1**). In order to reliably define FACS gates, an irradiated control sample (exposed to 60 Gy of irradiation and collected at 1 day post-irradiation) as well as a nonirradiated control sample (which may consist of an experimental sample, such as a *control(RNAi)* sample) should be included with each experiment.

3.1 EdU Administration

1. Collect starved animals and gradually adapt them to high-salt medium for 2 days prior to EdU administration (*see* **Note 2**).
2. Prepare a 5 mg/mL working EdU solution by diluting 1 μL of EdU stock solution in 8.5 μL water plus 0.5 μL food coloring.
3. Position the cold plate under the dissecting microscope and set up the nanoliter injector next to the microscope. Secure a glass needle to the injector and backfill with mineral oil as per the manufacturer's instructions.
4. Fill the needle by drawing up the EdU working solution.

5. Place a small piece of filter paper on the cold plate and dampen with planarian water.
6. Transfer planarians onto the filter paper in as little water as possible. Position the animals with their ventral side facing up.
7. At a 45-degree angle, insert the needle through the ventral side and into the gut. Inject 64.4 nL of EdU working solution (*see* **Note 3** and Fig. 1a).
8. Transfer the injected planarians to a clean petri dish containing high-salt medium. Perform daily water changes for the duration of the chase period.

3.2 Cell Dissociation and Fluorescence-Activated Cell Sorting

1. Collect 30 worms per sample into 1.5 mL Eppendorf tubes (*see* **Note 4** and Fig. 1b).
2. Remove planarian water and rinse animals twice with 1 mL cold CMF (*see* **Note 5**), and then place them in 750 μL of fresh CMF.
3. Dounce animals in Eppendorf tubes with a sterile plastic pestle until a cloudy cell suspension without large pieces of tissue is achieved.
4. Place a 40 μM cell strainer over a 50 mL conical. Wet the filter with 750 μL CMF.
5. Using a cut p1000 pipet tip, pipet the 750 μL cell suspension through the pre-wet strainer.
6. Rinse the strainer with an additional 1500 μL CMF.
7. *Optional*: Anti-TSPAN-1 immunostaining can be performed at this point (*see* **Note 6**). Add anti-TSPAN-1 primary antibody to the cell suspension at a concentration of 1:500. Incubate for 1 h. Centrifuge the cell suspension at 300 × g for 5 min. Rinse the cell pellet with cold CMF, and then resuspend in cold CMF containing anti-rabbit secondary antibody conjugated to APC at a concentration of 1:500. Incubate in the dark for 10 min. Proceed through **steps 8–10** without removing the secondary antibody solution to achieve a total incubation time of 30 min.
8. Add 2.5 μL/mL Hoechst stock solution to the cell suspension and incubate for 10 min in the dark.
9. Without removing the Hoechst solution, add 50 μL/mL Pyronin Y stock solution (to achieve a 50 mM working concentration, *see* **Note 7** and Fig. 2a) and incubate for an additional 10 min in the dark.
10. During incubation, prepare cell collection tubes for each sample by placing 50 μL cold CMF into 1.5 mL Eppendorf tubes. Keep on ice until collection.
11. Centrifuge the cell suspension at 300 × g for 5 min (*see* **Note 8**). Resuspend the cell pellet in 500 μL cold CMF and immediately proceed to cell sorting.

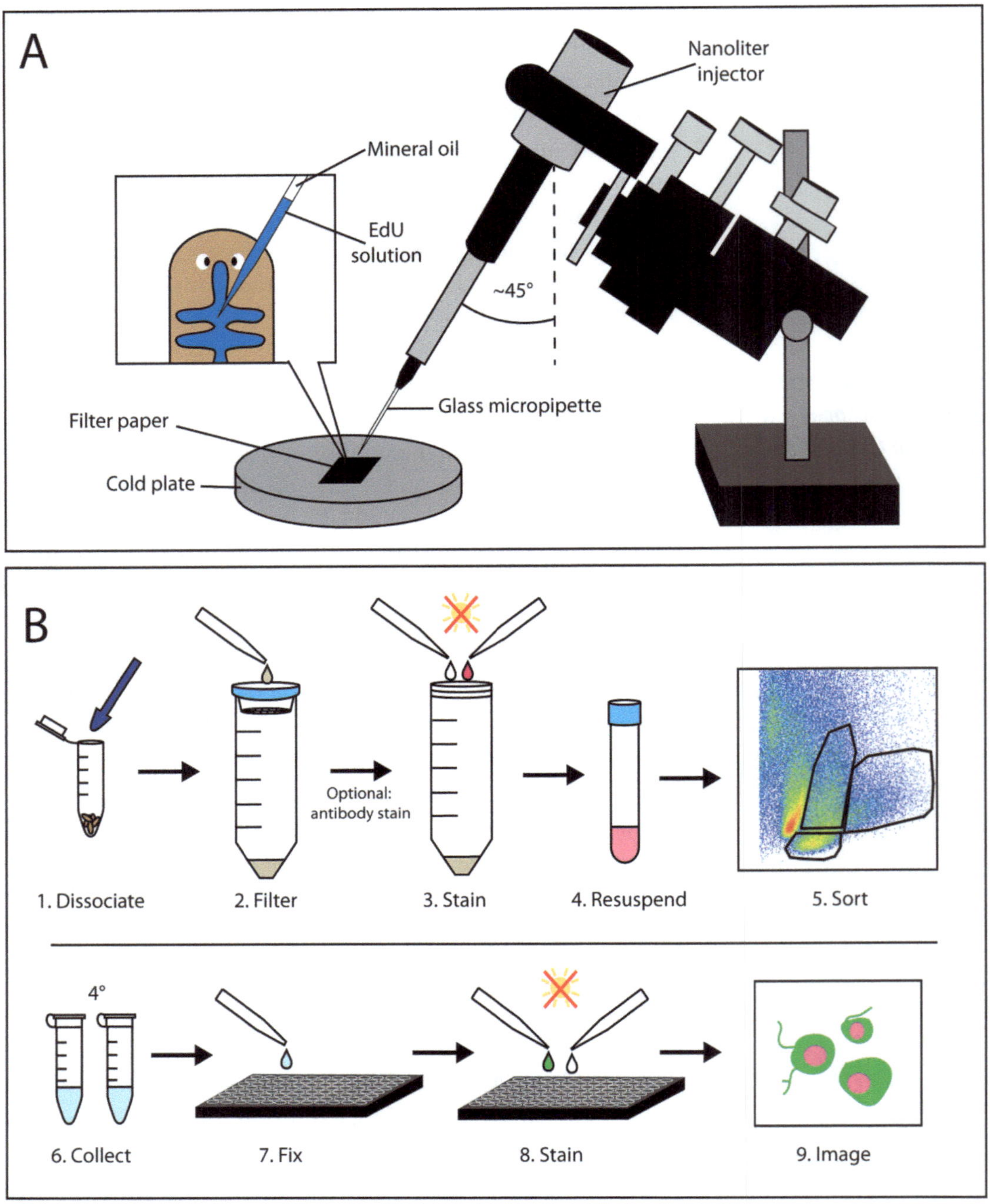

Fig. 1 Experimental schematics. Schematics of the (**a**) EdU injection strategy and (**b**) flow cytometry and cell staining protocols

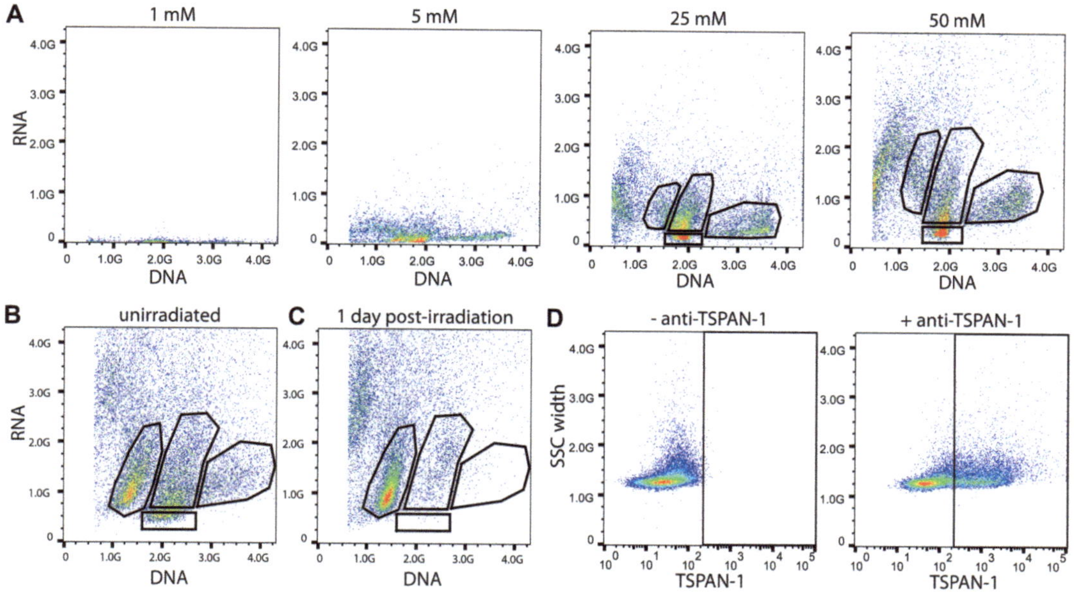

Fig. 2 Examples of flow cytometry plots representing different cell stains. (**a**) FACS plots showing gate separation with different concentrations of Pyronin Y. (**b**, **c**) Examples of FACS plots from (**b**) an unirradiated control sample and (**c**) an irradiated sample at 1 dpi. (**d**) FACS plots of TSPAN-1-negative and TSPAN-1-positive samples

12. To set FACS gates, select the appropriate laser inputs to construct a Hoechst blue (x-axis) × Pyronin Y (y-axis) FACS plot (*see* **Note 9**). Begin by running a nonirradiated control sample. Allow the sample to run until the distribution of events on the FACS plot stabilizes (*see* **Note 10**). Draw approximate gates on the FACS plot according to Fig. 2b. Switch to the irradiated control sample. Allow the sample to run until the distribution of events on the FACS plot stabilizes. The Xins and X1 gates should be populated, while the RNA^{low} and S/G2/M gates should be largely depleted. Adjust the gates to satisfy this expected phenotype (Fig. 2c).
13. If TSPAN-1 immunostaining was performed, also construct an APC (x-axis) x side scatter width (y-axis) FACS plot to visualize TSPAN-1 detection (*see* **Note 9** and Fig. 2d).
14. Proceed to run all experimental samples, being sure to allow each sample to run until the distribution of events on the FACS plot stabilizes before beginning collection. Collect the required number of cells from the desired gates into the prepared collection tubes (*see* **Note 11**). Store collected samples on ice until fixation.
15. Save resulting FACS data as .fcs files for future analysis if desired.

3.3 CellMask Staining and Cell Area Quantification

1. Immediately following cell sorting, proceed to fixation by using a cut pipet tip to transfer each collected sample into a well in a 96-well Cellview microplate. Top up each well with cold CMF to 200 μL and allow cells to settle to the bottom for 10 min.
2. Freshly prepare fixation solution (200 μL per well). Carefully hold the Cellview plate at a slight angle and, using a Pasteur pipet, gently aspirate the CMF (*see* **Note 12**). Pipet 200 μL of fixation solution into each well. Fix for 15 min.
3. Prepare CellMask staining solution. Remove the fixation solution and rinse each well with PBS, then add 200 μL of CellMask staining solution to each well, and incubate in the dark for 20 min.
4. Remove the CellMask solution and add 200 μL of DAPI staining solution to each well. Incubate in the dark for 20 min.
5. Remove the DAPI staining solution and add 200 μL of mounting medium to each well. Store the Cellview plate at 4 °C in the dark until ready to image.
6. Use a confocal microscope equipped with a 60× objective to obtain images of the CellMask and DAPI-stained cells. We recommend acquiring a z-stack spanning the height of the cells, with a maximum spacing of 2 μM between z-planes (*see* **Note 13**).
7. Z-stacks can be imported into ImageJ software for area quantification. Open the image file in ImageJ. Convert the z-stack into a maximum projection image by selecting Image > Stacks > Z Project. Use the Freehand Selection tool to trace the edges of a cell (*see* **Note 14** and Fig. 3). Then select Analyze > Measure to obtain the area of the selection.

3.4 EdU Detection

1. Transfer collected samples to a 96-well Cellview microplate and fix as described above (see **steps 1–2** in Subheading 3.3).
2. Freshly prepare the Click reaction solution. Following fixation, rinse each well with PBS, and then replace the PBS with 200 μL of Click reaction solution (*see* **Note 15**). Incubate in the dark for 30 min.
3. Gently aspirate the click reaction solution and rinse wells with PBS.
4. Aspirate the PBS and incubate the cells in DAPI staining solution in the dark for 20 min.
5. Replace the DAPI staining solution with 200 μL of mounting medium. Store at 4 °C in the dark until ready to image.
6. Use a confocal microscope equipped with a 20× objective to obtain images of the stained cells (*see* **Note 16** and Fig. 4).

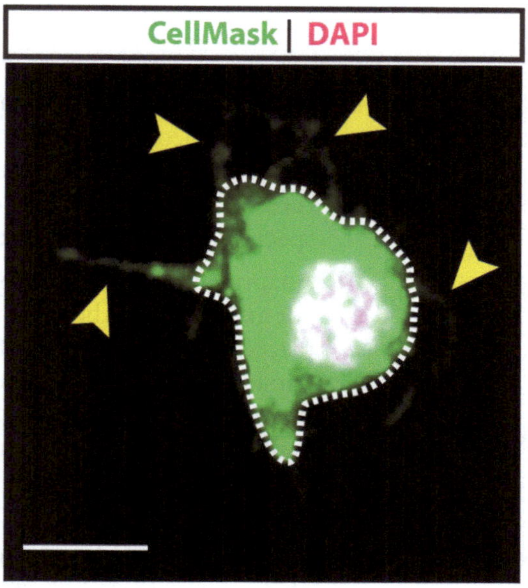

Fig. 3 Example of cell area quantification. The image shown is a maximum projection of an RNA^{low} cell at 48 h post-amputation. The white dashed line indicates the area measured. Cell projections (yellow arrow heads) are excluded from the area measurement. Scale bar, 5 μM

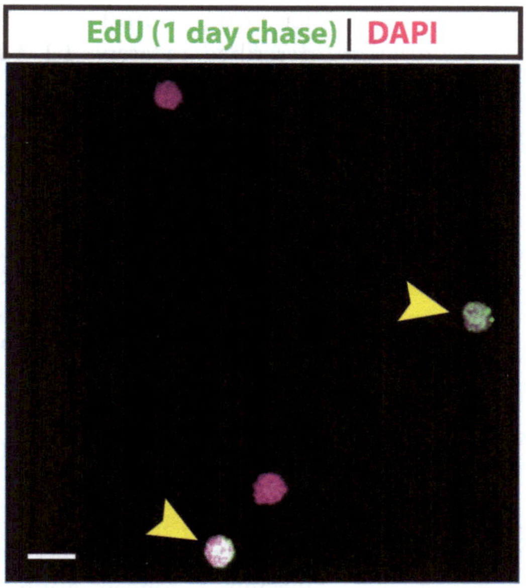

Fig. 4 Example of EdU-positive and EdU-negative cells. The image shown is a maximum projection of RNA^{low} cells at 48 h post-amputation. Yellow arrowheads indicate EdU^{+} cells. Scale bar, 5 μM

4 Notes

1. The Hoechst and Pyronin Y concentrations recommended in this protocol were optimized for samples consisting of 30, ~6-mm-long planarians. It may be necessary to adjust these concentrations for different-sized animals or different species, or to improve gate separation during cell sorting.
2. Adapting animals to high-salt medium helps to prevent or limit edema caused by the injection of EdU solution. Animals can be placed in a 1:1 mixture of high-salt medium and regular planarian water for 1 day before being transferred to high-salt medium for an additional day. Following EdU injection, animals should be maintained in high-salt medium for the duration of the chase period.
3. The gut is easiest to hit in the neck area, just anterior to the pharynx. Successful injection into the gut will result in an obvious accumulation of the colored EdU working solution. The volume of EdU solution injected need not be exact: for optimal labelling, continue to inject EdU until the gut appears full of the colored solution.
4. When comparing between experimental samples, it is imperative that all samples be prepared and sorted side by side to minimize technical variability. FACS gates should always be set using control samples (irradiated and nonirradiated, as described in **steps 12–14** of Subheading 3.2) and applied, without adjustment, to all samples within a given experiment.
5. Animals should be rinsed with CMF immediately prior to dissociation. Whole animals left for several minutes in cold CMF will release mucus, resulting in clumping of the cells during the staining steps.
6. If including TSPAN-1 detection, include a negative control sample in which the anti-TSPAN-1 primary antibody was either omitted (secondary antibody only) or, ideally, replaced with an isotype control antibody. This will be required to accurately define a TSPAN-1^+ FACS gate during data analysis.
7. As presented in Fig. 2a, 50 mM Pyronin Y yielded good separation of the various cell populations, making gating easier, but may also result in the loss of RNA^{high} cells out of the range of the y-axis due to very high pyronin uptake. Reducing the Pyronin Y concentration, as well as the voltage on the FACS instrument, will result in a lower Pyronin Y fluorescence reading and shift the events down toward the x-axis. Thus, more RNA^{high} cells will be visible at the expense of clear separation between populations.
8. If the option exists, reduce the brake speed on the centrifuge to prevent kicking up the cell pellet during deceleration.

9. Excitation and emission peaks: Hoechst, excitation at 361 nm and emission at 497 nm; Pyronin Y, excitation at 547 nm and emission at 566 nm; and APC, excitation at 650 nm and emission at 660 nm.
10. The final cell suspension will be bright pink in color because of Pyronin Y staining. As a result, it is necessary to run the sample through the sorter for several minutes prior to cell or data collection in order to equilibrate the system with the darker solution. During this time, the distribution of events on the live FACS plot may appear to migrate from the bottom right corner toward the top left of the plot. Once the system has equilibrated to the sample, the distribution of events will stabilize, and collection may commence. This process can take around 5 min. If the final sample was very concentrated, equilibration may take longer, and it is advised to dilute the sample further with CMF.
11. For computational analysis of the FACS plots, we recommend saving data for 500,000 events per sample. For cell imaging applications, we recommend collecting a minimum of 10,000 cells per gate per sample. Note that some cells will be lost with solution exchanges during staining.
12. Aspirate gently against the side of the well to avoid dislodging cells and minimize sample loss.
13. Overexposure of the cell body may be required to visualize cellular projections; however, for cell area quantifications, it is important to avoid overexposure as the blown-out signal will artificially increase the area measurement.
14. Due to their negligible contribution to the overall cell area, we recommend excluding cellular projections from cell area quantifications.
15. For EdU experiments, make sure to include a negative control sample collected from animals that were not injected with EdU and perform a click reaction on this alongside experimental samples. This will be helpful for distinguishing true EdU detection from background during imaging.
16. To quantify the proportion of EdU^+ cells in a given sample, it is necessary to randomly image the well by identifying cells based solely on the detection of $DAPI^+$ nuclei.

References

1. Robinson JP, Roederer M (2015) Flow cytometry strikes gold. Science 350(80):739–740
2. McKinnon KM (2018) Flow cytometry: an overview. Curr Protoc Immunol 2018: 5.1.1–5.1.11. https://doi.org/10.1002/cpim.40
3. Andreyev DS, Zybailov BL (2020) Integration of flow cytometry and single cell sequencing. Trends Biotechnol 38:133–136
4. Reddien PW (2018) The cellular and molecular basis for planarian regeneration. Cell 175:327–345

5. Plass M, Solana J, Alexander Wolf F et al (2018) Cell type atlas and lineage tree of a whole complex animal by single-cell transcriptomics. Science 360(80). https://doi.org/10.1126/science.aaq1723
6. Fincher CT, Wurtzel O, de Hoog T et al (2018) Cell type transcriptome atlas for the planarian Schmidtea mediterranea. Science eaaq1736. https://doi.org/10.1126/science.aaq1736
7. Forsthoefel DJ, Cejda NI, Khan UW, Newmark PA (2020) Cell-type diversity and regionalized gene expression in the planarian intestine. elife 9. https://doi.org/10.7554/eLife.52613
8. Zeng A, Li H, Guo L et al (2018) Prospectively isolated Tetraspanin+ Neoblasts are adult pluripotent stem cells underlying Planaria regeneration. Cell 173:1593–1608.e20. https://doi.org/10.1016/j.cell.2018.05.006
9. Raz AA, Wurtzel O, Reddien PW (2021) Planarian stem cells specify fate yet retain potency during the cell cycle. Cell Stem Cell. https://doi.org/10.1016/j.stem.2021.03.021
10. Hayashi T, Asami M, Higuchi S et al (2006) Isolation of planarian X-ray-sensitive stem cells by fluorescence-activated cell sorting. Develop Growth Differ 48:371–380. https://doi.org/10.1111/j.1440-169X.2006.00876.x
11. Reddien PW, Oviedo NJ, Jennings JR et al (2005) SMEDWI-2 is a PIWI-like protein that regulates planarian stem cells. Science 310(80):1327–1330. https://doi.org/10.1126/science.1116110
12. Mohamed Haroon M, Lakshmanan V, Sarkar SR et al (2021) Mitochondrial state determines functionally divergent stem cell population in planaria. Stem Cell Rep 16:1302–1316. https://doi.org/10.1016/j.stemcr.2021.03.022
13. Peiris TH, García-Ojeda ME, Oviedo NJ (2016) Alternative flow cytometry strategies to analyze stem cells and cell death in planarians. Regeneration 3:123–135. https://doi.org/10.1002/reg2.53
14. Wagner DE, Wang IE, Reddien PW (2011) Clonogenic neoblasts are pluripotent adult stem cells that underlie planarian regeneration. Science 332(80):811–816. https://doi.org/10.1126/science.1203983
15. van Wolfswinkel JC, Wagner DE, Reddien PW (2014) Single-cell analysis reveals functionally distinct classes within the planarian stem cell compartment. Cell Stem Cell 15:326–339. https://doi.org/10.1016/j.stem.2014.06.007
16. Molinaro AM, Pearson BJ (2016) In silico lineage tracing through single cell transcriptomics identifies a neural stem cell population in planarians. Genome Biol 17:87. https://doi.org/10.1186/s13059-016-0937-9
17. Eisenhoffer GT, Kang H, Alvarado AS (2008) Molecular analysis of stem cells and their descendants during cell turnover and regeneration in the planarian Schmidtea mediterranea. Cell Stem Cell 3:327–339. https://doi.org/10.1016/j.stem.2008.07.002
18. Zhu SJ, Hallows SE, Currie KW et al (2015) A *mex3* homolog is required for differentiation during planarian stem cell lineage development. elife 4:304–311. https://doi.org/10.7554/eLife.07025
19. Molinaro AM, Lindsay-Mosher N, Pearson BJ (2021) Identification of TOR-responsive slow-cycling neoblasts in planarians. EMBO Rep 22. https://doi.org/10.15252/embr.202050292

Chapter 10

ACME Dissociation-Fixation, Flow Cytometry, and Cell Sorting of Freshwater Planarian Cells

Helena García-Castro, Elena Emili, and Jordi Solana

Abstract

Planarian cell dissociation methods using enzymatic approaches are well established and have been widely used in the field. However, their use in transcriptomics and especially single-cell transcriptomics raises concerns as cells are dissociated alive, and this induces cellular stress responses. Here we describe a protocol for planarian cell dissociation using ACME, a dissociation-fixation approach based on acetic acid and methanol. ACME-dissociated cells are fixed, can be cryopreserved, and are amenable to modern methods of single-cell transcriptomics.

Key words Cell dissociation, Fixation, Flow cytometry, FACS, Planarian, Regeneration

1 Introduction

Cell dissociation is a fundamental method for cell biology and, in particular, for single-cell biology. In planarians [1], Romero and Baguñà used cell maceration with acetic acid as a dissociation method to observe the morphology of each cell type and quantify their abundance upon different conditions [2]. The maceration approach had been earlier introduced in cnidarians [3]. Briefly, when exposed to the acetic acid action, cells are fixed and dissociated. Romero and Baguñà added methanol to the original formula as it preserved better morphology.

Single-cell transcriptomic techniques were introduced later [4] and applied to planarians [5–8], using a trypsin dissociation approach developed by Hayashi and coworkers [9]. However, studies in other organisms have begun to indicate that enzymatic dissociation induces cellular stress, as the cells are dissociated while still metabolically alive. This in turn translates to changes at the gene expression levels [10, 11], which can affect single-cell expression studies.

Luca Gentile (ed.), *Schmidtea Mediterranea: Methods and Protocols*, Methods in Molecular Biology, vol. 2680,
https://doi.org/10.1007/978-1-0716-3275-8_10,
© The Author(s), under exclusive license to Springer Science+Business Media, LLC, part of Springer Nature 2023

To address this, we used the principle of acetic acid maceration to develop our ACME (acetic-methanol) dissociation protocol [12] for single-cell transcriptomic studies. ACME-dissociated cells are simultaneously fixed and therefore cannot develop the stress responses associated with enzymatic protocols. We introduced RNA-preserving measures to guarantee RNA integrity and implemented cryopreservation methods to make the use of ACME cells more flexible. We also developed a simple staining protocol for fixed and permeable cells, making imaging and cell sorting possible. Moreover, we showed that ACME works in many organisms beyond planarians. Finally, we proved that cells dissociated with ACME can be used in single-cell transcriptomics using both droplet-based methods and also more novel combinatorial barcoding methods.

In this chapter, we offer a detailed version of the ACME dissociation-fixation protocol, together with the cell cytometry imaging and sorting protocol for ACME-dissociated cells, in the planarian species *Schmidtea mediterranea*.

2 Materials

Prepare all solutions at room temperature, using molecular biology grade reagents and ultrapure water. Keep reagents and buffers at room temperature, unless otherwise stated. Use RNAse-free plasticware and reagents and maintain a general RNAse-free environment when handling samples.

2.1 ACME Dissociation-Fixation

1. 10× PBS buffer: 80 g of NaCl, 2 g of KCl, 2 g of monobasic KH_2PO_4, 11.5 g of Na_2HPO_4, up to 1 L of ultrapure water. Mix well, adjust pH to 7.4 with HCl, and autoclave. This stock solution can be aliquoted and stored for months.
2. Washing and resuspension buffer (1% BSA in 1× PBS): 1× PBS (diluted from 10× PBS buffer), 1% weight of BSA powder (store at 4 °C; *see* **Note 1**). Mix well with gentle agitation to avoid bubbles. Prepare and use on the same day. Prepare 10 mL of washing and resuspension buffer per sample (*see* **Note 2**).
3. NAC solution (7.5% N-acetyl cysteine in 1× PBS): 5 mL of 1× PBS (diluted from 10× PBS buffer), 0.375 g of N-acetyl cysteine powder (store at 4 °C). Agitate gently until N-acetyl cysteine dilutes. Prepare and use on the same day (*see* **Note 2**).
4. ACME solution: 6.5 mL of ultrapure water, 1.5 mL of methanol, 1 mL of glycerol, 1 mL of acetic acid. Shake vigorously to mix the reagents. Prepare and use on the same day. Prepare 10 mL of ACME solution per sample (*see* **Note 2**).
5. Pasteur pipettes.

6. Low-binding 1-mL pipette tips.
7. 1.5-mL tubes.
8. 15-mL tubes.
9. 50-mL tubes.
10. 50-μm cell strainers.
11. 40-μm cell strainers.
12. 40-μm strainer pipette tips (1000 μL).
13. Dimethyl sulfoxide (DMSO).
14. Seesaw rocker.
15. Refrigerated centrifuge for 1.5-mL tubes.
16. Refrigerated centrifuge for 15−/50-mL tubes.

2.2 ACME Cell Imaging and Sorting

1. Flow cytometer and/or cell sorter with red laser (780/60 nM filter) and yellow-green laser (525/40 nM filter).
2. 5 mM DRAQ5 fluorescence DNA dye (red laser with 780/60 nM filter).
3. 1 mg/mL concanavalin A conjugated with Alexa Fluor 488 dye (yellow-green laser with 525/40 nM filter).
4. Washing and resuspension buffer (1% BSA in 1× PBS) (see **step 2** in Subheading 2.1).
5. Refrigerated centrifuge for 1.5-mL tubes.
6. 50-μm cell strainers.

3 Methods

3.1 ACME Dissociation-Fixation

1. For each sample, prepare 100–300 μL of planarian biomass in a 15-mL tube (*see* **Note 3**).
2. Eliminate culture water carefully with a Pasteur pipette.
3. Add 100–300 μL of NAC solution, sufficient to cover the planarians, and incubate for 30 s at room temperature. Flick the tube gently to help cleaning planarian mucous (*see* **Note 4**).
4. Without removing the NAC solution, immediately add 10 mL of ACME solution to each sample. Incubate at room temperature for 40 min in a seesaw rocker (40–45 rpm) for dissociation-fixation (Fig. 1a). Orientate the tubes parallel to the direction of the movement (Fig. 1b).
5. Remove samples from the rocker and place them on ice (*see* **Note 5**).

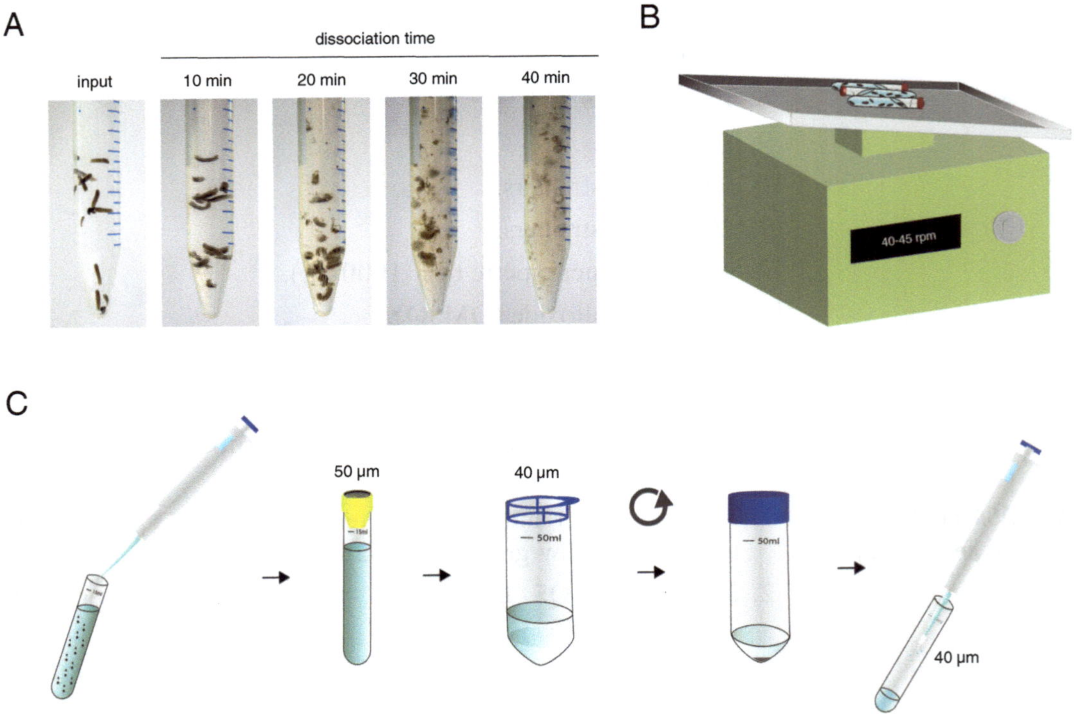

Fig. 1 ACME cell dissociation and fixation. (**a**) Dissociation-fixation process for the planarian *Schmidtea mediterranea*. From left to right: live worms used as input in water, ACME dissociation reaction after 10–40 min. (**b**) Incubation in a seesaw rocker. (**c**) Filtration steps. From left to right: pipetting up and down several times, filtering with a 50-µm cell strainer, filtering with a 40-µm cell strainer, centrifugation, discarding supernatant (leaving 1–2 µL), and filtering with a 40-µm cell strainer tip

6. Pipette each reaction up and down several times, until solution is clear of large tissue fragments, using low-binding 1-mL pipette tips (Fig. 1c).
7. To improve aggregate dissociation, filter each sample with a 50-µm cell strainer into a new 15-mL tube (Fig. 1c).
8. Filter samples one more time, with a 40-µm cell strainer, into a new 15-mL tube (Fig.1c).
9. Centrifuge at 1000 g for 5 min (4 °C).
10. Discard all the supernatant but 1–2 mL and resuspend the pellet in this volume (Fig. 1c).
11. Filter each sample using 40-µm cell strainer pipette tips (1000 µL) into a new 15-mL tube (Fig. 1c; *see* **Note 6**).
12. Add 8 mL of washing buffer (1% BSA in 1× PBS) and centrifuge at 1000 g for 5 min (4 °C) to remove the ACME solution.
13. Discard the supernatant (*see* **Note 7**).
14. Resuspend the pellet in 900 µL of resuspension buffer (1% BSA in 1× PBS) and transfer it to a 1.5-mL tube (*see* **Note 8**).

15. To cryopreserve cells, add 100 μL of DMSO per sample (10% volume) and freeze directly at −80 °C. Cells can be stored in these conditions for months.
16. When cells are going to be used, thaw them on ice, centrifuge at 1000 g for 5 min (4 °C), discard the supernatant, and resuspend in 1 mL of resuspension buffer.
17. Repeat the previous centrifugation and resuspend the pellet (*see* **Note 8**) in you desired volume of fresh resuspension buffer (*see* **Note 9**).

3.2 ACME Cell Imaging and Sorting

Dissociated cells can be used before or after cryopreservation for cell imaging and cell sorting. Samples must be kept on ice to prevent RNA degradation.

1. When using cryopreserved cells, thaw them on ice, centrifuge at 1000 g for 5 min (4 °C), discard the supernatant, and resuspend in 1 mL of fresh washing buffer (*see* **Note 10**).
2. Repeat the centrifugation and resuspend the pellet in your desired volume of fresh resuspension buffer (normally 400–1000 μL).
3. Filter ACME cells through a 50-μm cell strainer into a new 1.5-mL tube.
4. Stain the samples, or aliquoted dilutions (*see* **Note 11**), with 0.5–1 μL/mL of DRAQ5 fluorescence DNA dye (*see* **Note 12**) and 2 μL/mL of concanavalin A (Con-A) conjugated with Alexa Fluor 488 cytoplasm dye (*see* **Notes 13** and **14**).
5. Incubate cells in the dark for 30–40 min at room temperature.
6. Visualize cells using a regular cytometer or sort them using a cell sorter (*see* **Notes 15** and **16**).
7. Plot DRAQ5 vs Con-A (*see* **Note 17**) to have an overview of the sample content (Fig. 2a).
8. Gate from the FSC-H vs FSC-A plot by selecting the well-correlated events (Fig. 2b). This gating is used to remove aggregates.
9. Gate from the Con-A vs FSC-A plot by selecting concanavalin-A-positive events (*see* Fig. 2c). This gating is used to select events with cytoplasm.
10. Gate from the DRAQ5 vs FSC-A plot by selecting DRAQ5-positive events (Fig. 2d). This gating will select events with nucleus.
11. Gate from the DRAQ5-A vs DRAQ5-H plot by selecting well-correlated events (Fig. 2e). This gating is used for a further removal of aggregates.

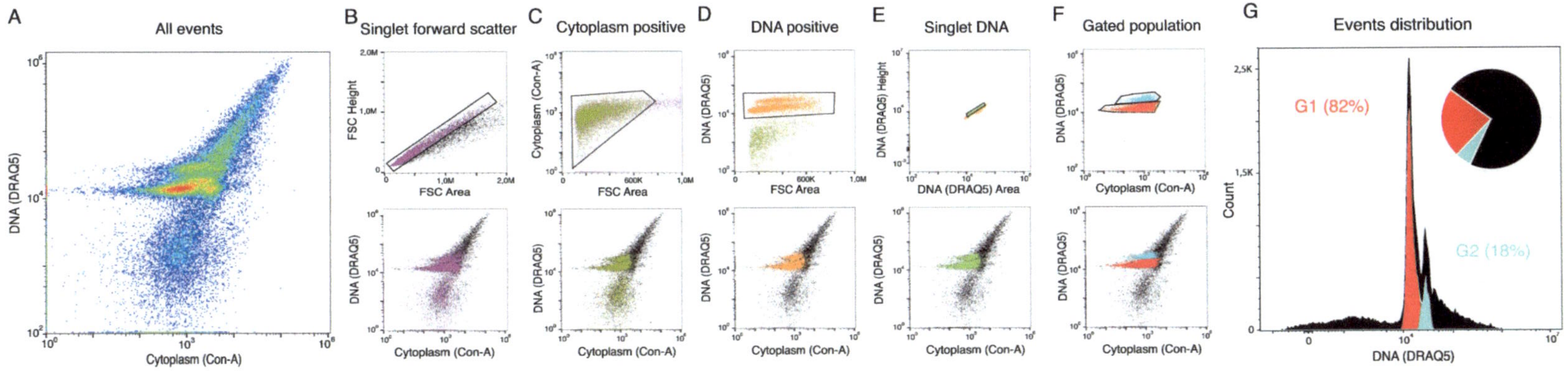

Fig. 2 Flow cytometry profiles and gating of ACME-dissociated cells. *Schmidtea mediterranea* ACME-dissociated cells stained with DRAQ5 (nucleus) and concanavalin A (cytoplasm). (**a**) Ungated populations, (**b**) forward scatter singlets gating, (**c**) cytoplasm-positive cells, (**d**) DNA-positive cells, (**e**) DRAQ5 singlet gating, (**f**) final gated populations, and (**g**) DRAQ5 (DNA content) histogram showing ungated populations (black) and gated single-cell populations, G1 and G2, with their relative proportions in a typical ACME dissociation for *S. mediterranea*

12. Use the DRAQ5 vs Con-A plot to visualize gated events, which correspond to DRAQ5 and Con-A-positive singlets, and to see the proportion of cells in G1 (1× DNA content) and G2 (2× DNA content). Single-cell populations can be counted or sorted from this final plot (Fig. 2f).
13. Alternatively, use the DRAQ5 histogram to visualize cell populations in G1 and G2 (Fig. 2g).
14. ACME cells can be cryopreserved after sorting. For this, collect the sorted cells in 1.5-mL tubes using 100–200 μL of resuspension buffer as collection buffer.
15. After sorting, centrifuge samples at 1000 g for 5 min (4 °C), discard the supernatant, and resuspend in 500–1000 μL of fresh resuspension buffer (*see* **Note 10**).
16. Add 10% volume of DMSO and store samples at −80 °C.

4 Notes

1. RNAse-free BSA powder can be obtained commercially but it is very expensive. We have used commercial non-RNAse free BSA with good results, but always checking RNAse activity in advance and aliquoting the powder to prevent contamination and bacterial growth. BSA is a potential source of contamination and RNA degradation in our protocol. We use 1% BSA in our washing and resuspension buffer to avoid cell clumps but, in case of persistent RNA degradation, we recommend reducing this percentage up to 0.2% or use a new BSA reagent.
2. We prepare the washing and resuspension buffer, NAC solution, and ACME solution in disposable 15-mL or 50-mL tubes, to avoid RNAse and other reagent contamination. When more than 50 mL volume is required, we split the solutions in different tubes.
3. We prefer to calculate planarian biomass in volume per sample, instead of number of individuals per sample, due to the variable size of the worms or if worm fragments are used. When working with whole individuals, we add them alive to the sample tube. When working with fragments, like blastema, regenerating animals, heads, or tails, we add them alive or freshly cut to the sample tube.
4. N-acetyl cysteine (NAC) was initially added as a mucolytic agent to remove planarian mucous, but this washing step also results in a better dissociation of tissues and higher RNA integrity after dissociation. NAC acts as an acidic reducing agent, cleaning the mucous and creating RNA protective conditions. An incubation of a few seconds is sufficient. Incubations longer than 3–4 min can result on cell lysis.

5. From this point on, samples are kept on ice or at 4 °C during handling to avoid RNA degradation.
6. We use this three-step filtration approach, but we encourage further optimization of filtration, trying different combination of filters. Either way, some kind of filtration is highly recommended at this point, when samples are still dissociating in ACME. Posterior filtration steps, when sample is already resuspended in buffer, have been proved less efficient.
7. If pellet is not compact, discard supernatant carefully, resuspend in 8 mL of washing buffer and repeat the washing step one more time. A noncompact pellet indicates the ACME solution has not been completely removed.
8. In this step, we can add Trizol to part of the sample, either the dry pellet or after resuspending in resuspension buffer, to evaluate RNA integrity.
9. For most applications, we resuspend our samples in 400–1000 μL of resuspension buffer. When working with smaller pellets, we resuspend in 200–400 μL. Normally, we adjust this volume depending on the size of the pellet and the final concentration of cells required.
10. When we work with very small pellets, of a few thousand cells, we help cell precipitation by adding 10% Triton X-100 to a final concentration of 0.1% to our samples before centrifugation. In this case, we also increment the speed of centrifugation up to 1200 g.
11. For cell imaging and cell counting, we normally stain a 1:3–1:5 dilution of the sample. For cell sorting, we stain the whole sample undiluted.
12. Other nuclear dyes, as Hoechst or DAPI, are cheaper and can be used for the same purpose, but they will require a cytometer with UV laser. When using Hoechst, it is worth to replace concanavalin A-Alexa Fluor 488 by red fluorescence concanavalin A-Alexa Fluor 594, to avoid fluorescence overlapping.
13. Concanavalin A is a lectin that selectively binds to mannose residues and glycoproteins. In cell biology, it is used to monitor endocytosis as is impermeable to cellular membranes. Concanavalin A is endocytosed by live cells and stays in the vesicles. However, ACME cells are dead and permeabilized, so concanavalin A can cross the membrane and stain all the cytoplasm.
14. We have optimized dye concentrations for a regular sample of 100–300 μL of planarian biomass, dissociated and resuspended in a final volume of 1 mL of resuspension buffer. Staining will require optimization for any lab as it is dependant not only on cell concentration, but also on the FACS/cytometer device and settings used.

15. We sort our cells using an 85-μm nozzle, 2.0 filter, and moderate-pressure separation (45 Psi). The sorting is set in 4-way purity mode in the BD FACSDiva Software.
16. To avoid RNAse contamination, the FACS is cleaned with bleach before sorting. Moreover, the injection chamber and collection tubes are kept at 4 °C during the process.
17. To visualize DRAQ5 we use a red laser with a 780/60 nm filter (named APC-A750-A in some cytometers). To visualize concanavalin A we use a yellow-green laser with a 525/40 filter (usually named as FITC-A).

References

1. Ivankovic M, Haneckova R, Thommen A, Grohme MA, Vila-Farre M, Werner S et al (2019) Model systems for regeneration: planarians. Development 146:17. https://doi.org/10.1242/dev.167684
2. Baguna J, Romero R (1981) Quantitative-analysis of cell-types during growth, degrowth and regeneration in the planarians Dugesia-Mediterranea and Dugesia-Tigrina. Hydrobiologia 84(Oct):181–194. https://doi.org/10.1007/Bf00026179
3. David CN (1973) A quantitative method for maceration of hydra tissue. Wilhelm Roux Arch Entwickl Mech Org 171(4):259–268. https://doi.org/10.1007/bf00577724
4. Stuart T, Satija R (2019) Integrative single-cell analysis. Nat Rev Genet 20(5):257–272. https://doi.org/10.1038/s41576-019-0093-7
5. Fincher CT, Wurtzel O, de Hoog T, Kravarik KM, Reddien PW (2018) Cell type transcriptome atlas for the planarian Schmidtea mediterranea. Science 360(6391). https://doi.org/10.1126/science.aaq1736
6. Plass M, Solana J, Wolf FA, Ayoub S, Misios A, Glažar P et al (2018) Cell type atlas and lineage tree of a whole complex animal by single-cell transcriptomics. Science 360(6391):eaaq1723. https://doi.org/10.1126/science.aaq1723
7. van Wolfswinkel JC, Wagner DE, Reddien PW (2014) Single-cell analysis reveals functionally distinct classes within the planarian stem cell compartment. Cell Stem Cell 15(3):326–339. https://doi.org/10.1016/j.stem.2014.06.007
8. Wurtzel O, Cote LE, Poirier A, Satija R, Regev A, Reddien PW (2015) A generic and cell-type-specific wound response precedes regeneration in planarians. Dev Cell 35(5):632–645. https://doi.org/10.1016/j.devcel.2015.11.004
9. Hayashi T, Asami M, Higuchi S, Shibata N, Agata K (2006) Isolation of planarian X-ray-sensitive stem cells by fluorescence-activated cell sorting. Develop Growth Differ 48(6):371–380. https://doi.org/10.1111/j.1440-169X.2006.00876.x
10. Denisenko E, Guo BB, Jones M, Hou R, de Kock L, Lassmann T et al (2020) Systematic assessment of tissue dissociation and storage biases in single-cell and single-nucleus RNA-seq workflows. Genome Biol 21(1):130. https://doi.org/10.1186/s13059-020-02048-6
11. van den Brink SC, Sage F, Vertesy A, Spanjaard B, Peterson-Maduro J, Baron CS et al (2017) Single-cell sequencing reveals dissociation-induced gene expression in tissue subpopulations. Nat Methods 14(10):935–936. https://doi.org/10.1038/nmeth.4437
12. García-Castro H, Kenny NJ, Iglesias M, Álvarez-Campos P, Mason V, Elek A et al (2021) ACME dissociation: a versatile cell fixation-dissociation method for single-cell transcriptomics. Genome Biol 22(1):89. https://doi.org/10.1186/s13059-021-02302-5

Chapter 11

Papain-Based Dissociation of *Schmidtea mediterranea* Cells

Claudia Ortmeier and Luca Gentile

Abstract

The need of highly viable cells dissociated from *Schmidtea mediterranea* is constantly growing. In this chapter, we describe a cell dissociation method based on papain (papaya peptidase I). This enzyme, often used to dissociate cells with complex morphology, is a cysteine protease with a broad specificity and increases both the yield and the viability of the dissociated cell suspension. The papain dissociation is preceded by a pretreatment for mucus removal, as this was shown to greatly improve the yield of cell dissociation, regardless of the method used. Papain-dissociated cells are suitable for a variety of downstream applications, like live immunostaining, flow cytometry, cell sorting, transcriptomics, and cell transplantation, also at the single-cell level.

Key words *Schmidtea mediterranea*, Cell dissociation, Papain, High cell viability

1 Introduction

Planarian *Schmidtea mediterranea* is an animal model system widely used to study regeneration. Many authors use dissociated cells for their studies; therefore, it is important to adopt a dissociation protocol that allows high yield and viability, regardless of the downstream applications. Papain (papaya peptidase I; [1]) was shown to outperform other dissociation reagents (e.g., trypsin, TrypLE, Accutase) with cells that either have a complex morphology (e.g., neurons) or are tightly interconnected (e.g., cardiomyocytes), and to provide a single-cell suspension of highly viable cells [2–4].

The following protocol describes an efficient papain-based cell dissociation technique that preserves *S. mediterranea* cell viability for downstream applications, like live immunostaining, flow cytometry and cell sorting, single-cell transcriptomics, and cell transplantation.

Luca Gentile (ed.), *Schmidtea Mediterranea: Methods and Protocols*, Methods in Molecular Biology, vol. 2680, https://doi.org/10.1007/978-1-0716-3275-8_11,

© The Author(s), under exclusive license to Springer Science+Business Media, LLC, part of Springer Nature 2023

2 Materials

Unless otherwise indicated, all solutions are prepared at room temperature using 18,2 MΩ ultrapure water. When performing the experiment, solutions are usually kept on ice, unless otherwise indicated. All reagents and buffers are stored at +4 °C, unless otherwise indicated. It is advisable to prepare the solutions in advance, unless otherwise indicated.

2.1 Mucus Removal and Digestion Pretreatment

1. Four to six 8 mm-long *S. mediterranea* per sample, starved for at least 2 weeks (*see* **Note 1**).
2. 10× CMF (calcium/magnesium-free medium) stock solution: 25.6 mM $NaH_2PO_4 \cdot 2H_2O$, 142.8 mM NaCl, 102.1 mM KCl, 94.2 mM $NaHCO_3$. Aliquot the 10x stock solution and store refrigerated (*see* **Note 2**). It is stable for several months.
3. 1× CMFHE++ (calcium/magnesium-free medium, complete): 1× CMF, 15 mM HEPES, 3 mM EDTA, 0.5 μg/mL DNase I, 0.1% BSA, 5 mg/mL glucose; pH 7.2. Prepare fresh prior to use and keep on ice (except 300 μL/sample; *see* **Note 3**).
4. 2% L-cysteine solution, pH 7.0 (for removal of the mucus). Dissolve 0.5 g of L-cysteine powder in 25 mL of ultrapure water. This volume is enough to remove the mucus from up to 60 animals. Swirl gently the tube until L-cysteine is completely dissolved and adjust the pH to 7 with 750 μL of 5 N NaOH. Use the solution within a few hours.
5. Disposable plastic Pasteur pipettes.
6. 6 or 10 cm plastic petri dishes (e.g., those for bacterial culture).
7. Large-borehole 1000-μL pipette tips (e.g., 1000G Art-tip, MβP USA; *see* **Note 4**).
8. Surgical blades (e.g., no. 29).
9. Soft dry wipe tissue (e.g., Kimwipe).
10. Microscope glass slides or glass petri dishes.
11. (Optional) horizontal shaker.

2.2 Cell Dissociation

1. *S. mediterranea* fragments (*see* **Note 5**).
2. 10× CMF (calcium/magnesium-free medium) stock solution: 25.6 mM $NaH_2PO_4 \cdot 2H_2O$, 142.8 mM NaCl, 102.1 mM KCl, 94.2 mM $NaHCO_3$. Aliquot the 10× stock solution and store refrigerated (*see* **Note 2**). It is stable for several months.
3. 1× CMFHE++ (calcium/magnesium-free medium, complete): 1× CMF, 15 mM HEPES, 3 mM EDTA, 0.5 μg/mL DNase I, 0.1% BSA, 5 mg/mL glucose; pH 7.2. Prepare fresh prior to use and keep on ice.

4. 0.2 M L-cysteine-HCl (as papain activator stock solution): Dissolve 35 mg of L-cysteine-HCl powder in 1 mL of ultrapure water. Sterile-filter, aliquot in 0.2 mL tubes, and store at −20 °C.
5. 30 mg/mL trypsin inhibitor, ovomucoid (stock solution): weight 300 mg of trypsin inhibitor, ovomucoid, and dissolve in 10 mL of ultrapure water. Sterile-filter, aliquot in 0.5 mL tubes, and store at −20 °C.
6. 10 mg/mL DNase I (stock solution; 1500–3000 U/mg): weight 10 mg of DNase I and dissolve in 1 mL of ultrapure water. Sterile-filter, aliquot in 0.2 mL tubes, and store at −20 °C.
7. 2× digestion solution: 30 U/mL papain (Worthington Biochemical Corp., USA; *see* **Note 6**) in 1× CMFHE++. Incubate at 37 °C until clear, and then add 2 mM L-cysteine (the stock solution is a 100×). Prepare 250 μL for each sample, conveniently aliquoted in 1.5 mL protein low-binding tubes ready to use. Use the 2× digestion solution within the day.
8. 3× stop solution: 1.5 mg/mL trypsin inhibitor, ovomucoid (the stock solution is a 20×), 60 μg/mL DNaseI (the stock solution is a 167×) in CMFHE++. Consider 250 μL for each sample. Use the 3x stop solution within the day.
9. 30-μm cell strainers, autoclaved.
10. Beveled 1000 μL pipette tips (e.g., StarLab 1000 XL, StarLab GmbH, Germany; *see* **Note 7**).
11. 1.5 mL tubes, protein low-binding.
12. 15 mL conical tubes, pre-chilled on ice.
13. Thermal block with adapter for 1.5 mL tubes, or water bath.
14. Refrigerated centrifuge with buckets and adapters for 15 mL tubes.
15. Hemocytometer.
16. 1 mg/mL propidium iodide stock solution: dissolve 1 mg of propidium iodide (powder) in 1 mL of ultrapure water. Store at +4 °C; the dye is stable for several months.

3 Methods

This protocol is carried out mostly at room temperature, except for the incubation with papain (25 °C). However, it is important that the dissociated cells are kept on ice or at 4 °C, as their viability quickly reduces at higher temperatures. The initial treatment for the removal of the mucus takes approximately 5 min, and it is recommended that it is performed on small batches of planarians—

ideally, those needed for one specimen. Therefore, the enzymatic digestion starts and ends sequentially for each sample, and the total incubation time depends on the overall number of samples (*see* **Note 5**).

3.1 Mucus Removal and Digestion Pretreatment

1. For each sample, transfer four to six 8 mm-long planarians to a petri dish containing 2% L-cysteine, pH 7, for max 2 min at room temperature, in order to remove the mucus (*see* Fig. 1a, b) (*see* **Note 8**).
2. Transfer the planarians on a glass surface (e.g., a microscope slide or a petri dish).
3. Rinse the planarians with a few drops of CMFHE++ at room temperature.
4. Grossly remove the excess of CMFHE++, with a pipette.
5. Carefully remove any residual CMFHE++ with the help of a dry wipe (*see* **Note 9**).
6. Roughly cut the planarians in small pieces using a surgical blade (*see* Fig. 1c) (*see* **Note 10**).
7. Use 250 μL of CMFHE++ to immediately wet the fragments and gently transfer them to the 1.5 mL protein low-binding tube that were filled with 250 μL of 2× digestion solution (*see* **Note 5**).

3.2 Cell Dissociation

1. Transfer the tube containing the fragments from **step** 7 of Subheading 3.1 on the thermal block and incubate at 25 °C for 60 min.
2. Remove the samples from the thermal block, one at a time (*see* Fig. 2a, b) (*see* **Note 5**).
3. Add 250 μL of 3× stop solution at room temperature.
4. Using a beveled 1000 μL tip, gently pipette up and down for 10–20 times, until the cell suspension is turbid and mostly devoid of visible tissue fragments (*see* **Note 11**).
5. Place a 30 μm cell strainer on a 15 mL tube placed on ice. Rinse the strainer with 1 mL of ice-cold CMFHE++ and apply the triturated sample onto it. Rinse the strainer with 1 additional mL of ice cold CMFHE++ (*see* **Note 12**).
6. Repeat **steps 2–5** for each of the samples planned for the experiment. Once passed through the strainer and kept on ice, the cells in the suspension maintain high viability. Therefore, from this step onwards, all samples could be processed simultaneously after the trituration of the last processed sample.
7. Transfer the tubes in a refrigerated centrifuge pre-cooled at +4 °C and spin for 5 min at 350 × g.

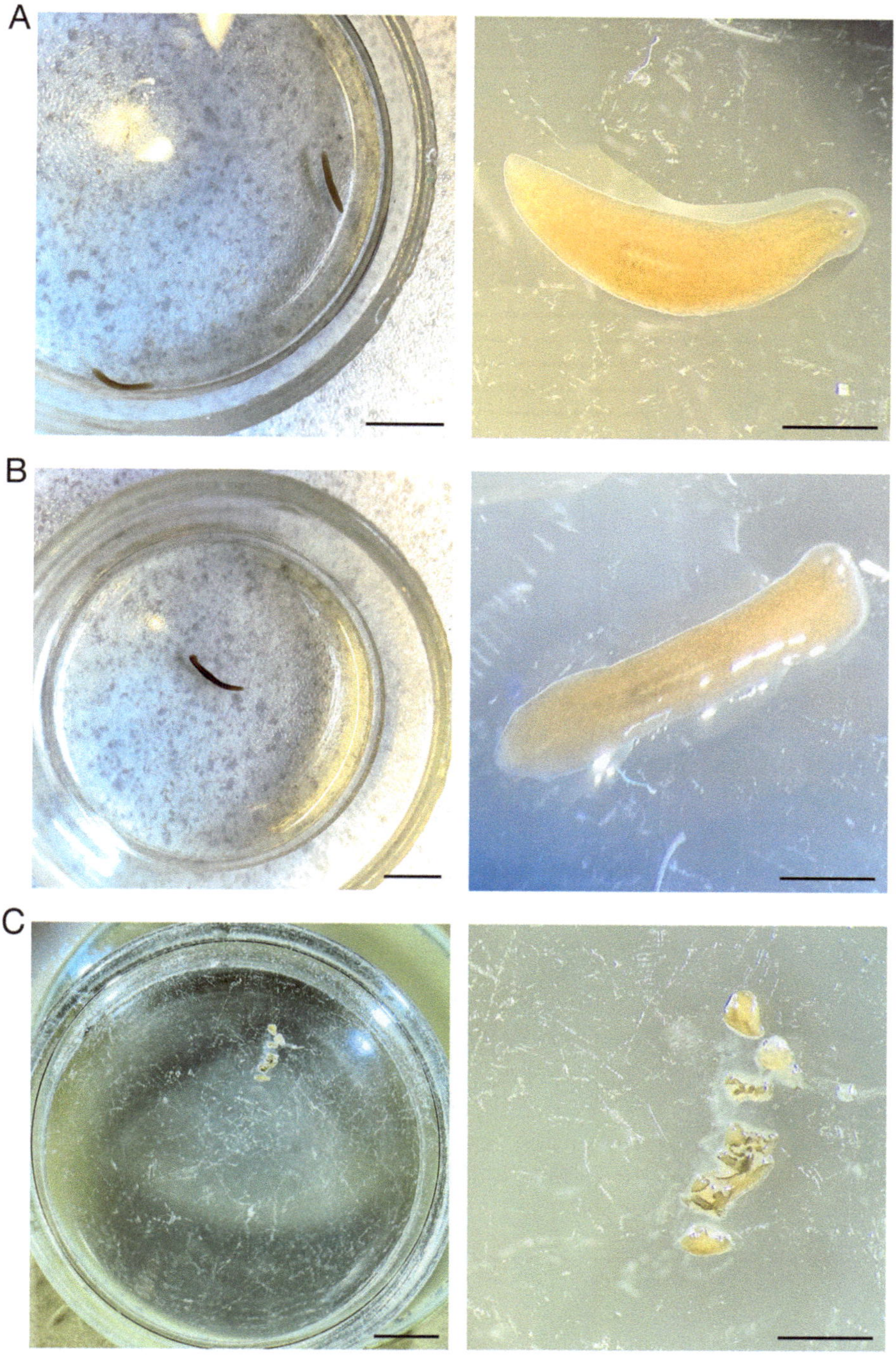

Fig. 1 Mucus removal and digestion pretreatment. (**a**, **b**) Planarians in planarian artificial medium (**a**) and in 2% L-cysteine solution, pH 7.0, (**b**) are shown for comparison as they appear to the naked eye (left panels) or under a stereomicroscope (right panels). (**c**) The amputation of an 8 mm-long, L-cysteine-treated planarian is shown as it appears to the naked eye (left panel) and under a stereomicroscope (right panel). Scale bar: 1 cm (left panels); 2 mm (right panels)

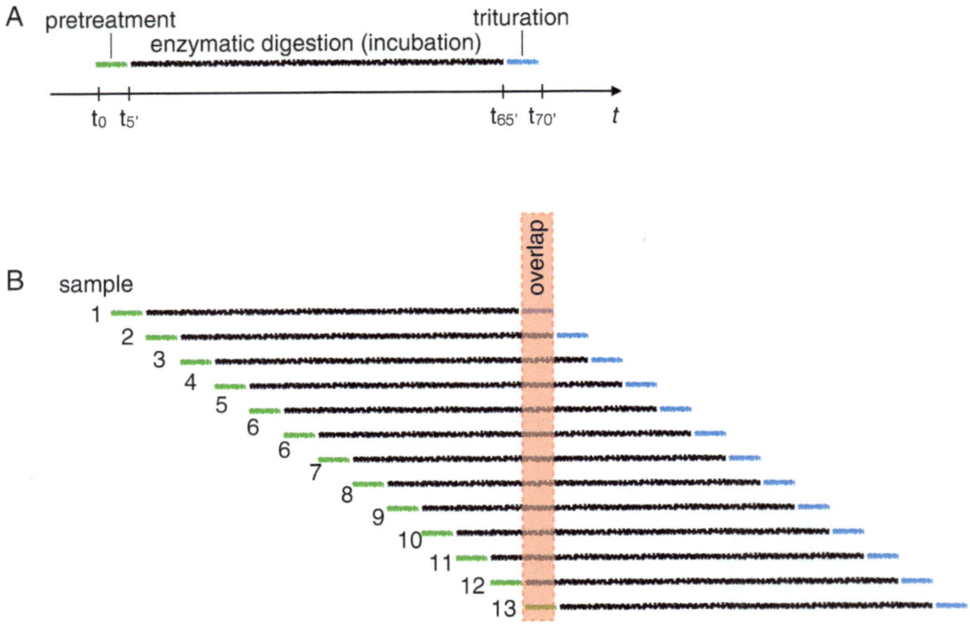

Fig. 2 Sequential processing of up to 12 samples. (**a**) The papain-based enzymatic dissociation consists of the pretreatment (5 min), the papain incubation (60 min), and the trituration (5 min). (**b**) The sequential processing of up to 12 samples is graphically depicted. In case of >12 samples, it is recommended to split the samples in batches of 12, so that the pretreatment of one sample does not overlap with the trituration of another one

8. Transfer the tubes on ice and carefully remove the supernatant.
9. Gently resuspend the pellet in 4 mL CMFHE++.
10. Repeat **steps 7–8**, then proceed to **step 11**.
11. Gently resuspend the pellet in 1 mL CMFHE++.
12. Transfer to pre-chilled 1.5 mL tubes and keep on ice.
13. Count the cells and adjust their concentration according to the downstream application. Approximately, 2.0×10^6 cells/mL are expected per sample (*see* **Note 1**). The viability of the cell suspension could be assessed via staining with 1 μg/mL of propidium iodide (*see* Fig. 3).

4 Notes

1. Planarian of virtually any size could be used. The expected number of cells will vary according to the size and number of animals used. The figures given in the Material section should result in approximately 2.0×10^6 cells/specimen. The number of specimens depends also on the downstream applications, as in the case of the live immunostaining of dissociated cells (*see* Chap. 11). The starvation time of 2 weeks is generally

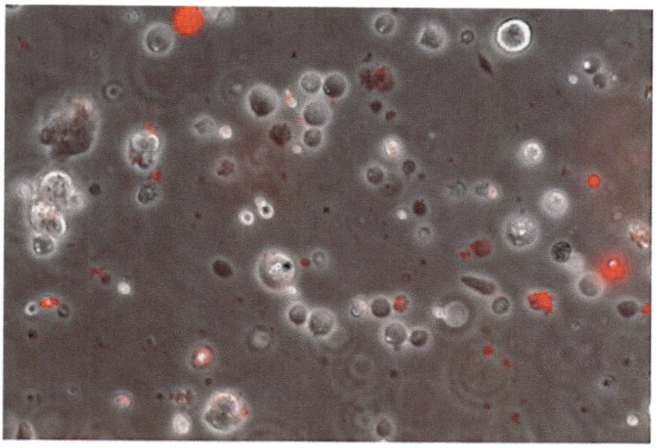

Fig. 3 Papain-dissociated cells after staining with propidium iodide. On average, a cell viability above 65% is expected. Depending on the skills of the operator, that could reach 95%

considered long enough to prevent food-related artifacts. For sensitive applications like single-cell transcriptomics, it is also important that the animals used for the experiments have no wounds and/or blastema, as this could alter the cell-type composition.

2. The 10× CMF is aliquoted in either 15 or 50 mL sterile disposable tubes, to prevent the risk of contamination, which increases if larger vessels (e.g., 1 or 2 L bottles) are used.
3. The 1× CMFHE++ at room temperature is both used to rinse the planarians after the L-cysteine treatment and to transfer the freshly cut fragments from the glass slide to the tube where the enzymatic dissociation is carried out.
4. The large borehole tips—like the ones used to transfer precipitated DNA—allow to collect effortlessly the animal fragments from the glass slide where animals were cut. Conventional 1000-μL tips could also be used, provided that the narrow tip is removed with a precise cut (e.g., using the appropriate tool, or sharp scissors or a blade). Alternatively, disposable plastic Pasteur pipettes could also be used, although this is not recommended, as controlling effectively the in- and outflow of as little as 250 μL that contains several small planarian fragments is very challenging.
5. It is recommended that the freshly cut fragments are immediately transferred to the digestion solution so that the enzymatic digestion could begin. Since both the pretreatment (mucus removal and cut; Subheading 3.1) and the post-digestion trituration step (Subheading 3.2, **steps 2–4**) last approximately 5 min each, the total time for the enzymatic digestion of one

sample is 70 min, approximately (*see* Fig. 2a). It is recommended that the samples are incubated in a sequential way, every 5 min (*see* Fig. 2b). Therefore, it is advisable that the number of samples processed in one experiment does not exceed $n = 12$, or the pretreatment of sample 13 would overlap with the trituration of sample 1 (Fi. 2B). The overall enzymatic digestion time is 60 + 5 x n min long, where n = number of samples (*see* Fig. 2b).

6. Papain quality is one of the bottlenecks of this protocol. We found that papain from Worthington Biochemical offered very consistent results; however, other papain reagents (either in powder or in suspension) could also be used, following protocol optimization.
7. The beveled edge of the 1000-μL tips reduces the shear stress applied to the dissociating cells, improving the cell viability, thus increasing the cell yield. Conventional 1000-μL tips could also be used, provided that the narrowest part of the tip is removed with a precise cut (e.g., using sharp scissors or a blade).
8. Process a small batch of planarians at time. Recommended is to use the number of animals necessary for a single sample. As a vessel, a plastic petri dish is ideal, although a 1.5 mL tube could also be used. The mucus removal treatment is size- and temperature-dependent; therefore, the actual incubation time should be empirically determined. In general, for animals 6 to 8 mm long, 2 min is enough. Moving the petri dish at regular intervals also helps in the effective removal of the mucus without increasing the incubation time. This could be achieved either using a horizontal shaker or swirling or flicking the dish at the side—as one does for cultured cells during trypsin incubation—or flushing, e.g., with a p1000 pipette. Notice that planarians will behave differently once placed in the L-cysteine solution. They no longer slide, as they do under normal conditions (*see* Fig. 1a), rather they swim, twisting and stretching their body (*see* Fig. 1b). This is a normal response to the dissolving of the mucus layer surrounding the animals.
9. The more medium is removed, the easier is the cutting in small fragments. A good strategy is to get close to the animal with the corner of the dry wipe, paying attention not to touch the animal itself, or it will stick to the paper tissue.
10. Critical step. If the fragments are too small, then the yield in terms of viable cells will be reduced. If the fragments are too large, then the yield in terms of total cells will be reduced. In general, fragments with a width 0.5–0.8 mm, transversely cut in rapid sequence, grant the best results (*see* Fig. 1c).

11. Critical step. The proper way should be empirically determined, adjusting the number, the strength, and the speed of the pipetting according to the results. The most important readout here is the number of viable cells.
12. This step effectively removes aggregates. However, cells larger than 30 μm could easily squeeze through the strainer without damage. Therefore, this filtration step helps in reducing cell aggregates (which could, for example, clog the nozzle of a flow cytometer, especially in the case a small nozzle is used), without introducing cell size-dependent biases.

References

1. Kimmel JR, Smith EL (1954) Crystalline papain. I. Preparation, specificity, and activation. J Biol Chem 207(2):515–531
2. Moritz S et al (2008) An induction gene trap screen in neural stem cells reveals an instructive function of the niche and identifies the splicing regulator sam68 as a tenascin-C-regulated target gene. Stem Cells 26(9):2321–2331
3. Moritz S et al (2012) Heterogeneity of planarian stem cells in the S/G2/M phase. Int J Dev Biol 56(1–3):117–125
4. Fischer B et al (2018) A complete workflow for the differentiation and the dissociation of hiPSC-derived cardiospheres. Stem Cell Res 32:65–72

Chapter 12

Live Immunostaining and Flow Cytometry of *Schmidtea Mediterranea* Cells

Claudia Ortmeier and Luca Gentile

Abstract

The use of flow cytometry and fluorescence-activated cell sorting to roughly separate subpopulations of cells in *Schmidtea mediterranea* is long established. In this chapter, we describe a method for the immunostaining—either single or double—of live planarian cells, using mouse monoclonal antibodies reactive against *S. mediterranea* plasma membrane antigens. This protocol allows to sort live cells according to their membrane signature, offering the possibility to further characterize the cell populations in *S. mediterranea* in a variety of downstream applications, like transcriptomics and cell transplantation, also at the single-cell level.

Key words Live immunostaining, Plasma membrane antigens, Cell subpopulations, Flow cytometry, FACS, *Schmidtea mediterranea*, Monoclonal antibody, Hybridoma library

1 Introduction

The possibility of using nongenetic markers to characterize the different cell populations that make an organism is an added value when transgenic lines are not available. Planarian *Schmidtea mediterranea* is long known to have remarkable regeneration capabilities [1–3], which hinge on the presence of adult pluripotent stem cells. Although in the last decade individual planarian stem cells have been molecularly or functionally characterized, e.g., via single-cell transplantation or single-cell whole transcriptome analysis (WTA) [4–9], the presence of a given antigen on the membrane of a stem cell and its ability to repopulate a stem cell-depleted individual could not be correlated, since using genetic reporters is currently not possible in planarian.

Fluorescence-activated cell sorting (FACS) and flow cytometry use the intensity of fluorescent reporters to distinguish among the different cell types within heterogeneous cell preparations, also in planarians [10–13]. In this protocol, we describe the

Luca Gentile (ed.), *Schmidtea Mediterranea: Methods and Protocols*, Methods in Molecular Biology, vol. 2680,
https://doi.org/10.1007/978-1-0716-3275-8_12,
© The Author(s), under exclusive license to Springer Science+Business Media, LLC, part of Springer Nature 2023

immunostaining of live *S. mediterranea* cells that allows to classify cells according to the presence of up to two plasma membrane antigens, following either simultaneous or sequential double immunostaining. The immunostained cells could be sorted and further characterized in downstream applications, like WTA, or used for cell transplantation, also as single-cells.

2 Materials

Unless otherwise indicated, all solutions are prepared at room temperature using 18,2 MΩ ultrapure water. When performing the experiment, solutions are kept on ice, unless otherwise indicated. The monoclonal antibodies used in this protocol are used in the form of unpurified hybridoma cells supernatants (as described in [13]). In case purified antibodies are used, the proper dilution should be experimentally determined. All reagents and buffers are stored at +4 °C, unless otherwise indicated.

2.1 Live Immunostaining

1. 1× CMFHE++ (calcium/magnesium-free medium, complete): 2.56 mM $NaH_2PO_4 \cdot 2H2O$, 14.28 mM NaCl, 10.21 mM KCl, 9.42 mM $NaHCO_3$, 15 mM HEPES (4-(2-hydroxyethyl)-1-piperazineethanesulfonic acid), 3 mM EDTA (ethylenediaminetetraacetic acid), 0.5 μg/mL DNase I, 0.1% BSA (bovine serum albumin, fraction V), 5 mg/mL glucose; pH 7.2. Prepare fresh prior to use and keep on ice.
2. 10 mg/mL Hoechst 33342 (2″-(4-ethoxyphenyl)-5-(4-methyl-1-piperazinyl)-2,5″-bi-1*H*-benzimidazol trihydrochloride) stock solution: dissolve 10 mg of Hoechst 33342 (powder) in 1 mL of ultrapure water. Store at +4 °C; the dye is stable for several months.
3. 1× CMFHE++/H (CMFHE++ added with Hoechst 33342): 1 μg/mL Hoechst 33342 in 1× CMFHE++, pH 7.2. Prepare fresh prior to use.
4. Antibody/ies of choice, directed against plasma membrane antigens, either in the form of hybridoma supernatant or purified.
5. 1:1000 isotype mouse IgG control, unconjugated in CMFHE++/H.
6. (Optional, in the case IgM antibodies are used): 1:1000 isotype mouse IgM control, unconjugated in CMFHE++/H.
7. One or two 1:1000 fluorescent conjugated secondary antibody (e.g., Alexa Fluor 647 conjugated rabbit anti-mouse IgG; Alexa Fluor conjugated rabbit anti-mouse IgM; Alexa Fluor sheep anti-mouse IgG) in CMFHE++/H. Both host and target species should be selected accordingly to the primary

antibody/ies used. Fluorophore(s) with excitation/emission spectra that overlap with Hoechst 33342 and PI must be avoided.

8. Single planarian cell suspension (2×10^6 cells/sample + up to 3×10^6 cells for the controls).
9. Horizontal shaker.
10. Refrigerated 1.5 mL tubes.
11. Refrigerated centrifuge for 1.5 mL tubes.
12. Optional: refrigerated centrifuge for 15 mL tubes.
13. Optional: refrigerated 15 mL tubes.
14. Optional: 30-μm cell strainers, autoclaved.

2.2 Flow Cytometry and Cell Sorting

1. 1× CMFHE++ (calcium/magnesium-free medium, complete): 2.56 mM $NaH_2PO_4 \cdot 2H2O$, 14.28 mM NaCl, 10.21 mM KCl, 9.42 mM $NaHCO_3$, 15 mM HEPES (4-(2-hydroxyethyl)-1-piperazineethanesulfonic acid), 3 mM EDTA, 0.5 μg/mL DNase I, 0.1% BSA, 5 mg/mL glucose; pH 7.2. Prepare fresh prior to use and keep on ice.
2. 10 mg/mL Hoechst 33342 (2″-(4-ethoxyphenyl)-5-(4-methyl-1- piperazinyl)-2,5″-bi-1*H*-benzimidazol trihydrochloride) stock solution: dissolve 10 mg of Hoechst 33342 (powder) in 1 mL of ultrapure water. Store at +4 °C; the dye is stable for several months.
3. 1× CMFHE++/H (CMFHE++ added with Hoechst 33342): 1 μg/mL Hoechst 33342 in 1× CMFHE++, pH 7.2. Prepare fresh prior to use.
4. 1 mg/mL propidium iodide stock solution: dissolve 1 mg of propidium iodide (powder) in 1 mL of ultrapure water. Store at +4 °C; the dye is stable for several months.
5. Live immunostained planarian cells from 3.1.
6. Flow cytometer and cell sorter (with 365 nm laser, e.g., BD FACSAria III; *see* **Note 1**).
7. 30 μm cell strainers, autoclaved.
8. Refrigerated 1.5 mL tubes.
9. (Optional, in the case cell sorting is required) collection tubes (e.g., 5 mL FACS collection tubes or 1.5 mL DNA low-binding tubes or 0.2 mL PCR tubes, depending on the application and the number of cells to be collected).

3 Methods

Schmidtea mediterranea cells should be stained, immunostained, and processed immediately after dissociation. Papain-dissociated cells kept in suspension at +4 °C show good (≥ 50%) viability even after 1 week, but regardless of the dissociation method used, data acquired from non-freshly dissociated cells might be biased.

Any cell dissociation method could be used, given that it provides a good cell viability (i.e., ≥ 60%).

Being n the number of samples considered for the experiment, a total of $n + 3$ (single immunostaining) or $n + 5$ (double immunostaining) tubes are necessary. The additional control samples are helpful to set the gating strategy and in case of troubleshooting.

Three different scenarios are given: single immunostaining (*see* Subheading 3.1.1), double immunostaining using non-cross-reacting secondary antibodies (e.g., rabbit anti-mouse IgG Alexa Fluor 647 and rabbit anti-mouse IgM Alexa Fluor 488; *see* Subheading 3.1.2), and double immunostaining using cross-reacting secondary antibodies (e.g., rabbit anti-mouse IgG Alexa Fluor 647 and rabbit anti-mouse IgG Alexa Fluor 488; *see* Subheading 3.1.3), for which the pre-absorption of the secondary antibodies on the primary antibodies is required. A schematic is provided in Fig. 1 where the number of control samples is calculated based on the number of antigens to detect and the cross-reactivity between the secondary antibodies. In spite of the differences among the three protocols, the initial steps are in common (*see* Subheading 3.1).

Unless indicated differently, incubations are carried out at room temperature, while centrifugations are done at +4 °C, to preserve cell viability. All the solutions needed should be prepared in advance.

3.1 Live Immunostaining

1. Name one of the 1.5 mL tubes "unstained control" and transfer 500 μL of cell suspension (approx. 1×10^6 cells) in it. Keep on ice; this sample is not subject to any of the following steps.
2. Name one of the 1.5 mL tubes "Hoechst/PI control" and transfer 500 μL of cell suspension (approx. 1×10^6 cells) in it. This sample is processed together with the other samples until **step 5**.
3. In the case of single immunostaining, name one of the control tubes "isotype control" (*see* **Note 2**) and transfer 500 μL of cell suspension (approx. 1×10^6 cells) in it. In the case of double immunostaining with non-cross-reacting secondary antibodies, name one control tube "isotype controls," one "isotype control 1," and one "isotype control 2" and transfer 500 μL of cell suspension (approx. 1×10^6 cells) in each of them. In the case of double immunostaining with cross-reacting secondary

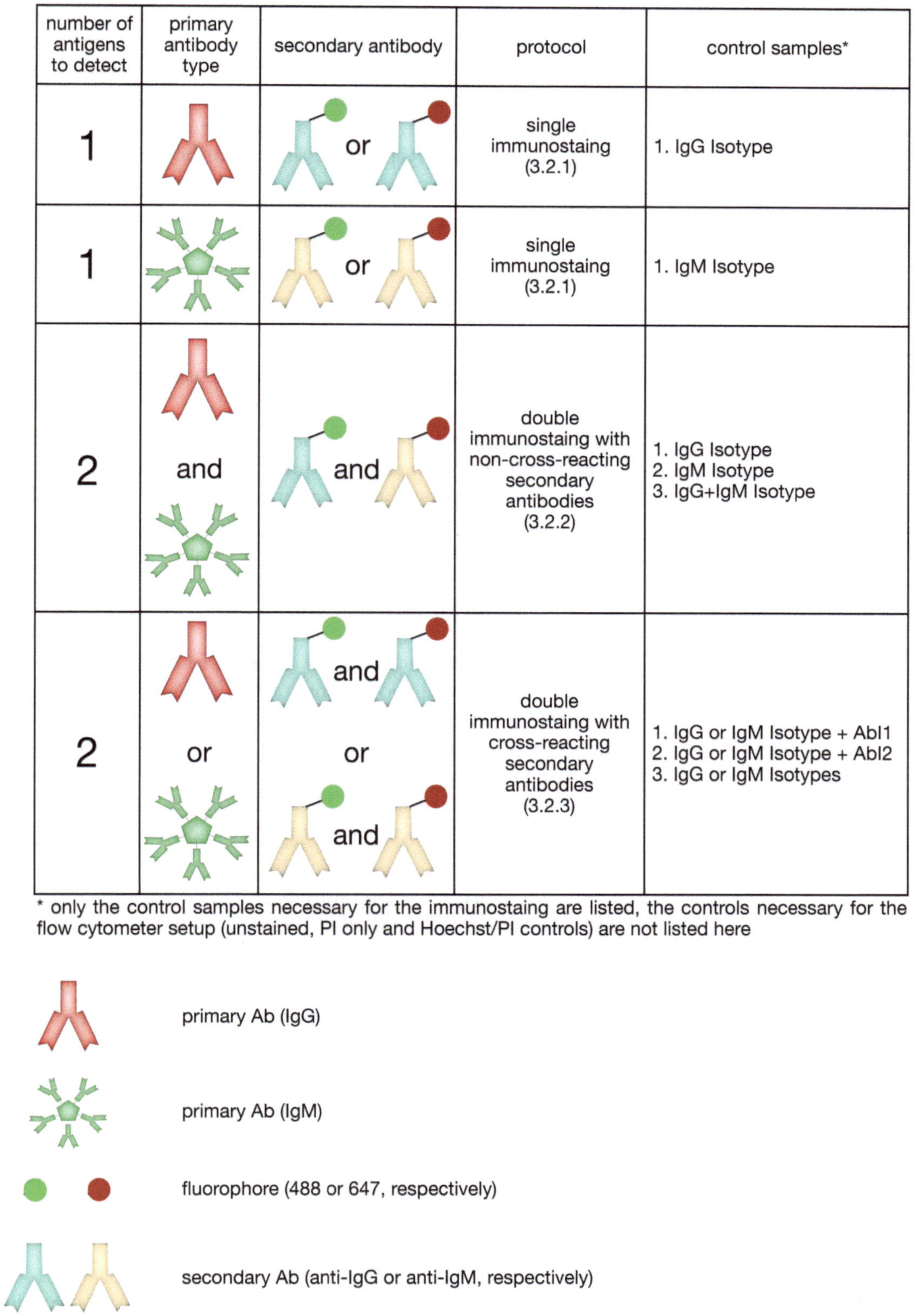

Fig. 1 Summary of the methodology and number of control samples. Secondary antibodies and number of control samples needed for the study are given according to the number and type of the primary antibodies used

antibodies, name one control tube "isotype control 1" and one "isotype control 2" and transfer 500 μL of cell suspension (approx. 1×10^6 cells) in each of them. The isotype control samples are processed together with the other samples, except that isotype IgGs and/or IgMs are used during the incubation steps instead of the respective primary antibodies.

4. Add 7.5 μg/mL Hoechst 33342 to all cell suspensions except to the "unstained control" sample and incubate for 90 min under slow agitation (e.g., 10 rpm) on a shaker, at room temperature.
5. Transfer the tubes in a refrigerated centrifuge pre-cooled at +4 °C and spin for 5 min at 500 × *g*.
6. Transfer the tubes to a rack kept on ice and carefully remove the supernatant.
7. In case of single live immunostaining, proceed to **step 1** in Subheading 3.1.1; in case of double live immunostaining with non-cross-reacting secondary antibodies, proceed to **step 1** in Subheading 3.1.2; in case of double live immunostaining with cross-reacting secondary antibodies, proceed to **step 1** in Subheading 3.1.3 (differences among these protocols are shown at a glance in Fig. 1).

3.1.1 Single Live Immunostaining

1. Dilute the primary antibody of choice in CMFHE++/H (*see* **Note 3**) and keep at room temperature. Prepare *n* milliliters of primary antibody solution, where *n* is the number of samples in the study (all control samples are *not* included in this count). If purified antibodies are used, dilute them to the recommended working concentration in CMFHE++/H. In case hybridoma supernatants are used, dilute them 1:4 (*see* **Note 4**) in CMFHE ++/H.
2. Dilute the isotype control antibody of choice (*see* **Note 2**) to the recommended working concentration in 500 μL of CMFHE++/H and keep at room temperature.
3. Gently resuspend the pellet in the tube named "isotype control" with the solution prepared in the previous step. Skip **step 4** for this sample.
4. Gently resuspend the pellets of the remaining samples in 1 mL (*see* **Note 5**) of the primary antibody solution prepared in **step 1**.
5. Incubate for 20 min at room temperature, under static conditions.
6. Transfer the tubes in a refrigerated centrifuge pre-cooled at +4 °C and spin for 5 min at 500 × *g*.
7. Transfer the tubes in a rack at room temperature and carefully remove the supernatant.

8. Gently resuspend the cell pellets in 1 mL of CMFHE++/H.
9. Repeat **steps 6–7**, then proceed to **step 10**.
10. Dilute the secondary antibody of choice 1:1000 in CMFHE++/H (*see* **Note 6**). Prepare *n* + 1 milliliters of secondary antibody solution, where *n* is the number of samples in the study (the "unstained control" and the "Hoechst/PI control" samples are *not* included in this count).
11. Gently resuspend the cell pellets of all the samples (including the "isotype control" sample) in the secondary antibody solution prepared in the previous step. Use 1 mL of solution for each sample and 500 μL for the "isotype control" sample.
12. Incubate for 15 min at room temperature, under static conditions.
13. Place a 30-μm cell strainer on a 15 mL tube placed on ice. Rinse the strainer with 1 mL of ice-cold CMFHE++ and apply the immunostained cells (*see* **Note 7**).
14. Rinse the filter with 1 mL of CMFHE++.
15. Transfer the tubes in a refrigerated centrifuge pre-cooled at +4 °C and spin for 5 min at 500 × *g*.
16. Transfer the tubes on ice and carefully remove the supernatant.
17. Gently resuspend the pellet in 1 mL (0.5 mL for the "isotype control" sample) of CMFHE++ at +4 °C.
18. Proceed with **step 1** in Subheading 3.2.

3.1.2 Double Live Immunostaining (Non-Cross-Reacting Secondary Antibodies)

1. Dilute each of the two primary antibodies of choice in two tubes of CMFHE++/H (*see* **Notes 3** and **8**) and keep at room temperature. Prepare *n* + 1 milliliters of the first primary antibody solution and *n* + 1 milliliters of the second primary antibody solution, where *n* is the number of samples. Label the vials "AbI1" and "AbI2," respectively. If hybridoma supernatants are used, dilute them 1:4 in CMFHE++/H; if purified antibodies are used, dilute them to the recommended working concentration in CMFHE++/H (*see* **Note 4**).
2. Dilute each of the two isotype control antibodies of choice (*see* **Note 2**) in two tubes with 1 mL of CMFHE++/H, at the concentration recommended by the manufacturer. Name the tubes "Iso1" and "Iso2" and keep at room temperature.
3. Gently resuspend the cell pellet in the tubes named "isotype controls" and "isotype control 1" in 500 μL of the "Iso1" solution. Skip **step 4** for these samples.
4. Gently resuspend the pellets of all remaining samples in the appropriate volume (*see* **Note 5**) of "AbI1" solution, at room temperature (*see* **Note 8**).

5. Incubate for 20 min at room temperature, under static conditions.
6. Transfer the tubes in a refrigerated centrifuge pre-cooled at +4 °C and spin for 5 min at 500 × *g*.
7. Transfer the tubes in a rack at room temperature and carefully remove the supernatant.
8. Gently resuspend the pellet in ≥1 mL of CMFHE++/H.
9. Repeat **steps 6–7**, then proceed to **step 10**.
10. Gently resuspend the cell pellet in the tubes named "isotype controls" and "isotype control 2" in 500 μL of the "Iso2" solution. Skip **step 11** for these samples.
11. Gently resuspend the pellets of the remaining samples in 1 mL of "AbI2" solution, at room temperature.
12. Incubate for 20 min at room temperature, under static conditions.
13. Transfer the tubes in a refrigerated centrifuge pre-cooled at +4 °C and spin for 5 min at 500 × *g*.
14. Transfer the tubes in a rack at room temperature and carefully remove the supernatant.
15. Gently resuspend the pellet in ≥1 mL of CMFHE++/H.
16. Repeat **steps 13–14**, then proceed to **step 17**.
17. Dilute each of the two non-cross-reacting secondary antibodies of choice (e.g., rabbit anti-mouse IgG Alexa Fluor 647 conjugated and rabbit anti-mouse IgM Alexa Fluor 488 conjugated) 1:1000 (*see* **Note 6**) in CMFHE++/H. Label this vial as "Ab-II-1 + 2." Prepare *n* + 2 milliliters of this solution.
18. Gently resuspend the pellet of all the samples in the appropriate volume of "Ab-II-1 + 2" solution.
19. Incubate for 15 min at room temperature, under static conditions.
20. Place a 30 μm cell strainer on a 15 mL tube placed on ice. Rinse the strainer with 1 mL of ice-cold CMFHE++ and apply the immunostained cells (*see* **Note 7**).
21. Transfer the tubes in a refrigerated centrifuge pre-cooled at +4 °C and spin for 5 min at 500 × *g*.
22. Transfer the tubes on ice and carefully remove the supernatant.
23. Gently resuspend the pellet in 1 mL (0.5 mL for the "isotype control" samples) of CMFHE++ at +4 °C.
24. Proceed with **step 1** in Subheading 3.2.

3.1.3 Double Live Immunostaining (Cross-Reacting Secondary Antibodies)

1. Mix the first primary antibody with the appropriate secondary antibody in CMFHE++/H. The final concentrations of primary and secondary antibodies are 1:1250 and 1:250, respectively, in case the primary antibody is a purified one and 1:5 and 1:250 in case the primary antibody is a hybridoma supernatant (*see* **Note 9**). Prepare $n + 1$ milliliters of this solution where n is the number of samples. Label this mixture as "Ab-mix 1" and keep at room temperature until use.
2. Mix the second primary antibody with the appropriate secondary antibody in CMFHE++/H. The final concentrations of primary and secondary antibodies are 1:1250 and 1:250, respectively, in case the primary antibody is a purified one and 1:5 and 1:250 in case the primary antibody is a hybridoma supernatant (*see* **Note 9**). Prepare $n + 1$ milliliters of this solution where n is the number of samples. Label this mixture as "Ab-mix 2" and keep at room temperature until use.
3. Mix the isotype control antibody I with the first secondary antibody in 750 μL of CMFHE++/H. The final concentration of the isotype control immunoglobulins is according to the manufacturer; the final concentration of the secondary antibody is 1:250. Label this mixture as "Iso-mix 1" and keep at room temperature until use.
4. Mix the isotype control antibody II with the second secondary antibody in 750 μL of CMFHE++/H. The final concentration of the isotype control immunoglobulins is according to the manufacturer; the final concentration of the secondary antibody is 1:250. Label this mixture as "Iso-mix 2" and keep at room temperature until use.
5. Gently resuspend the pellet of the sample labeled "isotype control 1" in 500 μL of a 1:1 mixture of "Iso-mix 1" and "Ab-mix 2." Skip **steps 6–8** for this sample.
6. Gently resuspend the pellet of the sample labeled "isotype control 2" in 500 μL of a 1:1 mixture of "Iso-mix 2" and "Ab-mix 1." Skip **steps 7–8** for this sample.
7. Gently resuspend the pellet of the sample labeled "isotype controls" in 500 μL of a 1:1 mixture of "Iso-mix 1" and "Iso-mix 2." Skip **step 8** for this sample.
8. Gently resuspend the pellets of all the remaining samples in the appropriate volume of a 1:1 mix of "Ab-mix 1" and "Ab-mix 2," at room temperature.
9. Incubate all samples for 20 min at room temperature, under static conditions.
10. Transfer the tubes in a refrigerated centrifuge pre-cooled at +4 °C and spin for 5 min at 500 × *g*.

Table 1
Lasers and filters (optical ways) used on the flow cytometer/cell sorter

Channel	Laser (nm)	Mirror (nm)	Filter (nm)
FSC	488	N/A	2.0 ND
SSC	488	N/A	488/10 BP
Hoechst $_{BLUE}$	375	N/A	450/20 BP
Hoechst $_{RED}$	375	610 LP	670 LP
Alexa Fluor-488	488	502 LP	530/30 BP
PI	561	600 LP	610/20 BP
Alexa Fluor-647	633	N/A	660/20 BP

11. Transfer the tubes in a rack at room temperature and carefully remove the supernatant.
12. Gently resuspend the pellet in ≥1 mL of CMFHE++/H.
13. Repeat **steps 10–12**, then proceed to **step 14**
14. Place a 30-μm cell strainer on a 15 mL tube placed on ice. Rinse the strainer with 1 mL of ice-cold CMFHE++ and apply the immunostained cells (*see* **Note 7**).
15. Transfer the tubes in a refrigerated centrifuge pre-cooled at +4 °C and spin for 5 min at 500 × g.
16. Transfer the tubes on ice and carefully remove the supernatant.
17. Gently resuspend the pellet in the appropriate volume of CMFHE++ at +4 °C.
18. Proceed with **step 1** in Subheading 3.2.

3.2 Flow Cytometry and Cell Sorting

The following applies to any flow cytometers/cell sorters; the data shown in the figures were acquired using a BD FACSAria II upgraded with a 365 nm laser. The nozzle had a size of 85 μm, and the pressure separation was set to 45 Psi. The injection chamber and the analysis and collection tubes were kept at +4 °C. Lasers and filter sets used were as indicated in Table 1.

1. Transfer 250 μL from the tube named "unstained control" into a new tube. Name it "PI control."
2. Add 1 μg/mL of propidium iodide to all samples except to the "unstained control" and incubate for 1 min on ice.
3. Start loading the "unstained control" sample on the flow cytometer. Keep the flow speed at the minimum. Adjust the laser intensity and plot the data according to FSC/SSC (*see* Fig. 2a). Exclude debris (events plotted in the left portion of the plot) as well as large cell aggregates (events plotted in the top/right portion of the plot) and name the central gate as "cells" (*see* Fig. 2b, thick black polygon; blue dots in Fig. 2c; blue dots in Fig. 2d).

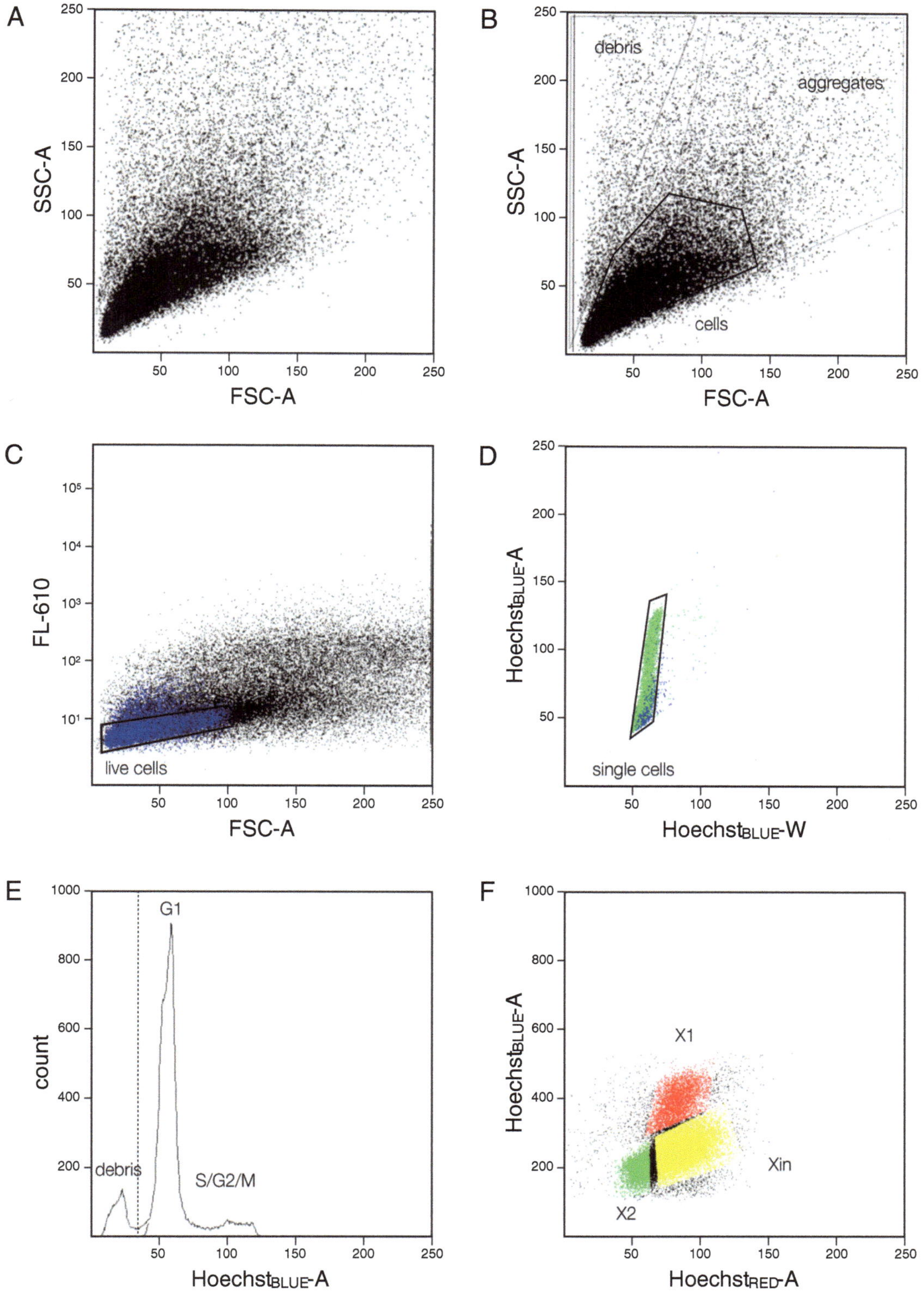

Fig. 2 Gating strategy for dissociated planarian cells stained with Hoechst 33342 and propidium Iodide. (**a**) Forward (FSC) and sideward (SSC) scatterplot of the *S. mediterranea* cell suspension. (**b**) Debris and large aggregates, which show up to the left and upper/right portions of the plot, are gated out, where bona fide

4. Load the "PI control" sample on the flow cytometer. Gate for FSC/SSC according to the "unstained control." No differences between the plots of these two controls are expected. Plot the "cells" gate according to FSC/PI-610. Two populations show up, one PI^- (live cells, lower portion of the plot) and one PI^+ (dead cells, upper portion of the plot). Draw a new gate around the PI^- events and name it "live cells" (*see* Fig. 2c, thick black polygon; green dots in Fig. 2d).
5. Load the "Hoechst/PI control" sample on the flow cytometer. Plot the "live cells" gate according to $Hoechst_{BLUE}$ intensity/ $Hoechst_{BLUE}$ width. This defines the single cells, excluding droplets (*see* Fig. 2d). Apply the "single-cell" gate according to Fig. 2d.
6. Plot the "single-cell" gate according to the fluorescence intensity of Hoechst. Two peaks show up (*see* Fig. 2e): the peak to the right is made of cells in G2/M phase, the peak to the left (the prevalent peak) is made of cells in G1. In case a third peak appears to the far left of the plot, it owes to the presence of residual debris and should be gated out, applying a threshold to the Hoechst intensity channel (dashed line in Fig. 2e).
7. Go back to the "single-cell" gate and plot it according to $Hoechst_{Blue}$/$Hoechst_{Red}$ fluorescence intensity. This will result in the three so-called planarian FACS populations (*see* Fig. 3): X1, X2 (X-ray-sensitive populations 1, 2, respectively) and Xin (X-ray-insensitive population). Set the three gates accordingly (*see* Fig. 2f) (*see* **Note 10**).
8. Load in turn the isotype control samples (if more than 1) and gate as for the "Hoechst/PI control" (**steps 5–7**).
9. In the case of single immunostaining, display the events count for the "isotype control" sample as a function of the secondary antibody fluorescence intensity (use individual plots for each of the "X1," "X2," and "Xin" gates) (*see* Fig. 4a).

Fig. 2 (continued) single cells (named "cells") are gated in with the use of a polygon gate, shown as a thick black line. (**c**) When stained with 1 μg/mL of propidium iodide, dead or dying cells emit at 610 nm and could be visualized (for example, as a function of FSC). Bona fide single cells in Fig. 2b are here shown as blue events. (**d**) Live single cells are gated according to $Hoechst_{BLUE}$ width (-W) as a function of $Hoechst_{BLUE}$ intensity (–A), as very small cell aggregates could not be gated out according to FSC/SSC. (**e**) The $Hoechst_{BLUE}$ intensity in function of the event count. In the case some debris was still present in the single-cell gate, this could be easily gated out at this point, for example, introducing a threshold (dashed line in (**e**). (**f**). Live single cells visualized according to the fluorescent intensity of Hoechst in the red ($Hoechst_{RED}$) in function of the fluorescent intensity of Hoechst in the blue ($Hoechst_{BLUE}$). Three populations of cells could be visualized: stem cells in S/G2/M phase (X1: low in $Hoechst_{RED}$ and high in $Hoechst_{BLUE}$); stem cells in G1, early progeny cells, and small differentiated cells (X2: very low in $Hoechst_{RED}$ and low in $Hoechst_{BLUE}$), and large differentiated cells (Xin: High in $Hoechst_{RED}$ and low in $Hoechst_{BLUE}$)

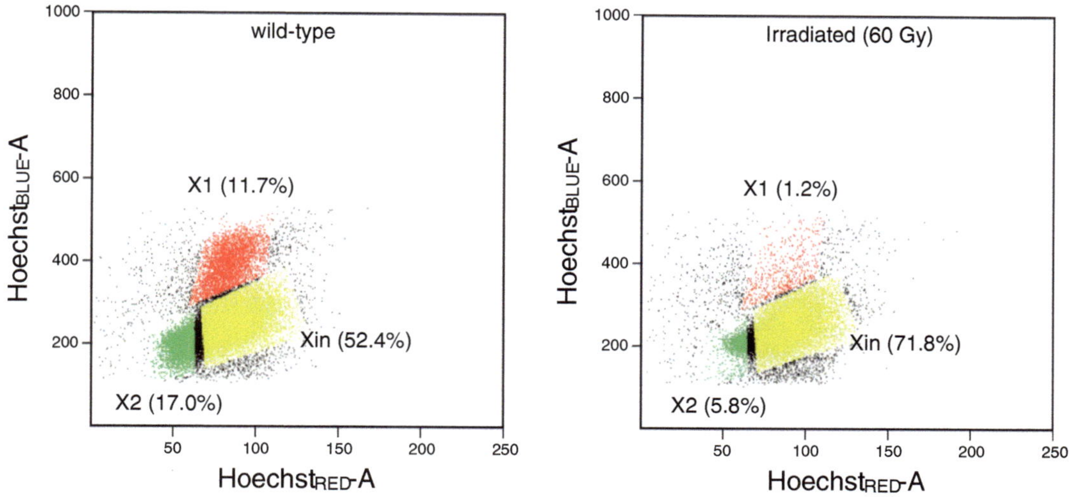

Fig. 3 The three conventional FACS populations in wild-type and lethally (60 Gy) irradiated *S. mediterranea*, 3 days after irradiation. Comparison between the Hoechst$_{BLUE}$/Hoechst$_{RED}$ fluorescence intensities of cells dissociated either from wild-type or lethally irradiated planarians. The main differences are as follows: (1) a 90% reduction of the X1 population (from 11.7% to 1.2%), owing to the fact that irradiation selectively kills cycling cells; (2) a strong reduction in the X2 population (from 17.0% to 5.8%), owing to the fact that the X2 is in part made of cycling cells in G1, which are selectively killed in irradiated animals, and postmitotic small progeny cells, which are irradiation-tolerant; and (3) a mild increase of the Xin population (from 52.4% to 71.8%), owing to the fact that the given percentages are relative, and not absolute; therefore, a decrease in two out of three populations results in the apparent increase of the third population, which is irradiation-tolerant

10. In the case of double immunostaining, display the events for the "isotype controls" sample as a density plot with quadrants (use individual plots for each of the "X1," "X2," and "Xin" gates) (*see* Fig. 4b; only X1 is shown). Set the negative gates accordingly (*see* **Note 11**).

11. In the case of double immunostaining, display in turn the events for the "isotsype control 1" and "isotype control 2" samples as a density plot with quadrants (use individual plots for each of the "X1," "X2," and "Xin" gates) (*see* Fig. 4c; only X1 is shown) (*see* **Note 12**).

12. Load in turn all the experimental samples. Analyze an equal number of events from each sample (e.g., 3×10^4, 1×10^5, 5×10^5, depending on the number of viable cells available). In the following figure, both a representative single immunostaining of X1, X2, and Xin cells using 6/9.2 antibody [13] (*see* Fig. 5a) and a double immunostaining of the X1 population using both 6/9.2 and 7/22.2 antibodies [13] (*see* Fig. 5b) are shown. In the latter, four subpopulations are visible, for which the relative percentages are given.

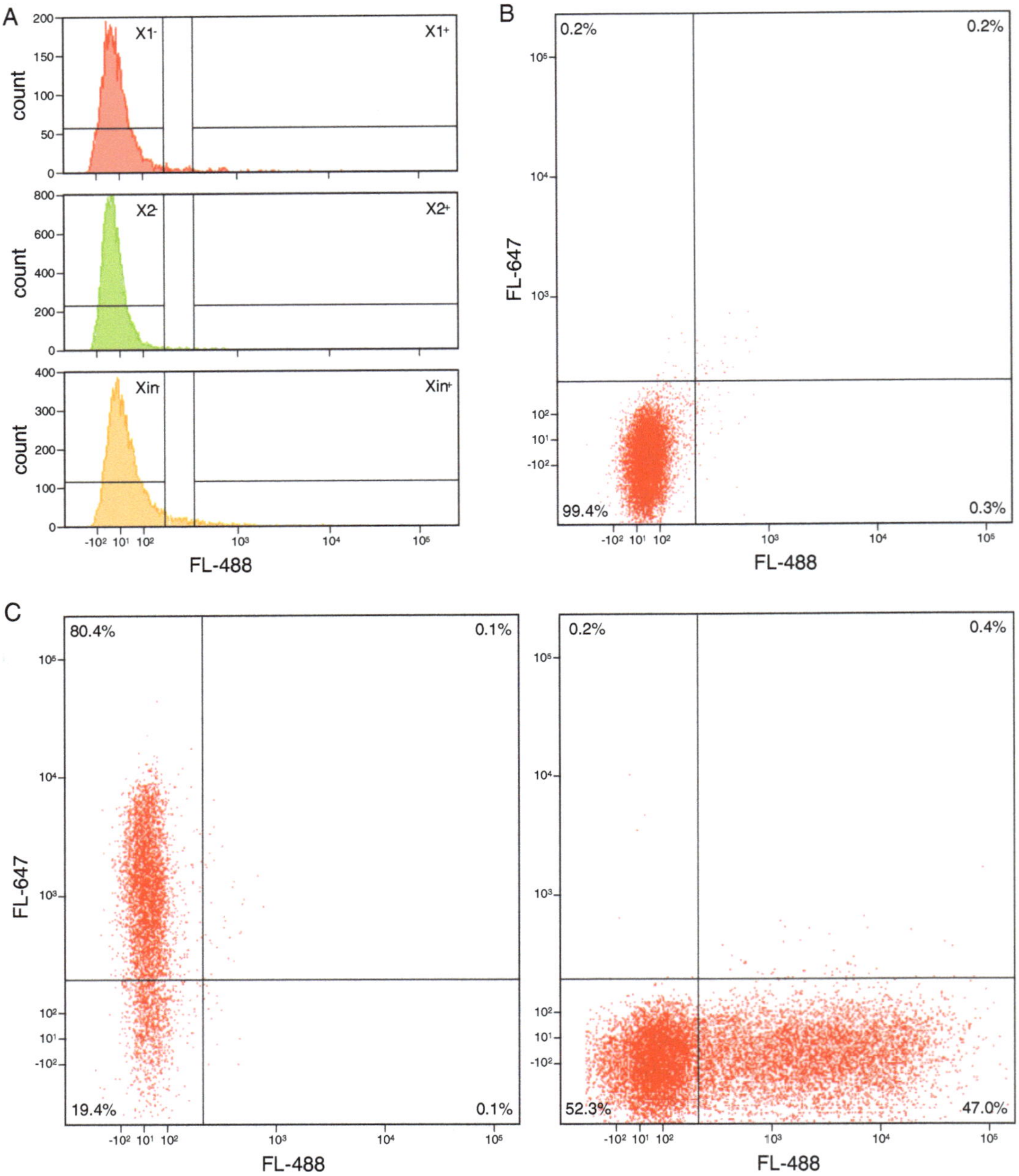

Fig. 4 Isotype control(s). (**a**) Dissociated *S. mediterranea* cells were incubated with a single isotype control (mouse IgGs) and then incubated with Alexa Fluor 488-conjugated rabbit anti-mouse IgG secondary antibody. The fluorescence intensity was plotted as a function of the events count for the three FACS populations, X1, X2, and Xin. (**b**) Dissociated planarian cells were incubated with two isotype control immunoglobulins (mouse IgMs and mouse IgGs) and then incubated with both Alexa Fluor 647-conjugated rabbit anti-mouse IgG and Alexa Fluor 488-conjugated sheep anti-mouse IgM secondary antibodies. The fluorescence intensities were plotted for the three FACS populations (only X1 is shown), with a quadrant gating strategy. (**c**) Dissociated planarian cells were incubated either with one IgG primary antibody (hybridoma clone 7/22.2) and mouse IgM isotype control immunoglobulins (left panel) or with one IgM primary antibody (hybridoma clone 6/9.2) and mouse IgG isotype control immunoglobulins (right panel). The resulting fluorescence intensities were assessed after simultaneous incubation with the two non-cross-reacting secondary antibodies used in (**b**) and plotted for the three FACS populations (only X1 is shown), with a quadrant gating strategy

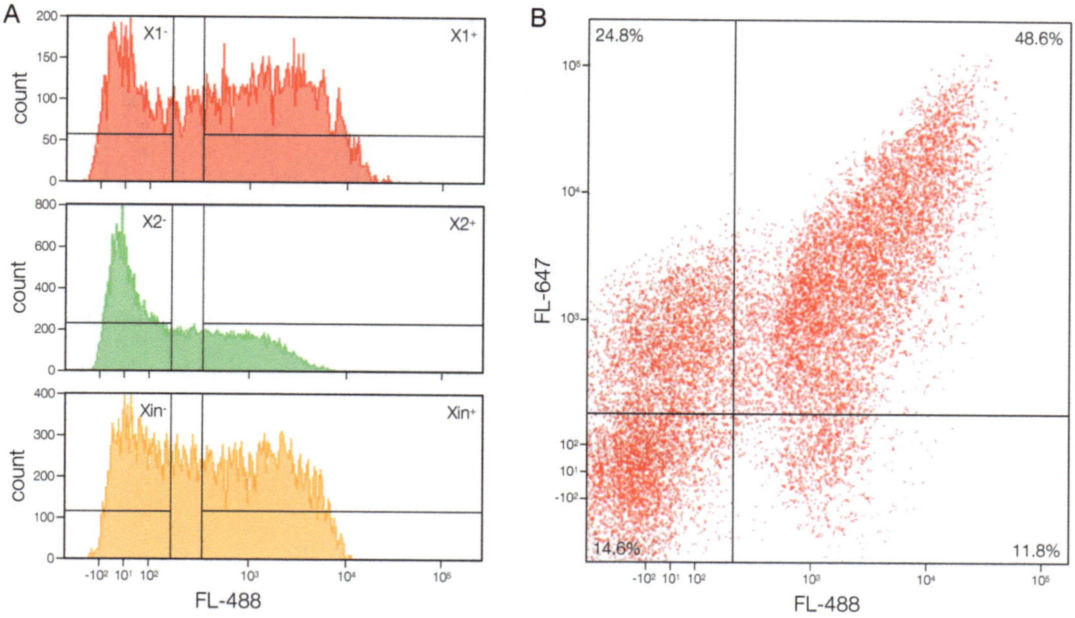

Fig. 5 Live immunostaining flow cytometry results. (**a**) Dissociated cells immunostained with 6/9.2 hybridoma clone supernatant and cross-reacting Alexa Fluor 488-conjugated rabbit anti-mouse IgM were plotted according to the resulting fluorescence intensities for the three FACS populations, in function of the events count. (**b**) Dissociated cells immunostained with both 7/22.2 IgGs and 6/9.2 IgMs and their respective non-cross-reacting secondary antibodies (Alexa Fluor 647-conjugated rabbit anti-mouse IgG and Alexa Fluor 488-conjugated sheep anti-mouse IgM) were plotted according to the resulting fluorescence intensities for the three FACS populations (only X1 is shown), with a quadrant gating strategy. Double-negative (bottom-left), double-positive (top/right) and single-positive (either top/left or bottom/right) cells are present in different percentage

13. If sorting is required, select the "4-way purity mode" (on the BD FACSAriaII/III, with FACSDiva software) or equivalent and collect up to four subpopulations in CMFHE++. If more than four subpopulations are needed, split the sample into two (for collecting up to eight subpopulations) or more tubes, apply the different gating strategies and sort in turn.
14. Live collected cells could be spread on a poly-L-ornithine-coated glass slide, allowed to attach and fixed, e.g., with 4% PFA for imaging (*see* Fig. 6a, b), or used immediately for transplantation. In the case of molecular analyses (e.g., qPCR, WTA), the sorted cells could be collected directly in the appropriate lysis buffer (*see* **Note 13**) and tube.

4 Notes

1. Violet (405 nm) laser also excites Hoechst 33342, although with much lower efficiency. It could be used, but it requires additional setup.

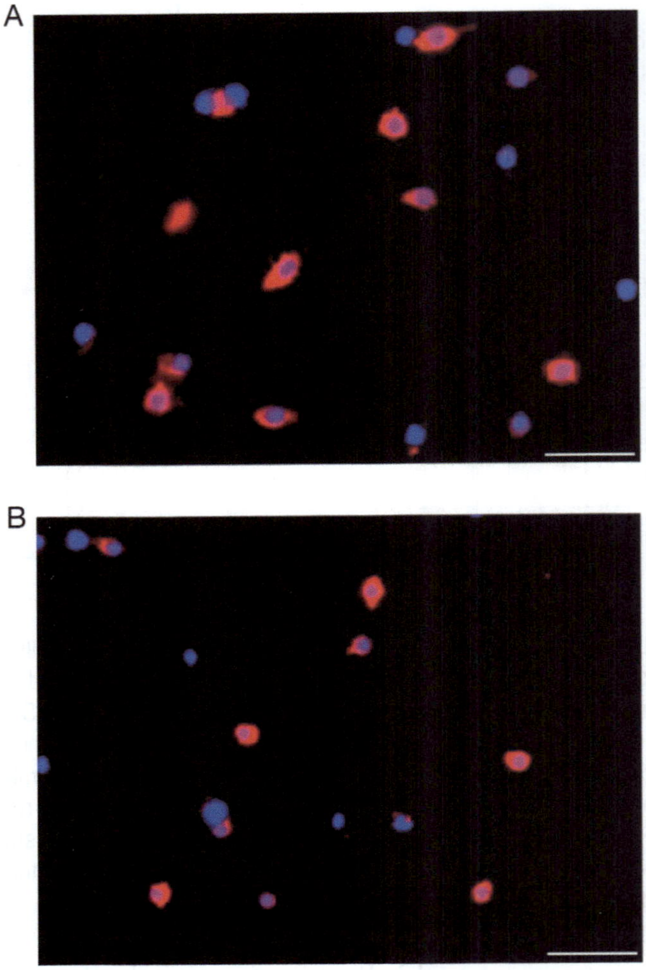

Fig. 6 Immunostained stem cells of the X1 and X2 populations fixed and imaged. (**a**, **b**) Dissociated cells were live immunostained with 6/9.2 hybridoma cell supernatant and cross-reacting Alexa Fluor 647-conjugated rabbit anti-mouse IgM. The X1 (**a**) and X2 (**b**) gates were sorted on a BD FACSAria II in individual tubes, spread on poly-ornithine-coated glass slides, fixed with 4% PFA and imaged on an inverted fluorescence microscope (Zeiss LSM780). Cells expressing and not expressing the 6/9.2 antigen could be observed. Scale bar: 25 μm

2. Isotype control antibodies are primary antibodies that do not possess any specific reactivity towards the target molecule (the antigen). They come from the same host species of the specific primary antibody used to conduct the experiment and belong to the same class of antibody (typically, either IgG or IgM; in the case that the subclass of the antibody is known, then an isotype control antibody of the same subclass should be used).

3. Hoechst 33342 is added to CMFHE++ (and therefore named CMFHE++/H) for every incubation and washing steps before

flow cytometry. This is because the immunostaining is carried out at room temperature, and eukaryotic cells actively pump out the Hoechst dye if not kept at +4 °C. Since the clearance activity depends on the cell type, the continuous presence of Hoechst 33342 reduces cell type-induced intensity changes in the Hoechst 33342 channels.

4. The 1:4 dilution of the hybridoma supernatants was empirically determined to be the most suitable for the immunostaining. However, it is a rule of thumb indication, as the antibody concentration in the supernatant depends on several factors (e.g., the number of hybridoma cells seeded, the growth rate, the time in culture, the volume of medium used). Therefore, we recommend a titration of the supernatants (usually between 1:20 and 1:3). As supernatant concentrations higher than 1:3 showed toxic effects to planarian cells, the use of affinity-purified antibodies could be a suitable alternative.
5. For consistency within the experiment and among different experiments, we recommend resuspending the samples in a fixed volume of medium, which is kept constant in all the steps. The cell density will gradually reduce, owing to partial cell loss during the centrifugation steps, but it will reduce consistently in all the samples used. Such a fixed volume (referred to as "the appropriate volume") is 1 mL in case of the study samples and 500 μL in the case of control samples. Exception to this rule are the wash steps, where the volume could be increased to up to 1.4 mL.
6. The suggested dilution of 1:1000 should work with most secondary antibodies; however, the recommended dilution is usually indicated by the manufacturer, depending on the application. Given that three channels of the flow cytometer are used to assess the fluorescent signals of Hoechst 33342 (both blue and red) and PI, the fluorophore conjugated to the secondary antibody should be chosen so that its excitation/emission spectra do not overlap with the fluorophores used. In the protocol, we use green (488) and far red (647) fluorophores.
7. This step is useful to remove residual/newly formed aggregates that could potentially clog the nozzle of the flow cell, especially in the case a small nozzle is used. CMFHE++, and not CMFHE ++/H is used, because from this step onwards, the cell suspensions are kept on ice, and therefore, the cell-specific ability to actively pump the Hoechst 33342 outside of the cell could no longer introduce a bias to the detection of the fluorescence intensity of Hoechst 33342.
8. The simultaneous incubation with both primary antibodies is also possible. In a few cases, the simultaneous incubation

slightly changed the intensity of the fluorescent signals from the secondary antibodies, but the main problem is that if two hybridoma supernatants are used at 1:4 dilution, the cumulative concentration of the hybridoma medium is too high (1:2) and kills the cells.

9. These dilutions were empirically tested. The general recommendation is to use a lower-than-usual concentration of primary antibody together with a higher-than-usual concentration of secondary antibody, so that the secondary antibody is in large excess compared to the primary antibody. This decreases the event of primary antibodies unbound to the secondary antibody of choice and could therefore introduce biases in the flow cytometry analysis.
10. The best way to define the three FACS populations is analyzing wild-type and lethally irradiated (e.g., with 60 Gy) samples side by side. The direct comparison of the plots of these two specimens allows to define the stem cell populations more precisely, as shown in Fig. 3.
11. It is important that the negative gate is set using the isotype controls, rather than using a generic control, like the one where the cells were incubated with the secondary antibody without the primary one, or the Hoechst/PI control, because the immunostaining conditions are more comparable.
12. It is recommended to run these controls while setting up and optimizing the protocol, as they are useful to understand if the concomitant presence of two primary and two secondary antibodies in the incubation steps introduces biases in the immunostaining, e.g., altering the boundary between negative and positive events. After the protocol has been established for a given pair of primary antibodies, these controls could be omitted.
13. Attention should be paid to the inevitable dilution of the lysis buffer due to the sheath fluid contained with the cells in the droplets. Ad hoc experiments are needed to quantify the effects of the dilution on downstream applications.

References

1. Aboobaker AA (2011) Planarian stem cells: a simple paradigm for regeneration. Trends Cell Biol 21(5):304–311
2. Gentile L, Cebria F, Bartscherer K (2011) The planarian flatworm: an in vivo model for stem cell biology and nervous system regeneration. Dis Model Mech 4(1):12–19
3. Rink JC (2013) Stem cell systems and regeneration in planaria. Dev Genes Evol 223(1–2): 67–84
4. Wagner DE, Wang IE, Reddien PW (2011) Clonogenic neoblasts are pluripotent adult stem cells that underlie planarian regeneration. Science 332(6031):811–816
5. Guedelhoefer OCT, Alvarado AS (2012) Amputation induces stem cell mobilization to sites of injury during planarian regeneration. Development 139(19):3510–3520
6. van Wolfswinkel JC, Wagner DE, Reddien PW (2014) Single-cell analysis reveals functionally

distinct classes within the planarian stem cell compartment. Cell Stem Cell 15(3):326–339

7. Abnave P et al (2017) Epithelial-mesenchymal transition transcription factors control pluripotent adult stem cell migration in vivo in planarians. Development 144(19):3440–3453
8. Fincher CT et al (2018) Cell type transcriptome atlas for the planarian Schmidtea mediterranea. Science 360(6391)
9. Plass M et al (2018) Cell type atlas and lineage tree of a whole complex animal by single-cell transcriptomics. Science 360(6391)
10. Asami M et al (2002) Cultivation and characterization of planarian neuronal cells isolated by fluorescence activated cell sorting (FACS). Zool Sci 19(11):1257–1265
11. Reddien PW et al (2005) SMEDWI-2 is a PIWI-like protein that regulates planarian stem cells. Science 310(5752):1327–1330
12. Hayashi T et al (2006) Isolation of planarian X-ray-sensitive stem cells by fluorescence-activated cell sorting. Develop Growth Differ 48(6):371–380
13. Moritz S et al (2012) Heterogeneity of planarian stem cells in the S/G2/M phase. Int J Dev Biol 56(1–3):117–125

Chapter 13

Live Imaging in Planarians: Immobilization and Real-Time Visualization of Reactive Oxygen Species

Vincent Jaenen, Karolien Bijnens, Martijn Heleven, Tom Artois, and Karen Smeets

Abstract

Imaging of living animals allows the study of metabolic processes in relation to cellular structures or larger functional entities. To enable in vivo imaging during long-term time-lapses in planarians, we combined and optimized existing protocols, resulting in an easily reproducible and inexpensive procedure. Immobilization with low-melting-point agarose eliminates the use of anesthetics, avoids interfering with the animal during imaging—functionally or physically—and allows recovering the organisms after the imaging procedure. As an example, we used the immobilization workflow to image the highly dynamic and fast-changing reactive oxygen species (ROS) in living animals. These reactive signaling molecules can only be studied in vivo and mapping their location and dynamics during different physiological conditions is crucial to understand their role in developmental processes and regeneration. In the current protocol, we describe both the immobilization and ROS detection procedure. We used the intensity of the signals together with pharmacological inhibitors to validate the signal specificity and to distinguish it from the autofluorescent nature of the planarian.

Key words In vivo studies, Live imaging, Immobilization, Planarians, Reactive oxygen species, ROS, Developmental dynamics, Autofluorescence

1 Introduction

Imaging of fixed specimens limits our ability to understand cellular dynamics. Physical and chemical fixation methods induce changes in the composition and appearance of in vivo structures and only reflect the status of the organism at a specific point in the past [1]. In addition, fixation does not allow imaging of highly reactive molecules, such as reactive oxygen species (ROS).

Real-time imaging of living organisms, on the other hand, is challenging, Even small movements result in unfocussed and blurry images. Current immobilization techniques include physical or pharmacological interference with either motility or consciousness. In small animals such as planarians, zebrafish, *C. elegans*, annelids,

Luca Gentile (ed.), *Schmidtea Mediterranea: Methods and Protocols*, Methods in Molecular Biology, vol. 2680,
https://doi.org/10.1007/978-1-0716-3275-8_13,
© The Author(s), under exclusive license to Springer Science+Business Media, LLC, part of Springer Nature 2023

and hydra, the use of sedatives, including ethanol, chloretone (1,1,1-trichloro-2-methyl-2-propanol), linalool, tetrodotoxin (TTX), sodium azide, and tricaine mesylate (MS-222), is widespread [2–8]. However, biological side effects, such as neurotoxicity and damage to the epidermis or epidermal appendages (e.g., cilia), are often induced [3, 9]. Alternatives include physical restraining methods, such as embedding in hydrogels [10], methyl cellulose [11], or agarose [9, 11]. Polystyrene nanobeads are then added to increase friction [12] or a (poly)dimethylsiloxane membrane is used to restrict animal movement [9, 13, 14]. As these steps also interfere with the physiological state of the animals, optimization procedures are challenging, and regularly specialized devices are required.

Here, we describe a protocol to perform live imaging of the freshwater invertebrate *Schmidtea mediterranea*, a commonly used model organism in regeneration and stem cell research [15, 16] and an upcoming (eco)toxicological model organism [17–23]. Its relatively large size (up to 25 mm), easily deformed and sensitive body, optimal temperature and humidity requirements , strong motility, and negative phototactic behavior make *S. mediterranea* a challenging target for live imaging [9]. Dexter et al. (2014) previously described the construction of an agarose-based chip in which they used an additional membrane to increase pressure for better immobilization. The present workflow combines elements of this protocol and other immobilization procedures [5, 9, 24, 25]. It is a cheap, easy, reproducible, and reliable method to image living planarians that can be performed with commonly available lab materials. The use of low-melting-point (LMP) agarose allows long-term in vivo imaging, without compromising the planarian's survival rate or regenerative ability during and after imaging.

These advantages are especially important when imaging redox molecules, such as ROS. ROS are indispensable in all living organisms and regulate cellular events in developmental processes such as cell proliferation, differentiation, patterning, and apoptosis [26–30]. Their high reactivity and related short half-life, make the visualization process extremely difficult. Most researchers use indirect measurements to map redox dynamics, for example by measuring oxidative damage at proteins, lipids, and nucleic acids or by determining total antioxidant capacity [31]. Direct In vivo electrochemical measurements [32] and electron spin resonance-based detection [33] often require specialized equipment. More commonly used visualization methods are fluorescent reporter-based methods, including genetically encoded biosensors and synthetic ROS indicators [34]. [35–37].

In this manuscript, we describe a protocol to image living planarians, using three different ROS detection methods: the Image-IT LIVE Green ROS detection kit (Thermo Fisher—I36007) for detecting general ROS production and the ROS-ID superoxide detection kit (Enzo—51,012) and the dye Peroxy Orange 1, PO1 (Sigma Aldrich—SML0688), to visualize superoxide and hydrogen peroxide, respectively. We used the intensity of the signal, together with pharmacological inhibitors, to validate signal specificity and to distinguish it from the the autofluorescent nature of the planarian. Although the current manuscript takes ROS live imaging as a specific example, similar principles and concepts can be applied for visualizing other molecules, or for autofluorescence imaging.

2 Materials

Prepare and store all reagents at room temperature (unless indicated otherwise). Diligently follow all waste disposal regulations when disposing waste materials.

2.1 Detection of Reactive Oxygen Species

1. ROS detection solution: Dissolve the dye (for general ROS and hydrogen peroxide labeling) in DMSO (dimethyl sulfoxide), resulting in a final stock concentration of 10 mM carboxy-- H_2DCFDA or 2 mM Peroxy Orange 1. Dissolve the superoxide detection reagent in DMF (N,N-dimethylformamide) in order to obtain a 5 mM stock solution. Prepare aliquots following the manufacturer's instructions (*see* **Notes 1–4**).
2. Working solution: Dilute all dyes in planarian medium, up to final concentrations of 25 μM carboxy-H_2DCFDA, 20 μM Peroxy Orange 1, or 0.5 μM superoxide detection reagent.
3. 24-well plate (*see* **Note 5**).
4. Planarian medium: Make a solution of Milli-Q water containing the following ingredients: 1.2 mM $NaHCO_3$, 1.6 mM NaCl, 1.0 mM $CaCl_2$, 1.0 mM $MgSO_4$, 0.1 mM $MgCl_2$, and 0.1 mM KCl. Precool the planarian medium at 4 °C.
5. Black box to allow the planarians to reside in the dark.
6. Temperature-controlled room at 20 °C.
7. Plastic Pasteur pipettes with a big (± 2 mm) and a small (± 1 mm) tip opening.

2.2 Immobilization

1. Make a 2.5% low-melting-point (LMP) agarose solution in planarian medium (*see* **Note 6**).
2. Make a 7% LMP agarose solution in planarian medium (*see* **Note 6**).
3. Water bath or thermomixer at 80 °C , and an ability to cool to a temperature of 40 °C.
4. Precooled, plastic, reusable freezer block at a temperature of −20 °C.

5. Paper towel or filter paper.
6. Plastic Pasteur pipette with a big (± 2 mm) and a small (± 1 mm) tip opening and a glass Pasteur pipette (± 2 mm) with a big tip opening (*see* **Note 7**).
7. Planarian medium: Make a solution of Milli-Q water containing the following ingredients: 1.2 mM $NaHCO_3$, 1.6 mM NaCl, 1.0 mM $CaCl_2$, 1.0 mM $MgSO_4$, 0.1 mM $MgCl_2$, and 0.1 mM KCl.
8. 3-well chamber slide, preferably with a glass bottom (*see* **Note 8**).

2.3 Live Imaging

1. Ice cubes and a microcentrifuge tube rack (*see* Subheading 3.2, **step 11**).
2. A fluorescent microscope, preferably in an inverted setup, with the appropriate filters and a high-resolution and high-sensitivity camera. (*see* **Note 9**) .
3. Plastic Pasteur pipettes with a small (± 1 mm) tip opening.
4. Serrated lab tweezers.
5. ROS-inhibiting compounds: Prepare a stock solution of 3 mM diphenyleneiodonium chloride (DPI, Sigma Aldrich, D2926) and 4 M apocynin (APO, 4-hydroxy-3 -methoxy-acetophenone, Sigma Aldrich, A10809) in DMSO (*see* **Note 4**). Make the exposure solutions in planarian medium with a final concentration of 3 μM DPI or 400 μM APO. Exposure occurs 5 h prior to in vivo ROS staining (*see* **Note 10**).
6. The Drummond Nanoject II: This device is used for functional or physical manipulation during imaging.
7. A generic 3-axes micromanipulator with adaptor for the Drummond.
8. Glass capillary needles compatible with the Drummond Nanoject II: The needles have an outer diameter of 1.1 mm, an inner diameter of 0.53 mm, and a length of 8.5 cm. The glass capillaries are hand-pulled into needles above a Bunsen burner, but can also be bought pre-pulled.
9. Utility knife: Used for physical manipulation during imaging.

2.4 Recovery of the planarian

1. Curved lab tweezers.
2. Glass plate.
3. Petri dish (60 mm x 15 mm).
4. Planarian medium at room temperature: Make a solution of Milli-Q water containing the following ingredients: 1.2 mM $NaHCO_3$, 1.6 mM NaCl, 1.0 mM $CaCl_2$, 1.0 mM $MgSO_4$, 0.1 mM $MgCl_2$, and 0.1 mM KCl.

3 Methods

The combined workflow is visualized in Fig. 1. To distinguish the signal from the autofluorescent background of the planarian, the intensity of the signal and the effects of ROS-inhibiting treatments were used as a validation (*see* **Note 10**; Fig. 2).

3.1 Detection of Reactive Oxygen Species

An overview of the labelling step is represented in Fig. 1.1 (*see* **Notes 2, 9,** and **10**). All steps can be performed with multiple planarians simultaneously. The number of planarians during the subsequent immobilization steps depends on the number of wells on the chamber slide.

1. Select planarian(s) of approximately 2–3 mm in size and starve it/them for at least 7 days (*see* **Note 11**). We recommend using no more than three organisms at a time.
2. Transfer each individual planarian to a single well of a 24-well plate (*see* **Notes 5** and **12**).
3. Remove as much medium as possible and add the working solution of the specific fluorophore (*see* **Note 13**).
4. Incubate the planarian(s) for 1 h, in a black box (dark) and at 20 °C in a temperature-controlled room. To capture the fast-changing redox dynamics, it is important to ensure uptake of the dye before the start of the experiment (*see* **Notes 11** and **13**).
5. *Optional:* Induce the condition of interest, e.g., regeneration by artificially cutting or wounding the animal. After this intervention, continue with another 15 min of incubation in the same labeling solution (dark, 20 °C).
6. Gently transfer the planarian(s) to a new well and wash it/them at least three times with precooled planarian medium, with a new well each time (*see* **Note 14**).
7. The planarian(s) is/are now ready to immobilize.

3.2 Immobilization

The immobilization workflow is visualized in Fig. 1.2. For short-term imaging, the immobilization method described from **steps 1** to **10** is sufficient. For a long-term time-lapse, include **steps 11–14** (*see* Fig. 1.3). In both cases, it is important to minimize experimental variation (*see* **Note 15**).

1. Heat both the 2.5% and 7% LMP agarose up to 80 °C until it is completely liquified. Set the temperature of the water bath or thermomixer at 40 °C and allow the agarose to cool down. Flick regularly to keep the agarose fully dissolved, do not vortex (*see* **Note 6**).

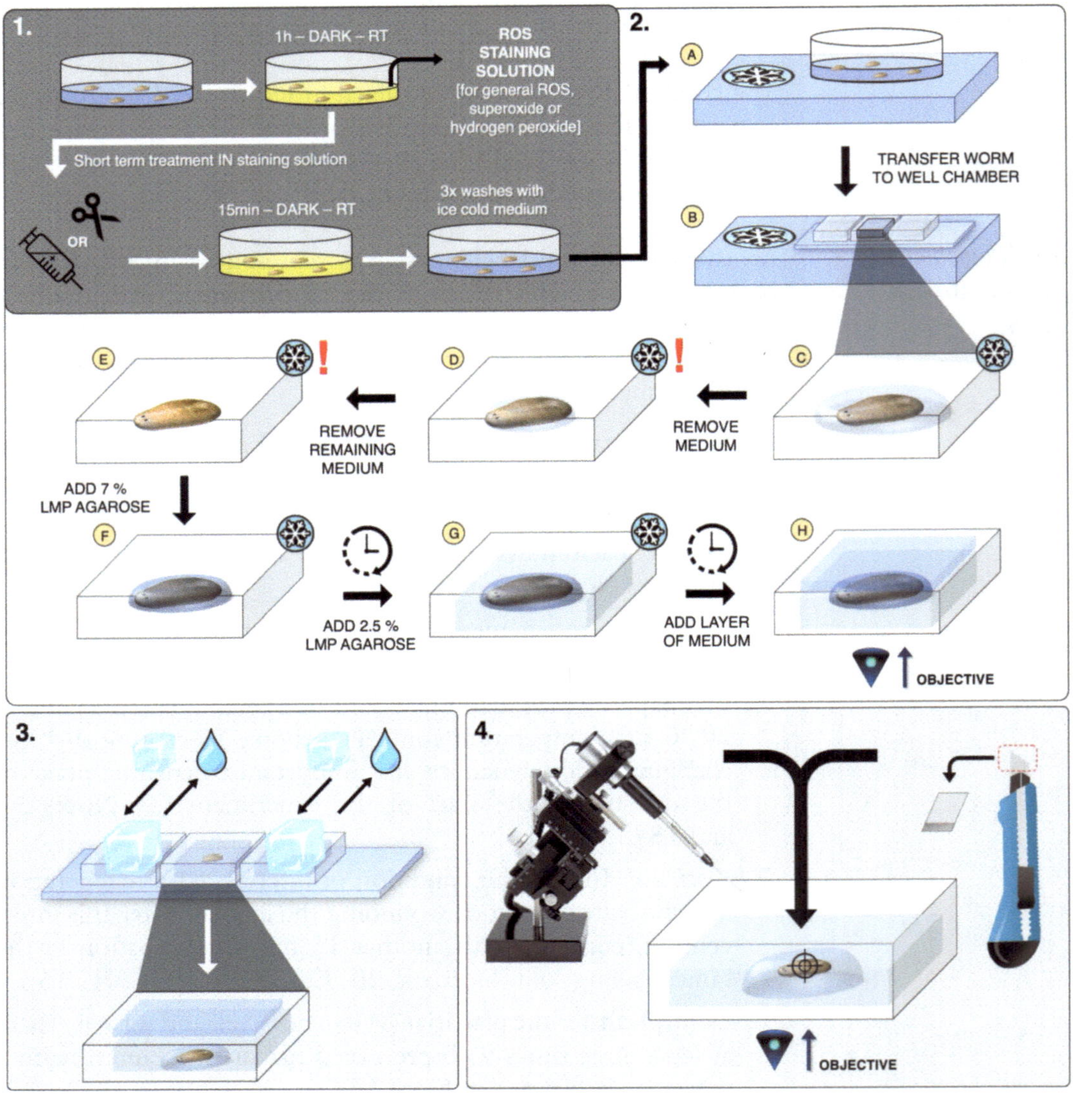

Fig. 1 Graphical representation of the in vivo ROS detection procedure. (**1**) Staining procedure for visualizing in vivo production of ROS in general, superoxide and hydrogen peroxide. (**2**) Workflow to immobilize a living organism, using a glass-bottom well chamber slide and with the intention to use inverted microscopy for imaging. Procedure steps are numbered from **a** to **h**. In short; place the planarian in a petri dish on a precooled freeze block to reduce motility (**a**), precool the well chamber slide on the precooled freeze block (**b**), and transfer the planarian in a droplet of medium to the middle of the well chamber (**c**). Remove the medium around the planarian, first with a Pasteur pipette (small tip opening) (**d**) and then with a paper towel or filter paper (**e**). Add one droplet of 7% LMP agarose on top of the planarian (**f**). After the droplet is solidified, cover the whole chamber with 2.5% LMP agarose (**g**). When all LMP agarose is solidified, add a thin layer of medium on top (**h**). The sample is now ready for imaging. (**3**) Setup for long-term imaging (1–3 h), using a simple cooling method in combination with the immobilization addressed in Fig. 1.2. **4**). Representation of how live manipulation can be performed during long-term in vivo imaging. A microinjector can be used to [1] administer fluorophores, pharmacological inhibitors, RNAi probes, or toxicants, or [2] to inflict a healing wound (H-wound) by the glass capillary needle of the microinjector, while a blade of a utility knife can be used to inflict a regenerative wound (R-wound) (*see* **Note 29**)

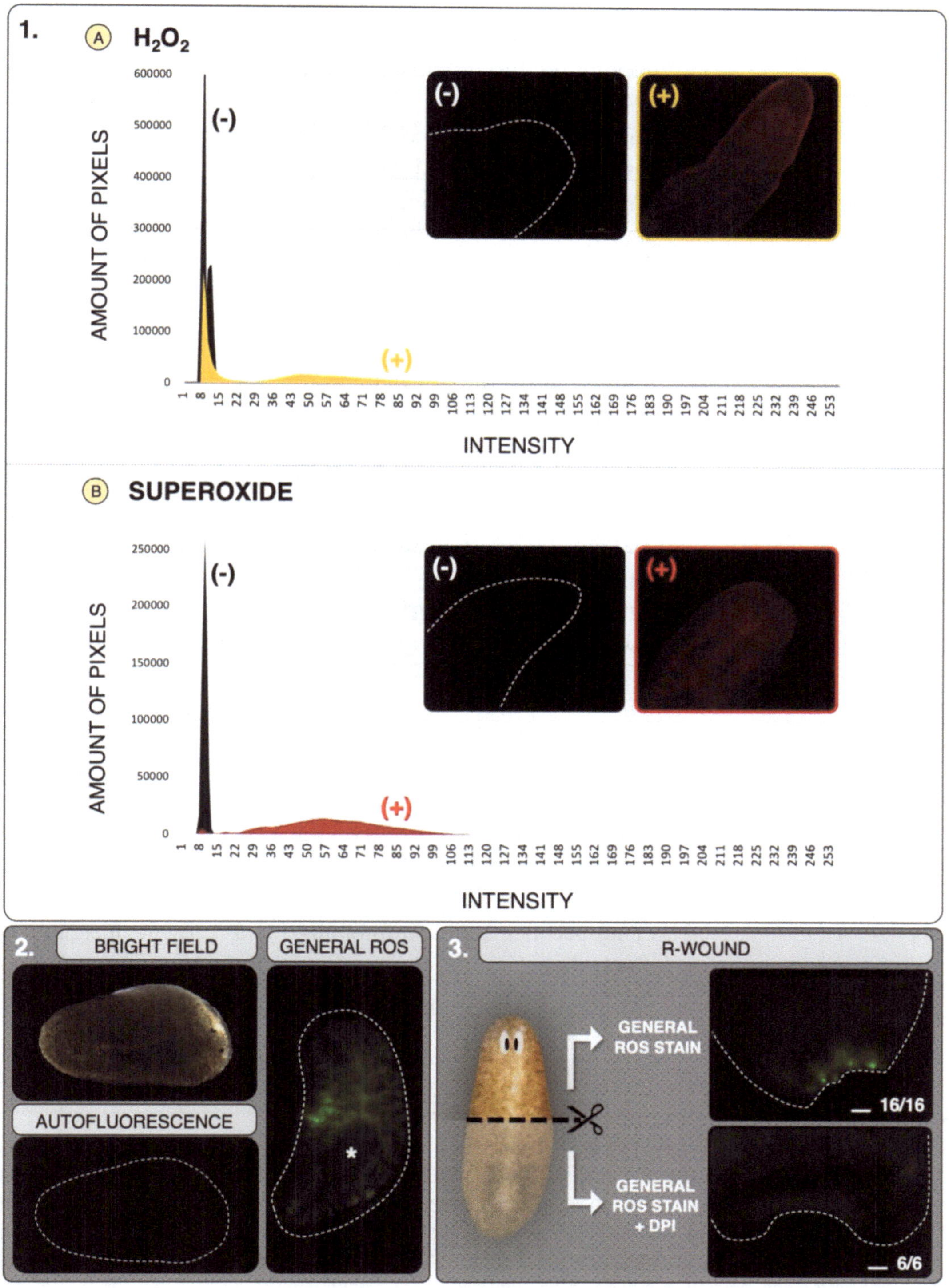

Fig. 2 Autofluorescence versus ROS signals . (**1**) Fluorescence intensity per pixel observed in images of a planarian with (+) and without (−) the in vivo ROS stain, for (**a**) hydrogen peroxide (Ex/Em 543/545–750) and (**b**) superoxide (Ex/Em 550/620). The data represented in the graphs are based on an average of six individual measurements (specimens) per condition. The fluorescence intensities of the condition without the staining solution (control) are a measure for the level of autofluorescence and indicated in black. A representative

2. Place the planarian in a petri dish filled with plenty of medium and let it cool down on a cooling element. Wait until the motility of the planarian is strongly reduced (*see* Fig. 1.2.a) (*see* **Note 16**).
3. In the meantime, precool the well chamber slide on a cooling element (*see* Fig. 1.2.b) (*see* **Note 17**).
4. Transfer the planarian to the well chamber. Place the planarian in one droplet of medium in the middle of the well chamber. Wait until the planarian is fully stretched out, stops moving, and lays in the preferred imaging position (*see* Fig. 1.2.c) (*see* **Notes 18** and **19**). This position depends on the structures you aim to study. In our experimental setup, we imaged at the ventral side of the planarian.
5. Carefully remove the medium around the planarian using a Pasteur pipette with a small tip opening (*see* Fig. 1.2.d).
6. Take the well chamber slide off the cooling element. Remove the remaining medium by means of a paper towel or filter paper (*see* Fig. 1.2.e) (*see* **Note 20**).
7. Place the slide on the cooling element. Immediately add one droplet of 7% LMP agarose on top of the planarian (*see* **Note 21**). Wait until the droplet is completely solidified (*see* Fig. 1.2.f) (*see* **Note 22**).
8. Fill the rest of the well chamber slide with 2.5% LMP agarose until the whole chamber is covered with a 3–4 mm layer of 2.5% LMP agarose and wait for this layer to solidify completely (*see* Fig. 1.2.g) (*see* **Note 23**).
9. Add a thin layer of medium on top of the 2.5% LMP agarose to prevent dehydration of the agarose and of the organism (*see* Fig. 1.2.h) (*see* **Note 24**).
10. The specimen is now immobilized and ready for imaging (*see* **Note 25**).

Additional steps for long-term imaging:

Fig. 2 (continued) picture is shown on the left, labeled "(−)." The yellow/red graph shows the fluorescence intensities of stained planarians, and a representative picture is shown on the right, labeled "(+)." The white dotted line indicates the edge of the specimen as the specimen itself is not visible. The pictures were made using an inverted microscope with identical camera and device settings for all conditions. (**2**) General ROS production in intact animals, including a negative control showing limited autofluorescence (Ex/Em 495/529). The white dotted line indicates the edge of the planarian as the planarian itself is not visible and the asterisk represents the presence of a pharynx. (**3**) ROS signal specificity via a representative close-up image of general ROS production induced at the wound site after amputation (16/16), without and with exposure to the ROS inhibitor DPI (6/6). The white dotted line indicates the edge of the amputation site of the planarian. Scale bars represent 100 μm

11. Prepare the ice cubes at least one day in advance. Fill the holes of the microcentrifuge rack with tap water and store overnight at −20 °C. When the water is frozen, place the microcentrifuge tube rack in warm water to heat up the bottom of the rack. The ice cubes will slightly move upwards. Pull them out using lab tweezers.
12. Prepare the immobilized specimen in a well chamber glass slide as described above (*see* **Note 26**).
13. Place an ice cube in each well chamber on either side of the immobilized sample (*see* **Note 27**).

3.3 Live Imaging

1. First check the fluorescent signal of the negative controls, being the samples without ROS labeling or samples treated with ROS inhibitors such as DPI or APO. Focus the sample under bright-field using a low magnification (4–5×) and low intensity. Once the image is focused, switch to fluorescence imaging at the wavelength corresponding to the fluorophore you are using. Be as quick as possible as prolonged, strong light exposure will induce ROS production.
2. Next, the samples with the ROS labeling can be imaged. Repeat the first step to bring the specimen into focus.
3. Scan the specimen for clear fluorescent signals. If the image remains blurry, dry the bottom of the well chamber slide before loading it under the microscope as the cold slide condensates at the bottom.
4. The fluorescent signal attributed to ROS can be distinguished from the autofluorescent background by looking at (1) the intensity and spectra of the signals, (2) specific locations, and (3) delineated structures showing signals and comparing this to the negative controls (*see* **Note 9**) (*see* Fig. 2).
5. Adapt the gain and intensity of the fluorescence lamp/laser to create an image with as little background as possible. When a background signal is observed outside the planarian, reduce the gain and the intensity of the lamp/laser. If you still experience difficulties, carefully read **Notes 8–12, 14, 23,** and **25** (*see* **Note 28**). Pay attention to use the same settings in all conditions within a single experiment.
6. Via microinjection, fluorophores, pharmacological inhibitors, RNAi probes, or drugs can be administered while imaging. The microinjector also allows to inflict a healing wound (H-wound) by the glass capillary needle and a regenerative wound (R-wound) by attaching a blade of a utility knife to the injector (*see* Fig. 1.4.; *see* **Note 29**).

3.4 Recovery of the Planarian

Make sure the specimen is at room temperature before starting the following recovery steps, visualized in Fig. 3.

1. Use a lab tweezer to remove the 2.5% LMP agarose square containing the 7% LMP agarose droplet including the planarian from the well chamber slide (*see* Fig. 3.1.a) (*see* **Note 30**).
2. Place the whole on a glass plate with the specimen facing down (*see* Fig. 3.1.b).
3. Turn the specimen upside down using the lab tweezers. Use the tweezers to separate the 7% LMP agarose droplet, containing the planarian, from the 2.5% LMP agarose square (*see* Fig. 3.1.c).
4. Gently transfer the 7% LMP agarose droplet, with the planarian, to the middle of an empty petri dish. Pay attention that the droplet is laying upside down with the side of the planarian pointed upwards.
5. Gently fill the petri dish with planarian medium (RT) until just above the 7% LMP agarose droplet with the planarian and wait until the planarian is acclimatized and ready to move out of the droplet into the medium (*see* Fig. 3.1.d) (*see* **Note 31**). After a few minutes the planarian releases itself, which is visualized in Fig. 3.2 (*see* **Note 32**).

4 Notes

1. Depending on the type of ROS, different detection solutions exist. To study general ROS production, the Image-IT LIVE Green ROS detection kit (Thermo Fisher—I36007), based on carboxy-H_2DCFDA, is a commonly used application. To detect superoxide or hydrogen peroxide, the ROS-ID superoxide detection kit (Enzo—51,012) or Peroxy Orange 1, PO1 (Sigma Aldrich—SML0688), can be used. All three dyes are cell-permeable and display a fluorescent signal in response to an intracellular reaction with the specific ROS, depending on the dye of your choice. Other fluorophores or detection systems are not yet tested or optimized for their use in planarians.
2. Work in the dark throughout the procedure. Light and temperature changes will deteriorate the fluorophores. Planarians are negatively phototactic and will experience less stress in the dark.
3. The components of the ROS detection kits are oxygen-sensitive. It is essential to store the opened and stocked vials following the manufacturer's instructions, preferably together with the included oxygen scavenger pouch.

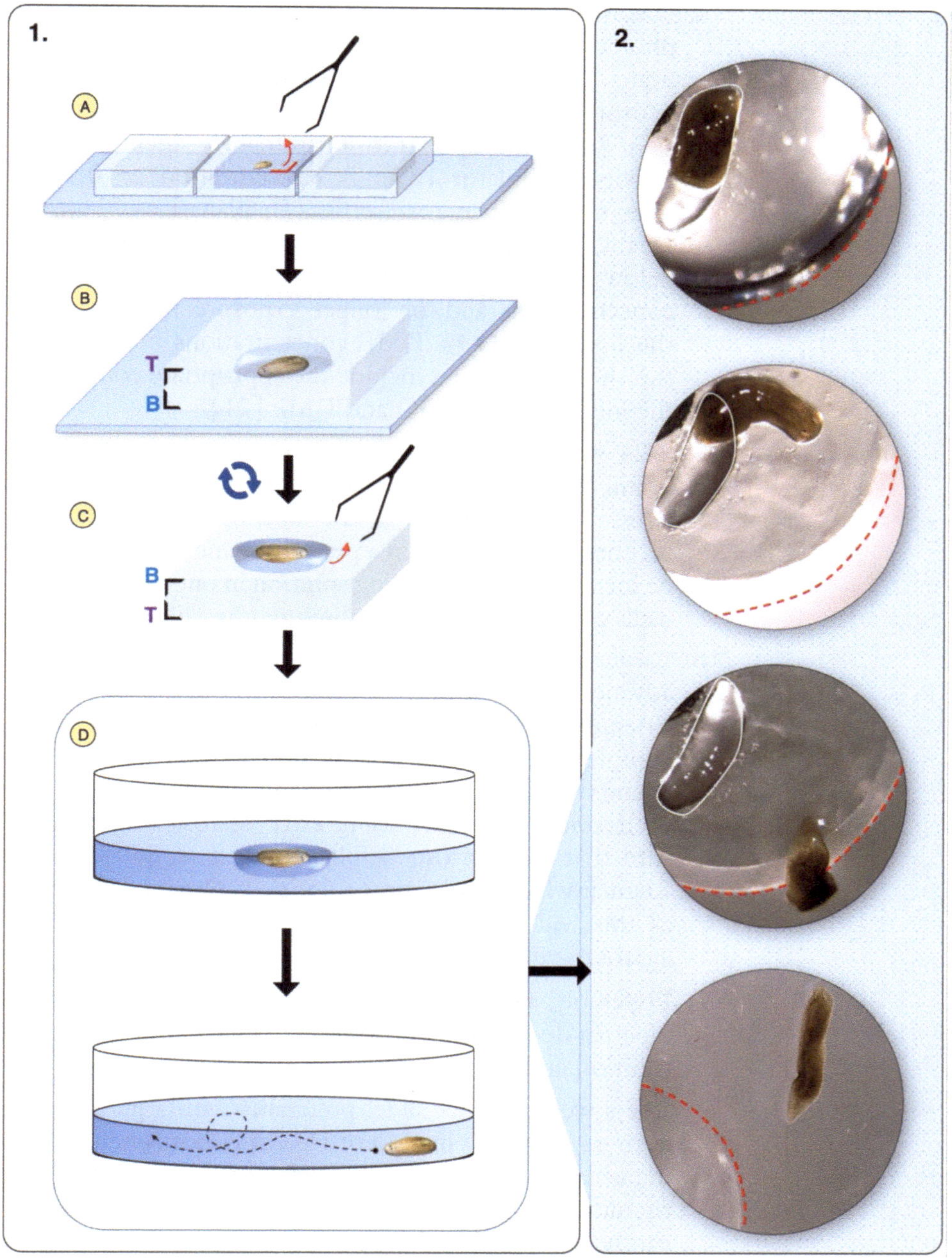

Fig. 3 Visualization of the procedure to recover the planarian after immobilization. (**1**) Procedure steps are numbered from **a** to **d.** Briefly, use lab tweezers to remove the 2.5% LMP agarose square from the well chamber slide (**a**). Place the agarose cube on a glass plate with the sample facing down (T = top and B = bottom) (**b**). Turn the specimen and use a lab tweezer to pull out the 7% LMP agarose droplet with the planarian inside (**c**). Gently transfer the droplet including the planarian to a petri dish and fill it with medium. Make sure the droplet remains upside down with the planarian facing up and allow the planarian to detach (**d**). (**2**) Brightfield images to illustrate how the planarian releases itself from the LMP agarose, allowing an easy recovery without damaging the organism. The white line indicates the area of the planarian when it was stuck in the 7% LMP agarose droplet, while the red dotted line points to the edge of the 7% LMP agarose droplet

4. DMSO and DMF are used to enhance the solubility of hydrophobic compounds in aqueous solutions [38]. In both vertebrates and invertebrates, DMSO is known to induce neurotoxicity and can interfere with the redox status by acting as a ROS scavenger, anti-inflammatory agent, and cytokine activity modulator [39]. A concentration of 0.1% DMSO does not result in toxic effects in planarians [21], and final DMSO concentrations in the staining solutions were kept below this threshold. DMF is toxic for planarians at higher concentrations and/or longer exposure times [40]. Reduce the final DMSO or DMF concentrations below the toxicity threshold and always include the appropriate controls to assess the influence of DMSO and DMF [21].
5. A 24-well plate is convenient to work with as (1) the planarian is still able to move freely and does not experience additional stress, (2) the volume of the labeling solution necessary for incubation is less than in a petri dish, and (3) the organism can be incubated in the labeling solution in one well while the other wells can be used for the subsequent washing steps.
6. Larger amounts of LMP agarose solutions can be prepared in advance by dissolving the agarose in planarian medium in an Erlenmeyer flask using a microwave. When the LMP agarose is completely dissolved, it can be transferred to 15 mL or 50 mL falcon tubes or aliquoted in Eppendorf tubes and stored at 4 °C up to one week. When using LMP agarose after storage, make sure that the agarose is fully liquified at 80 °C before use. Regularly flick until a clear solution is obtained. Do not vortex as this will induce air bubbles. Continue mixing the LMP agarose as it cools to 40 °C.
7. Preferably, use glass pipettes to handle LMP agarose. Glass pipettes do not deform when in contact with the hot agarose solution and can be washed and reused.
8. Use a well chamber slide, preferably with a glass bottom and high walls. A glass instead of a polymer bottom increases imaging quality and allows a more efficient cleaning and recovery of the slide afterwards. A regular glass bottom slide of 1 mm thick can be used for imaging at lower magnification (4, 5, or 10×). High-magnification imaging requires a shorter working distance between the objective and the specimen; a 1-mm-thick slide will negatively influence the focus point. Therefore, a well chamber slide with a glass coverslip bottom and a thickness of 1.7 μm is advised in order to obtain high-quality images. High walls around the well make it possible to apply more LMP agarose (2.5%) on top of the specimen, to increase the pressure on the planarian. They facilitate long-term storage as a layer of medium can be easily applied on top (*see* Fig. 4.6). After

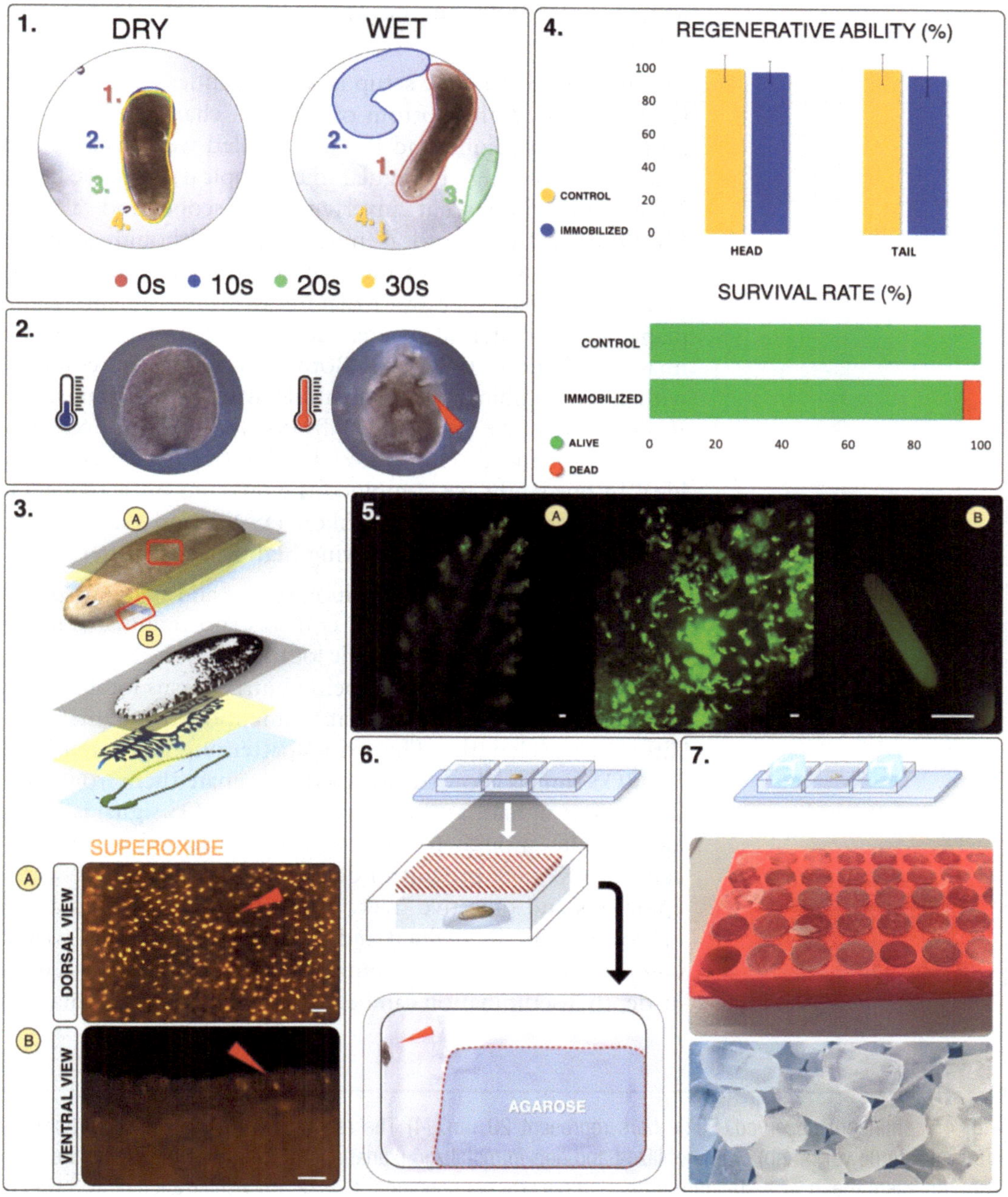

Fig. 4 Details of the in vivo ROS staining and immobilization procedure. (**1**) Immobilization with and without complete removal of the medium. **Steps 1–4** represent different time points after immobilization (0 sec, 10 sec, 20 sec, and 30 sec) and show that the motility of a dry planarian (left) is significantly more reduced compared to a situation where the medium is not completely removed (right). (**2**) After cooling the 7% LMP agarose to 40 °C, additional cooling inside the Pasteur pipette is required. This figure shows the induction of tissue lysis without additional cooling (right; approximately 40 °C) versus no effect with additional cooling (left; approximately 30 °C) (Fig. 1.2.F) (*see* **Note 21**). (**3**) The tissue types as well as the location of imaging (being anterior vs posterior or dorsal vs ventral) strongly influence the character of the observed fluorescent ROS signals. The lower panel represents a dorsal and a ventral view after in vivo superoxide staining (location of the pictures taken Is indicated on the model, red squares, labeled with **a** and **b**). Red arrowheads indicate signals of superoxide production in epidermal cells. A clear difference in distribution and quantity of superoxide can be

imaging, the slide can be reused if properly cleaned with 70% ethanol.

9. An inverted microscopy setup is recommended, as the planarian is located at the bottom of the well chamber. The LMP agarose, the medium, and the high-walled well chamber will complicate focusing in a regular microscopical setup. We used a Nikon Ts2 fluorescent inverted microscope and a Zeiss LSM900 KMAT confocal microscope. Although confocal imaging greatly increases the resolution, quality, and level of detail of the images, be aware that miniscule movements (of the planarian or of internal systems such as the gut) will disturb the image stack along the z-axis. Both microscopes are equipped with filter combinations compatible with the stainings, i.e., Ex/Em 495/529 (for general ROS), Ex/Em 550/620 (for superoxide), and Ex/Em 543/545–750 (for hydrogen peroxide). In our study, we combined the inverted microscope and confocal microscope with a Nikon DS-Fi3 and Photometrics 95B camera, respectively, allowing high-quality images.

10. To discriminate between autofluorescence and staining-specific fluorescence, include negative controls (no staining) and pay close attention to (1) the specific location, (2) the intensity, and (3) the nature of the signals (being blurry versus localized and delineated) [41–43]. Check abovementioned parameters at different wavelengths. The spatial patterning and intensity of the ROS signals allow to visually discriminate them from autofluorescent signals, as the intensity of the ROS signal is strong and specific enough to overcome the observed autofluorescence. ROS were present in delineated structures such as the epidermis and digestive tract. The autofluorescent signal was (1) less specifically defined and (2) five to seven times weaker compared to the signal attributed to ROS (*see* Fig. 2.1). A more in-depth discrimination can be carried out via spectral analyses.

Fig. 4 (continued) observed. Scale bars represent 20 μm. (**4**) The regenerative ability (upper graph) and survival rate (lower graph) of immobilized and non-immobilized planarians, 7 days after the procedure. Error bars represent mean +/− standard error. Sample size per condition: $n = 12$. (**5**) Visualization of ROS with (**a**) and without (**b**) washing steps (Fig. 1.1) (*see* **Note 14**). 5.**a** shows a clear ROS signal in the intestines without the presence of a background signal. 5.**b** displays a strong signal outside the planarian, when the planarian is not properly rinsed after the in vivo ROS staining procedure. These rod-shaped structures (protists) negatively influence the imaging (5.**b**, left image). The right image shows a magnification of one protist. ROS are stained by means of carboxy-H_2DCFDA, but similar effects were noticed with the other protocols. Scale bars are 10 μm. (**6**) Dehydration and shrinking of the agarose can be avoided by applying a layer of medium on top, especially for long-term imaging. The red dotted line indicates the edge of the LMP agarose (2.5%) and the red arrowhead indicates the position of the planarian after shrinking. (**7**) With common lab equipment such as a microcentrifuge tube rack, small (high-density) ice cubes can be made to aid cooling the sample during longer imaging procedures (1–3 h)

To verify signal specificity, we used two compounds, i.e., apocynin (APO) and diphenyleneiodonium (DPI) [44, 45], that are widely used to inhibit ROS production in several in vitro and in vivo regeneration models [46–49]. Based on spectra analyses and as no visible fluorescent signal was observed after DPI (*see* Fig. 2.3) or APO exposure [28], we concluded that the observed fluorescent signals are ROS-specific.

11. Use small planarians of approximately 2–3 mm in size as they are easier to immobilize and image. In larger planarians uptake of the fluorescent dye might be compromised. The variability in dye uptake and distribution must be considered in each experiment. It is important to use planarians that are starved for at least 7 days, as the food (here we used bovine liver) itself can be a source of autofluorescence in planarians [50]. In addition, it is known that feeding induces ROS production as a by-product of the absorption process [50, 51].
12. Be gentle when handling the planarian because each wound or additional stress will induce ROS production. Avoid variation between experiments, given the highly dynamic nature of ROS.
13. Make sure the planarian is completely submerged in the labeling solution. Different incubation times were studied and optimized.
14. These washing steps are essential when visualizing ROS in vivo. Non-planarian materials or life forms present in the medium (probably protists) can take up the dye as well, as illustrated in Fig. 4.5, resulting in reduced imaging clarity.
15. When performing real-time imaging of redox molecules, additional stress should be avoided and experiments should be performed using exactly the same settings to limit experimental variation. Limit light exposure as much as possible by operating the shutter manually or take this into account in the settings by automatically closing the shutter between images in time-lapses.
16. Fill the petri dish with plenty of precooled 4 °C medium, to just below the rim, to rapidly reduce motility. Keep in mind that a large, thick precooled freezing block will extract the heat much faster than a smaller, thin one, although care must be taken to not avoid damage the planarian by cooling it too fast. This procedure allows to (almost) completely reduce the motility of the planarian before it is further immobilized it in agarose.
17. Put one layer of paper towel around the precooled freezing block in order to (1) prevent scratches at the bottom of the slide which might hinder imaging and (2) to prevent the planarian from freezing to death.

18. The nature, quantity, and quality of the observed fluorescent ROS signal strongly depend on the orientation of the planarian relative to the objective, as well as on the target structure that is imaged. Figure 4.3 visualizes in vivo superoxide production in ventral and dorsal epidermal cells, whereas in Fig. 4.5.a, intestinal ROS are represented. There is a clear difference between the quantity, distribution, and nature of the fluorescent signal between the different tissues. Hence, it is important to determine the target of interest in advance and adapt the positioning of the specimen accordingly.

19. To avoid condensation induced by temperature differences, dry the well chamber slide with a paper towel or a cotton swab. To avoid stress induced by a cold shock, use a larger volume in the pipet to transfer the planarian. However, when positioning the planarian in the middle of the well chamber, a small droplet of medium is preferred. Otherwise, the planarian will (1) move to the edge of the well chamber or (2) make a cavity in the LMP agarose during the following steps. These events will disturb imaging.

20. Take the following aspects into account when using a filter paper or paper towel (1). Remove as much medium as possible beforehand. This way, the planarian cannot be accidentally absorbed by the paper (2). Avoid direct contact between the specimen and filter paper, as planarians tend to stick to the paper (3). Ensure that all the medium is removed before continuing to the next step. Even a small volume of medium will create a cavity in the 7% LMP agarose droplet, allowing the planarian to move. This is illustrated in Fig. 4.1.

21. Be careful and work fast. The planarian is not covered with a buffer and can easily freeze to death.

22. Precool the slide before adding the 7% LMP agarose droplet. This fastens the solidification process and reduces the possibility that the planarian will make a cavity by moving (*see* Fig. 4.1). Cool the LMP agarose below 40 °C inside the pipet by gently pipetting up and down. If the temperature of the agarose is above 40 °C, the planarian will lyse as illustrated in Fig. 4.2. Try to avoid air bubbles in the droplet of LMP agarose as this will affect imaging. A needle can be used to pop and remove air bubbles.

23. Hold the slide with the specimen on the precooled freezing block when applying the layer of 2.5% LMP agarose on top of the specimen. The cooling will increase the rate of solidification. The extra layer of 2.5% LMP agarose on top of the 7% LMP agarose elevates pressure and minimizes the motility of the planarian. Try to reduce air bubble formation as this can affect imaging.

24. To avoid dehydration and shrinking of the LMP agarose, add a thin layer of medium on top of the 2.5% LMP agarose square, especially when aiming for long-term imaging or storing. Figure 4.6 illustrates what happens when no medium is applied: the agarose shrinks (red dotted line); the planarian detaches and can again move around (red arrowhead).
25. During the optimization of the protocol, we tried to immobilize the planarian using polystyrene nanoparticles and agarose pads, as described by Kim and colleagues [12]. They demonstrated a successful immobilization of *C. elegans* without the use of anesthetics, by adding polystyrene nanoparticles to subsequently increase the friction between the animal and the agarose. However, instead of an expected decrease in motility, we observed the opposite in *S. mediterranea*. The planarian appeared stressed and showed strong uncontrollable movements, already immediately after administering the polystyrene nanobeads. The biocompatibility of nanobeads in this or other settings has to be investigated further.
26. Small ice cubes in the empty chambers on both sides of the well chamber slide reduce specimen heating and animal movement. Ice cubes can be made with standard lab equipment such as a microcentrifuge tube rack (*see* Figs. 1.3 and 4.7).
27. Be very gentle when replacing the melted ice cubes with new ones as even a small displacement of the specimen might negatively affect the imaging. Constantly pay attention to the state of the ice cubes while imaging. When the ice cubes are melted, remove the liquid with the aid of a Pasteur pipette with a small tip opening and put new ice cubes inside the well chambers using serrated lab tweezers.
28. Be sure to use the same imaging settings in all conditions within a single experiment. In addition, if possible, try not to process and image specimens of the same experiment in multiple separate sessions. Slight alterations in experimental conditions and imaging settings can easily induce a strong variation.
29. Interfering is possible as this is performed on top of the specimen, while the responses are inversely imaged. The microinjector with the glass capillary needle can be utilized for functional manipulation, administering the desired fluorophores or pharmacological compounds. The glass capillary needle can also be used to induce a healing wound (H-wound). When the intention is to make other types of wounds, a construction with a blade of a utility knife can be attached to the generic 3-axis micromanipulator with adaptor for the Drummond. To make the construction, make sure that the Drummond Nanoject II is removed from the adaptor. Fix a bar of any material in the adaptor where the injector is normally

located. Next, attach the blade of a utility knife to the bar with adhesive tape or glue. Now the construction to apply wounding, such as a regenerative (R) wound, is ready to use.

30. To remove the specimen from the slide, place the curved lab tweezer in a corner of the LMP agarose square and slide it underneath to move the agarose upwards, as indicated by the red arrow in Fig. 3.1.a.

    Do not put it on a paper surface as the planarian can get stuck and damaged.

31. Using planarian medium at room temperature speeds up the recovery process.

32. If the planarian does not free itself, it can be removed from the 2.5% LMP agarose droplet by gently shaking it with the tweezers. The planarian will be released without injury.

References

1. Li Y, Almassalha LM, Chandler JE, Zhou X, Stypula-Cyrus YE, Hujsak KA, Roth EW, Bleher R, Subramanian H, Szleifer I, Dravid VP, Backman V (2017) The effects of chemical fixation on the cellular nanostructure. Exp Cell Res 358(2):253–259. https://doi.org/10.1016/j.yexcr.2017.06.022
2. Talbot J, Schotz EM (2011) Quantitative characterization of planarian wild-type behavior as a platform for screening locomotion phenotypes. J Exp Biol 214(Pt 7):1063–1067. https://doi.org/10.1242/jeb.052290
3. Stevenson CG, Beane WS (2010) A low percent ethanol method for immobilizing planarians. PLoS One 5(12):e15310. https://doi.org/10.1371/journal.pone.0015310
4. Beane WS, Adams DS, Morokuma J, Levin M (2019) Live imaging of intracellular pH in planarians using the ratiometric fluorescent dye SNARF-5F-AM. Biol Methods Protoc 4: bpz005. https://doi.org/10.1093/biomethods/bpz005
5. Zattara EE, Turlington KW, Bely AE (2016) Long-term time-lapse live imaging reveals extensive cell migration during annelid regeneration. BMC Dev Biol 16:6. https://doi.org/10.1186/s12861-016-0104-2
6. Goel T, Wang R, Martin S, Lanphear E, Collins ES (2019) Linalool acts as a fast and reversible anesthetic in hydra. PLoS One 14(10): e0224221. https://doi.org/10.1371/journal.pone.0224221
7. Collymore C, Tolwani A, Lieggi C, Rasmussen S (2014) Efficacy and safety of 5 anesthetics in adult zebrafish (Danio rerio). J Am Assoc Lab Anim Sci 53(2):198–203
8. Leifer AM, Fang-Yen C, Gershow M, Alkema MJ, Samuel AD (2011) Optogenetic manipulation of neural activity in freely moving Caenorhabditis elegans. Nat Methods 8(2): 147–152. https://doi.org/10.1038/nmeth.1554
9. Dexter JP, Tamme MB, Lind CH, Collins EM (2014) On-chip immobilization of planarians for in vivo imaging. Sci Rep 4:6388. https://doi.org/10.1038/srep06388
10. Burnett K, Edsinger E, Albrecht DR (2018) Rapid and gentle hydrogel encapsulation of living organisms enables long-term microscopy over multiple hours. Commun Biol 1:73. https://doi.org/10.1038/s42003-018-0079-6
11. Renaud O, Herbomel P, Kissa K (2011) Studying cell behavior in whole zebrafish embryos by confocal live imaging: application to hematopoietic stem cells. Nat Protoc 6(12): 1897–1904. https://doi.org/10.1038/nprot.2011.408
12. Kim E, Sun L, Gabel CV, Fang-Yen C (2013) Long-term imaging of Caenorhabditis elegans using nanoparticle-mediated immobilization. PLoS One 8(1):e53419. https://doi.org/10.1371/journal.pone.0053419
13. Guo SX, Bourgeois F, Chokshi T, Durr NJ, Hilliard MA, Chronis N, Ben-Yakar A (2008) Femtosecond laser nanoaxotomy lab-on-a-chip for in vivo nerve regeneration studies. Nat Methods 5(6):531–533. https://doi.org/10.1038/nmeth.1203
14. Zeng F, Rohde CB, Yanik MF (2008) Sub-cellular precision on-chip small-animal immobilization, multi-photon imaging and

femtosecond-laser manipulation. Lab Chip 8(5):653–656. https://doi.org/10.1039/b804808h

15. Oviedo NJ, Nicolas CL, Adams DS, Levin M (2008) Planarians: a versatile and powerful model system for molecular studies of regeneration, adult stem cell regulation, aging, and behavior. CSH Protoc:pdb emo101. https://doi.org/10.1101/pdb.emo101
16. Aboobaker AA (2011) Planarian stem cells: a simple paradigm for regeneration. Trends Cell Biol 21(5):304–311. https://doi.org/10.1016/j.tcb.2011.01.005
17. Gentile L, Cebria F, Bartscherer K (2011) The planarian flatworm: an in vivo model for stem cell biology and nervous system regeneration. Dis Model Mech 4(1):12–19. https://doi.org/10.1242/dmm.006692
18. Hagstrom D, Cochet-Escartin O, Zhang S, Khuu C, Collins EM (2015) Freshwater planarians as an alternative animal model for Neurotoxicology. Toxicol Sci 147(1):270–285. https://doi.org/10.1093/toxsci/kfv129
19. Leynen N, Van Belleghem F, Wouters A, Bove H, Ploem JP, Thijssen E, Langie SAS, Carleer R, Ameloot M, Artois T, Smeets K (2019) In vivo toxicity assessment of silver nanoparticles in homeostatic versus regenerating planarians. Nanotoxicology 13(4):476–491. https://doi.org/10.1080/17435390.2018.1553252
20. Salvetti A, Gambino G, Rossi L, De Pasquale D, Pucci C, Linsalata S, Degl'Innocenti A, Nitti S, Prato M, Ippolito C, Ciofani G (2020) Stem cell and tissue regeneration analysis in low-dose irradiated planarians treated with cerium oxide nanoparticles. Mater Sci Eng C Mater Biol Appl 115:111113. https://doi.org/10.1016/j.msec.2020.111113
21. Stevens AS, Pirotte N, Plusquin M, Willems M, Neyens T, Artois T, Smeets K (2015) Toxicity profiles and solvent-toxicant interference in the planarian Schmidtea mediterranea after dimethylsulfoxide (DMSO) exposure. J Appl Toxicol 35(3):319–326. https://doi.org/10.1002/jat.3011
22. Stevens AS, Wouters A, Ploem JP, Pirotte N, Van Roten A, Willems M, Hellings N, Franken C, Koppen G, Artois T, Plusquin M, Smeets K (2018) Planarians customize their stem cell responses following genotoxic stress as a function of exposure time and regenerative state. Toxicol Sci 162(1):251–263. https://doi.org/10.1093/toxsci/kfx247
23. Lau AH, Knakievicz T, Pra D, Erdtmann B (2007) Freshwater planarians as novel organisms for genotoxicity testing: analysis of chromosome aberrations. Environ Mol Mutagen 48(6):475–482. https://doi.org/10.1002/em.20307
24. Upadhyay S, Vergara L, Shah P, Gustafsson JA, Kakadiaris I, Bondesson M (2020) A layered mounting method for extended time-lapse confocal microscopy of whole zebrafish embryos. J Vis Exp 155. https://doi.org/10.3791/60321
25. Berger S, Lattmann E, Aegerter-Wilmsen T, Hengartner M, Hajnal A, Demello A, Casadevall ISX (2018) Correction: Long-term C. elegans immobilization enables high resolution developmental studies in vivo. Lab Chip 18(12):1802. https://doi.org/10.1039/c8lc90050g
26. Bedard K, Krause KH (2007) The NOX family of ROS-generating NADPH oxidases: physiology and pathophysiology. Physiol Rev 87(1):245–313. https://doi.org/10.1152/physrev.00044.2005
27. Kennedy KA, Sandiford SD, Skerjanc IS, Li SS (2012) Reactive oxygen species and the neuronal fate. Cell Mol Life Sci 69(2):215–221. https://doi.org/10.1007/s00018-011-0807-2
28. Pirotte N, Stevens AS, Fraguas S, Plusquin M, Van Roten A, Van Belleghem F, Paesen R, Ameloot M, Cebria F, Artois T, Smeets K (2015) Reactive oxygen species in planarian regeneration: an upstream necessity for correct patterning and brain formation. Oxidative Med Cell Longev 2015:392476. https://doi.org/10.1155/2015/392476
29. Love NR, Chen Y, Ishibashi S, Kritsiligkou P, Lea R, Koh Y, Gallop JL, Dorey K, Amaya E (2013) Amputation-induced reactive oxygen species are required for successful Xenopus tadpole tail regeneration. Nat Cell Biol 15(2):222–228. https://doi.org/10.1038/ncb2659
30. Gauron C, Rampon C, Bouzaffour M, Ipendey E, Teillon J, Volovitch M, Vriz S (2013) Sustained production of ROS triggers compensatory proliferation and is required for regeneration to proceed. Sci Rep 3:2084. https://doi.org/10.1038/srep02084
31. Katerji M, Filippova M, Duerksen-Hughes P (2019) Approaches and methods to measure oxidative stress in clinical samples: research applications in the cancer field. Oxidative Med Cell Longev 2019:1279250. https://doi.org/10.1155/2019/1279250
32. Vaneev AN, Gorelkin PV, Garanina AS, Lopatukhina HV, Vodopyanov SS, Alova AV, Ryabaya OO, Akasov RA, Zhang Y, Novak P, Salikhov SV, Abakumov MA, Takahashi Y, Edwards CRW, Klyachko NL, Majouga AG, Korchev YE, Erofeev AS (2020) In vitro and

in vivo electrochemical measurement of reactive oxygen species after treatment with anticancer drugs. Anal Chem 92(12):8010–8014. https://doi.org/10.1021/acs.analchem.0c01256

33. Yamato M, Egashira T, Utsumi H (2003) Application of in vivo ESR spectroscopy to measurement of cerebrovascular ROS generation in stroke. Free Radic Biol Med 35(12): 1619–1631. https://doi.org/10.1016/j.freeradbiomed.2003.09.013
34. Belousov VV, Fradkov AF, Lukyanov KA, Staroverov DB, Shakhbazov KS, Terskikh AV, Lukyanov S (2006) Genetically encoded fluorescent indicator for intracellular hydrogen peroxide. Nat Methods 3(4):281–286. https://doi.org/10.1038/nmeth866
35. Miller EW, Tulyathan O, Isacoff EY, Chang CJ (2007) Molecular imaging of hydrogen peroxide produced for cell signaling. Nat Chem Biol 3(5):263–267. https://doi.org/10.1038/nchembio871
36. Freitas M, Lima JL, Fernandes E (2009) Optical probes for detection and quantification of neutrophils' oxidative burst. A review Anal Chim Acta 649(1):8–23. https://doi.org/10.1016/j.aca.2009.06.063
37. Rhee SG, Chang TS, Jeong W, Kang D (2010) Methods for detection and measurement of hydrogen peroxide inside and outside of cells. Mol Cells 29(6):539–549. https://doi.org/10.1007/s10059-010-0082-3
38. Pacheco S, Fung SY, Liu M (2017) Solubility of hydrophobic compounds in aqueous solution using combinations of self-assembling peptide and amino acid. J Vis Exp 127. https://doi.org/10.3791/56158
39. Santos NC, Figueira-Coelho J, Martins-Silva J, Saldanha C (2003) Multidisciplinary utilization of dimethyl sulfoxide: pharmacological, cellular, and molecular aspects. Biochem Pharmacol 65(7):1035–1041. https://doi.org/10.1016/s0006-2952(03)00002-9
40. Zhang J, Yuan Z, Zheng M, Sun Y, Wang Y, Yang S (2013) Effects of N, N-dimethylformamide on behaviour and regeneration of planarian Dugesia japonica. Toxicol Ind Health 29(8):753–760. https://doi.org/10.1177/0748233712443148
41. Lim Y, Shiver AL, Khariton M, Lane KM, Ng KM, Bray SR, Qin J, Huang KC, Wang B (2019) Mechanically resolved imaging of bacteria using expansion microscopy. PLoS Biol 17(10):e3000268. https://doi.org/10.1371/journal.pbio.3000268
42. Gonzalez-Estevez C, Momose T, Gehring WJ, Salo E (2003) Transgenic planarian lines obtained by electroporation using transposon-derived vectors and an eye-specific GFP marker. Proc Natl Acad Sci U S A 100(24): 14046–14051. https://doi.org/10.1073/pnas.2335980100
43. Forsthoefel DJ, James NP, Escobar DJ, Stary JM, Vieira AP, Waters FA, Newmark PA (2012) An RNAi screen reveals intestinal regulators of branching morphogenesis, differentiation, and stem cell proliferation in planarians. Dev Cell 23(4):691–704. https://doi.org/10.1016/j.devcel.2012.09.008
44. Hirano K, Chen WS, Chueng AL, Dunne AA, Seredenina T, Filippova A, Ramachandran S, Bridges A, Chaudry L, Pettman G, Allan C, Duncan S, Lee KC, Lim J, Ma MT, Ong AB, Ye NY, Nasir S, Mulyanidewi S, Aw CC, Oon PP, Liao S, Li D, Johns DG, Miller ND, Davies CH, Browne ER, Matsuoka Y, Chen DW, Jaquet V, Rutter AR (2015) Discovery of GSK2795039, a novel small molecule NADPH oxidase 2 inhibitor. Antioxid Redox Signal 23(5):358–374. https://doi.org/10.1089/ars.2014.6202
45. Holland PC, Sherratt HS (1972) Biochemical effects of the hypoglycaemic compound diphenyleneiodonnium. Catalysis of anion-hydroxyl ion exchange across the inner membrane of rat liver mitochondria and effects on oxygen uptake. Biochem J 129(1):39–54. https://doi.org/10.1042/bj1290039
46. Kucera J, Bino L, Stefkova K, Jaros J, Vasicek O, Vecera J, Kubala L, Pachernik J (2016) Apocynin and Diphenyleneiodonium induce oxidative stress and modulate PI3K/Akt and MAPK/Erk activity in mouse embryonic stem cells. Oxidative Med Cell Longev 2016:7409196. https://doi.org/10.1155/2016/7409196
47. Ortega-Villasante C, Buren S, Blazquez-Castro A, Baron-Sola A, Hernandez LE (2018) Fluorescent in vivo imaging of reactive oxygen species and redox potential in plants. Free Radic Biol Med 122:202–220. https://doi.org/10.1016/j.freeradbiomed.2018.04.005
48. Romero MMG, McCathie G, Jankun P, Roehl HH (2018) Damage-induced reactive oxygen species enable zebrafish tail regeneration by repositioning of hedgehog expressing cells. Nat Commun 9(1):4010. https://doi.org/10.1038/s41467-018-06460-2
49. Zhang Q, Wang Y, Man L, Zhu Z, Bai X, Wei S, Liu Y, Liu M, Wang X, Gu X, Wang Y (2016) Reactive oxygen species generated from skeletal muscles are required for gecko tail regeneration. Sci Rep 6:20752. https://doi.org/10.1038/srep20752

50. Bijnens K, Jaenen V, Wouters A, Leynen N, Pirotte N, Artois T, Smeets K (2021) A spatiotemporal characterisation of redox molecules in planarians, with a focus on the role of glutathione during regeneration. Biomol Ther 11(5). https://doi.org/10.3390/biom11050714

51. Bhattacharyya A, Chattopadhyay R, Mitra S, Crowe SE (2014) Oxidative stress: an essential factor in the pathogenesis of gastrointestinal mucosal diseases. Physiol Rev 94(2):329–354. https://doi.org/10.1152/physrev.00040.2012

Chapter 14

A Planarian Model System to Study Host-Pathogen Interactions

Eli Isael Maciel, Ashley Valle Arevalo, Clarissa J. Nobile, and Néstor J. Oviedo

Abstract

This protocol is focused on using the recently established planarian infection model system to study host-pathogen interactions during fungal infection. Here, we describe in detail the infection of the planarian *Schmidtea mediterranea* with the human fungal pathogen *Candida albicans*. This simple and reproducible model system allows for rapid visualization of tissue damage throughout different infection timepoints. We note that this model system has been optimized for use with *C. albicans*, but should also be applicable for use with other pathogens of interest.

Key words Planarians, *Candida albicans*, Host-pathogen interactions, Fungal infections, Infection model systems

1 Introduction

Planarians are flatworms commonly used to study fundamental mechanisms regulating cellular decisions in an adult body [1]. They possess a complex body plan consisting of several tissue types, including a nervous system, excretory system, and digestive system [2–5]. In addition, they have the capacity to regenerate complete tissues and maintain high rates of cellular turnover and tissue homeostasis, resulting from their large population of adult pluripotent stem cells called neoblasts [5–9].

Recently, the planarian immune system has garnered scientific interest for its ability to quickly eliminate (within 12 days) a wide range of microbial pathogens, such as the bacterial pathogens *Mycobacterium tuberculosis* and *Staphylococcus aureus*, and the fungal pathogen *Candida albicans* [10–15]. Remarkably, planarians rely on an evolutionarily conserved innate immune system to clear these infections [11].

Luca Gentile (ed.), *Schmidtea Mediterranea: Methods and Protocols*, Methods in Molecular Biology, vol. 2680, https://doi.org/10.1007/978-1-0716-3275-8_14,
© The Author(s), under exclusive license to Springer Science+Business Media, LLC, part of Springer Nature 2023

Three different methods (injection, feeding, and soaking) have been used in prior studies to infect planarians with microbial pathogens [10, 12, 14]. The infection by injection method introduces the pathogen inside the animal. In this case, a known concentration of the pathogen is injected directly into the animal at the body site of interest (*see* Fig. 1a). A notable advantage of this approach is that it provides an opportunity to initiate the infection at specific regions in the animal. However, a drawback is that injury is inflicted on the animal, which may trigger additional cellular responses, making it challenging to dissect the contributions of the immune and regenerative responses. The most popular method of infection is through feeding. Here, the planarians are offered food (e.g., liver paste) mixed with the pathogen for limited time periods (*see* Fig. 1b). Although this method does not involve injury, it is difficult to precisely determine the amount of pathogen ingested, which may lead to inconsistencies in infection levels across different animals. Another issue with the infection by feeding method is that planarians are natural scavengers and have evolved a large population of phagocytic cells within their digestive tracts that protect them against infection [11, 16]. Thus, immune responses to infection by feeding may differ compared to other methods since pathogens are exposed to phagocytic cells immediately after ingestion. These anatomical interactions with the pathogen suggest that the digestive route offers unique opportunities to assay pathogen clearance by the host, but it is more restrictive in evaluating pathogen virulence. Another method of infection involves soaking planarians in known concentrations of the pathogen for a limited time frame. Planarians live in water and adding pathogens to the media allows for the control of pathogen concentration and exposure time to the pathogen. In addition, with this method, pathogens can be easily and rapidly removed through the addition of fresh media. This "soaking method" is injury-free, simple, and consistent and allows for the visualization of the infection process accounting for the initial physical interactions between the host and the pathogen (i.e., pathogen adhesion) to the planarian epithelial surface and penetration to deeper host tissues [14]. The soaking method also facilitates the evaluation of pathogen clearance by the host, as well as virulence of the pathogen.

In this chapter, we present detailed protocols for the three planarian infection methods introduced above, with a primary focus on the soaking method. We use *C. albicans* as the pathogen example in our protocols, but note that *C. albicans* could be exchanged for other pathogens of interest. *C. albicans* is a common human fungal pathogen that can cause a variety of infections, ranging from superficial skin infections to severe systemic infections [17, 18]. *C. albicans* is multimorphic and can develop several different cellular morphologies depending on its environment [19, 20]. One notable *C. albicans* morphological transition that is

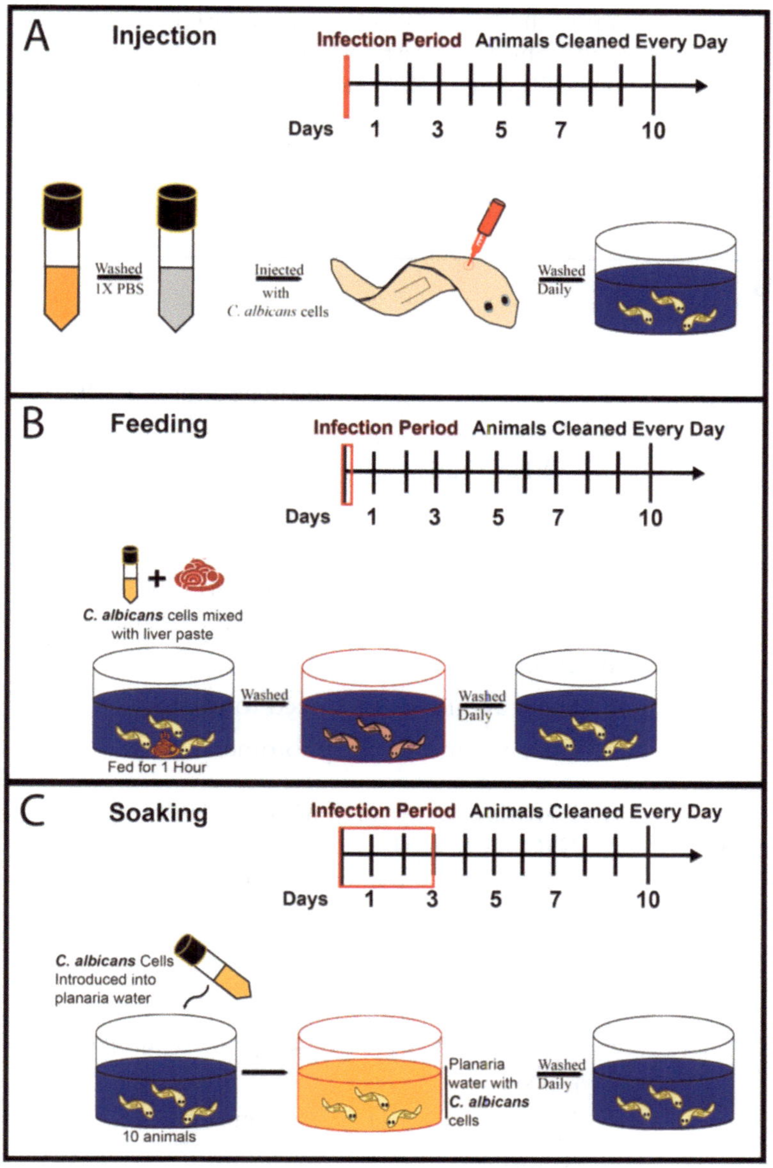

Fig. 1 Infection methods to study host-pathogen interactions in the planarian model system. (**a**) Injection. An overnight *C. albicans* culture is pelleted and resuspended in planarian water. A few microliters of the pathogen mixture are picked up in a glass capillary tube connected to a nano-injection needle and injected directly into the planarian at a body site of interest. The infected planarians are then placed in a well and observed and/or collected at specific timepoints for CFU enumeration and downstream experimentation. (**b**) Feeding. An overnight *C. albicans* culture is diluted to the desired concentration and pelleted. The pellet is mixed with liver paste and dispensed inside the planarian well. The planarians are left with the liver solution for one hour and then washed. Animals successfully infected will have a slightly red body tint indicative that they have recently eaten. The planarians are then directly observed and/or collected at specific timepoints for CFU enumeration and downstream experimentation. (**c**) Soaking. An overnight *C. albicans* culture is diluted to the desired concentration and dispensed into the well containing planarians in a 4 mL total volume of planarian water. The planarians are left in the well with the microbial cell mixture of planarian water for 3 days, after which the planarians are washed, and their water changed daily. The planarians are then directly observed and/or collected at specific timepoints for CFU enumeration and downstream experimentation

important for virulence is the yeast to hyphal transition, where round budding yeast-form cells transition into elongated filamentous cells (pseudohyphae and hyphae) and vice versa. This reversible morphological transition is important for *C. albicans* to adapt to multiple host environments as well as to cause host tissue damage [21].

2 Materials

2.1 Culturing of C. Albicans for Planarian Infection

1. *Schmidtea mediterranea* planarian asexual strain (CIW4).
2. *C. albicans* isogenic wild-type strain (SN250) [20].
3. *C. albicans nrg1/nrg1* hyper-filamentous mutant strain (TF125) [20].
4. *C. albicans efg1/efg1* nonfilamentous mutant strain (TF156) [20].
5. YPD (liquid and agar medium): 2% Bacto peptone, 2% dextrose, 1% yeast extract, 2% agar (for agar media only).
6. 20 mL test tubes.
7. Short and long autoclaved toothpicks.
8. Cuvettes and spectrophotometer.

2.2 Infecting Planarians Through Injection

1. Planarian water (1X Montjuic saltwater solution): 1.6 mM NaCl, 1.0 mM $CaCl_2$, 1.0 mM $MgSO_4$, 0.1 mM $MgCl_2$, 0.1 mM KCl, 1.2 mM $NaHCO_3$; using ultrapure water bring up to 13.5 L.
2. Overnight *C. albicans* culture.
3. Glass capillary tubes.
4. Glass needle puller (e.g., p-97 flaming/brown micropipette puller).
5. Nanoject II Injector (Drummond).
6. Kimwipes.
7. Parafilm.
8. Forceps.
9. 6-well polystyrene non-tissue culture plate.
10. 3 mL transfer pipette.
11. Petri dishes.

2.3 Infecting Planarians Through Feeding

1. Planarian water (1X Montjuic saltwater solution): 1.6 mM NaCl, 1.0 mM $CaCl_2$, 1.0 mM $MgSO_4$, 0.1 mM $MgCl_2$, 0.1 mM KCl, 1.2 mM $NaHCO_3$; using ultrapure water bring up to 13.5 L.
2. Overnight *C. albicans* culture.

3. Liver paste (derived fresh and organic from the butcher which is processed the same day with a blender, aliquoted, and stored at −80 °C).
4. 6-well polystyrene non-tissue culture plate.
5. 3 mL transfer pipette.
6. Petri dishes.

2.4 Infecting Planarians Through Soaking

1. Planarian water (1X Montjuic saltwater solution): 1.6 mM NaCl, 1.0 mM $CaCl_2$, 1.0 mM $MgSO_4$, 0.1 mM $MgCl_2$, 0.1 mM KCl, 1.2 mM $NaHCO_3$; using ultrapure water bring up to 13.5 L.
2. Overnight *C. albicans* culture.
3. 6-well polystyrene non-tissue culture plate.
4. 3 mL transfer pipette.
5. Petri dishes.
6. Dissecting microscope.

2.5 Fixation After Infection Via Feeding

1. 20 mL scintillation vials.
2. NAC solution: 7.5 g N-acetylcysteine; diluted in 10 mL of ultrapure water.
3. 1× PBS: 137 mM NaCl, 2.7 mM KCl, 4.3 mM Na2HPO4, 1.47 mM KH2PO4; using ultrapure water bring up to 1 L.
4. 4% fixative solution: 1.1 mL 36.5% formaldehyde; diluted in 1× PBS.
5. 1% SDS solution: 1 mL of 10% SDS and 9 mL of 1× PBS.
6. 0.3% PBSTx: 100 mL of 10× PBS, 3 mL of Triton X-100; using ultrapure water bring up to 1 L.

2.6 Fixation After Infection Via Injection and Soaking

1. 20 mL scintillation vials.
2. NAC solution: 9 g N-acetylcysteine; diluted in 10 mL ultrapure water.
3. 1× PBS: 137 mM NaCl, 2.7 mM KCl, 4.3 mM Na2HPO4, 1.47 mM KH2PO4; using ultrapure water, bring up to 1 L.
4. 4% fixative solution: 1.1 mL of 36.5% formaldehyde; diluted in 1× PBS.
5. Formamide solution: 300 μL of formamide, 400 μL of H_2O_2, and 9.3 mL of 1× PBS.
6. 0.3% PBSTx: 100 mL of 10× PBS, 3 mL of Triton X-100; using ultrapure water, bring up to 1 L.

2.7 Whole-Mount Immunohistochemistry

1. Anti-*Candida*, 1:500.
2. Goat anti-rabbit Alexa 568, 1:800.
3. HRP-conjugated goat anti-rabbit antibody, 1:1000.
4. PBSTB: 1.25 g of BSA; diluted in 50 mL of 0.3% PBSTx.
5. FITC tyramide solution: 1:1000 FITC, 500 μL of 1 M imidazole, and 50 mL of 0.3% PBSTx.
6. Quench solution: 600 μL of H_2O_2 in 6 mL of PBSTx.

2.8 Calculating C. albicans Concentration from Overnight Cultures

1. Glass hemacytometer.
2. Coverslips.
3. Handheld cell counter.
4. 70% ethanol.

3 Methods

3.1 Culturing C. albicans for Planarian Infections

1. *C. albicans* strain SN250 is the isogenic wild-type reference strain used to compare mutant strains like the hyperfilamentous strain TF125 (*nrg1/nrg1*) and nonfilamentous strain TF156 (*efg1/efg1*) to assess the effects of filamentation on the infection assays [20].
2. *C. albicans* strains are streaked and grown in yeast extract peptone dextrose (YPD) agar at 30 °C for 48 h (*see* **Note 1**).
3. Single-colony overnights are set up shaking at 30 °C for 12–16 h prior to the infection assay. This culturing condition is commonly used to grow *C. albicans* wild-type strains, mutant strains, and clinical isolates for experimentation [22].

3.2 Infecting Planarians Through Injection

1. Place planarians in a cold plate or ice block with a layer of parafilm and a damp Kimwipe for immobilization.
2. Spin down overnight *C. albicans* cultures grown for 12–16 h at 4596 g for 5 min.
3. Resuspend the pellet in planarian water.
4. Construct glass injection needles from glass capillary tubes after breaking the tip using forceps (*see* **Note 2**).
5. Place the glass needle in a Nanoject II Injector (Drummond) and backfill with mineral oil and a few microliters of the *C. albicans* cells.
6. Inject planarians with two pulses (*see* **Note 3**).
7. Place injected planarians into planarian water.

3.3 Infecting Planarians Through Feeding

1. Grow overnight culture of *C. albicans* for 12–16 h.
2. Allocate a concentration of ten million *C. albicans* cells to a tube and pellet using a centrifuge at 4596 g for 5 min.
3. Decant the supernatant and keep the pellet. Mix the pellet with 50 μL of liver paste, and pipette the mixture to a petri dish containing the planarians.
4. Allow the planarians to feed for 2–4 h. They will turn a slight pink/red color indicating that they have eaten.
5. Once they have eaten, clean the petri dishes or move the planarians to a clean petri dish.
6. The planarians will regurgitate some parts of the mixture eaten and will need to be cleaned again the next day.

3.4 Infecting Planarians Through Soaking

1. Grow overnight culture of *C. albicans* for 12–16 h.
2. Maintain ten planarians in 6 mL wells containing 3 mL of planarian water.
3. Add the desired concentration of *C. albicans* cells and adjust the total volume to 4 mL with planarian water (*see* **Note 4**).
4. Maintain planarians in the infected media for 3 days.
5. After the 3-day exposure, wash the planarians daily with fresh water and observe them under the microscope to record any behavioral or macroscopic defects until they are collected for specific downstream experiments at various timepoints.

3.5 Fixation After Infection Via Feeding

1. Collect planarians in 20 mL scintillation vials filled with 3 mL of planarian water.
2. Sacrifice planarians by placing in 7.5% of NAC solution for 5 min at room temperature with rocking.
3. Fix planarians by replacing liquid with 4% of formaldehyde in 1× PBS for 20 min at room temperature with rocking.
4. Remove the fixative and replace with a 1× PBS wash.
5. After the wash, permeabilize planarians using a formamide solution with 6% of H_2O_2 for 20 min (*see* **Note 5**).
6. Wash the animals three times with 1× PBS and store at 4 °C for up to 7 days.

3.6 Fixation After Infection Via Soaking or Injection

1. Place select planarians collected at specific timepoints of interest in a 20 mL scintillation vial filled with 3 mL of planarian water.
2. Sacrifice planarians by removing the planarian water and replacing it with 9% of NAC solution for 5 min at room temperature with rocking.
3. Remove the NAC solution and add 4% of fixative for 15–20 min at room temperature with rocking.

4. Remove the 4% of fixative and rinse two times with 1× PBS.
5. Remove the 1× PBS, and permeabilize the planarians with 1% of SDS for 20 min at room temperature with rocking.
6. Rinse the planarians three times with 1× PBS.
7. Bleach planarians in 6% of H_2O_2 in 1× PBS for 2–4 h under a light source (*see* **Note 5**).
8. Remove the bleaching solution, and rinse planarians with 1× PBS three times, and store at 4 °C for up 7 days.

3.7 Whole-Mount Immuno-histochemistry

1. Transfer planarians from 1× PBS solution to 0.3% of PBSTx solution (*see* **Note 6**).
2. After 20 min, replace the PBSTx with PSTB for 4 h at room temperature with rocking.
3. After the 4 h, transfer planarians to a 24-well plate.
4. Incubate planarians with the anti-*Candida* (1:500) antibody for 4 h at room temperature or 8 h at 4 °C overnight.
5. Wash planarians every 20 min for 2.5 h with PSTB.
6. After all the washes, add the Alexa 568 (1:800) secondary antibody, and incubate for 4 h at room temperature or 8 h at 4 °C.
7. Remove the secondary antibody, and wash planarians every 20 min for 2.5 h. Mount and observe in the microscope the stained planarians (*see* Fig. 2).

3.8 Calculating C. albicans Concentration from Overnight Cultures

1. Clean glass hemocytometer with alcohol and let dry. Moisten a coverslip with a small amount of water prior to adhering to the hemocytometer.
2. Using aseptic technique, make a 1:20 dilution of the overnight *C. albicans* culture and add 10 μL of the dilution to the hemocytometer. Allow cells to flow through and fill chamber(s).
3. Place the hemocytometer on a microscope, and using the 20× objective, focus on the grid lines of the hemocytometer.
4. Count the number of *C. albicans* cells in all four outer squares and divide by four. This is the mean number of cells per square.
5. Calculate the number of cells per square x 10^4 = the number of cells/mL of overnight culture. Multiply by 20 to account for the original 1:20 dilution. This will be the working concentration.
6. Clean the hemocytometer with 70% ethanol, and dispose of coverslip.

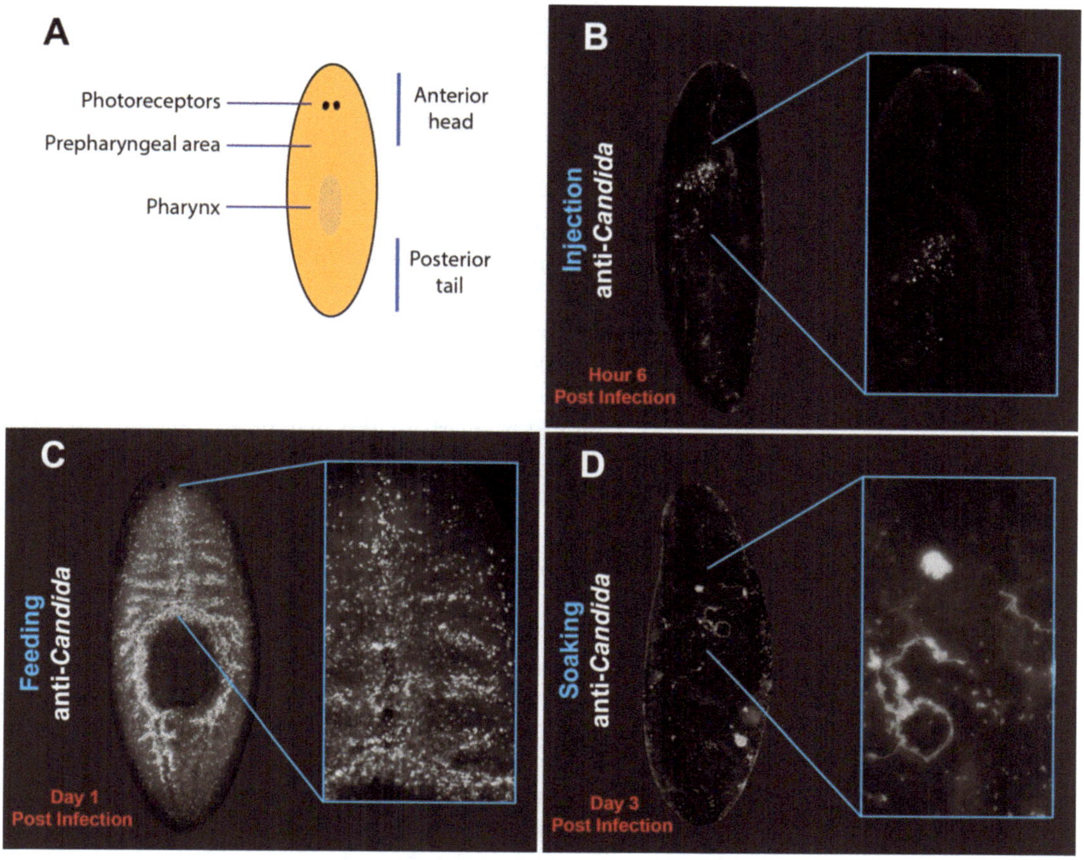

Fig. 2 Immunohistochemistry of C. albicans using different infection methods in the planarian model. (**a**) Outline of the planarian anatomy in its position in the subsequent images. (**b**) Injection. The planarian was injected in the parapharyngeal area with 6000 wild-type *C. albicans* cells, fixed, and stained 6 hours post infection. Using this infection method, a small number of *C. albicans* cells are shown to adhere to the indicated area. (**c**) Feeding. The planarian was fed ten million wild-type *C. albicans* cells/mL and stained 1 day post infection. Using this infection method, *C. albicans* cells can be observed throughout the digestive tract of the planarian. (**d**) Soaking. The planarian was soaked in 15 million wild-type *C. albicans* cells/mL and stained at 3 days post infection. Using this infection method, *C. albicans* cells can be observed to adhere throughout the planarian epithelial layer and different morphologies of *C. albicans* cells can be observed (*see* **Note 7**)

4 Notes

1. Recommendations for Growing *C. albicans* Strains. Do not use *C. albicans* colonies that are greater than 7 days old or store plates at 4 °C as *C. albicans* acquires aneuploidies under these conditions. Likewise, new plates should be streaked from glycerol stocks rather than re-streaked from existing plates. Alternative growth media can be used for these experiments, although the concentrations used in the protocol will need to be optimized for the specific media chosen. We recommend YPD media since it is a rich medium that supports the growth of all *C. albicans* strains and has been optimized for these infection assays.

2. Recommendation for Glass Capillaries. The dimensions of the pulled capillaries are as follows: outer diameter is ~1.15 mm, and inner diameter is ~0.5 mm.
3. Recommendations for Using the Different Planarian Infection Methods. When using the injection method, a few microliters can be used to infect all planarians using a nanoinjector dispensing two pulses ~32 nanoliters/pulse.
4. Recommendations on *C. albicans* Cell Concentrations to Use in the Soaking Infection Assay. We have tested a range of different infection concentrations from 5 to 20 million cells/mL for the three strains described in this chapter. For the hyper-filamentous strain, we determined that five million cells/mL cause 50% death of the planarians (*see* Fig. 3a). We determined that 15 million cells/mL are required of the wild-type and the nonfilamentous *C. albicans* strains to maintain survival of the planarians after 3 days of infection (*see* Fig. 3c). To comparatively visualize the different morphologies of these *C. albicans* strains during the infection process and to assess virulence, the use of the same number of *C. albicans* cells between strains is recommended. We typically use an infection concentration of 7.5×10^6 cells/mL for each strain to allow for enough planarians to survive throughout infection with any of the three *C. albicans* strains used in this chapter to conduct different downstream experiments, such as immunohistochemistry assays, transcriptional profiling experiments, and protein assays (*see* Fig. 3b). If the desired assay is simply to determine virulence via an endpoint survival assay, a concentration of 20 million *C. albicans* cells/mL and above is ideal (*see* Fig. 3d).

   When using the soaking infection method, once the *C. albicans* cells have been added, make sure to swirl the plate for 10 seconds. Lastly, when removing the infected media, make sure the animals are not left too long without water. Additionally, just transferring the animals to clean wells is a lot easier than cleaning the wells used for infection. We recommend using 3 mL transfer pipettes to move the planarians between wells. If a planarian gets stuck inside the transfer pipette, flick the region of the transfer pipette to detach the planarian. All procedures before, during, and after infection with *C. albicans* should be performed in water at room temperature, which is the ideal condition for planarians.
5. Recommendations on Soaking and Injection Fixation. During fixation, some planarians may start to float with many air bubbles surrounding them. Swirl the plate to help them to sink. If the planarians continue to float after the bleaching step, perform additional 1× PBS washes.

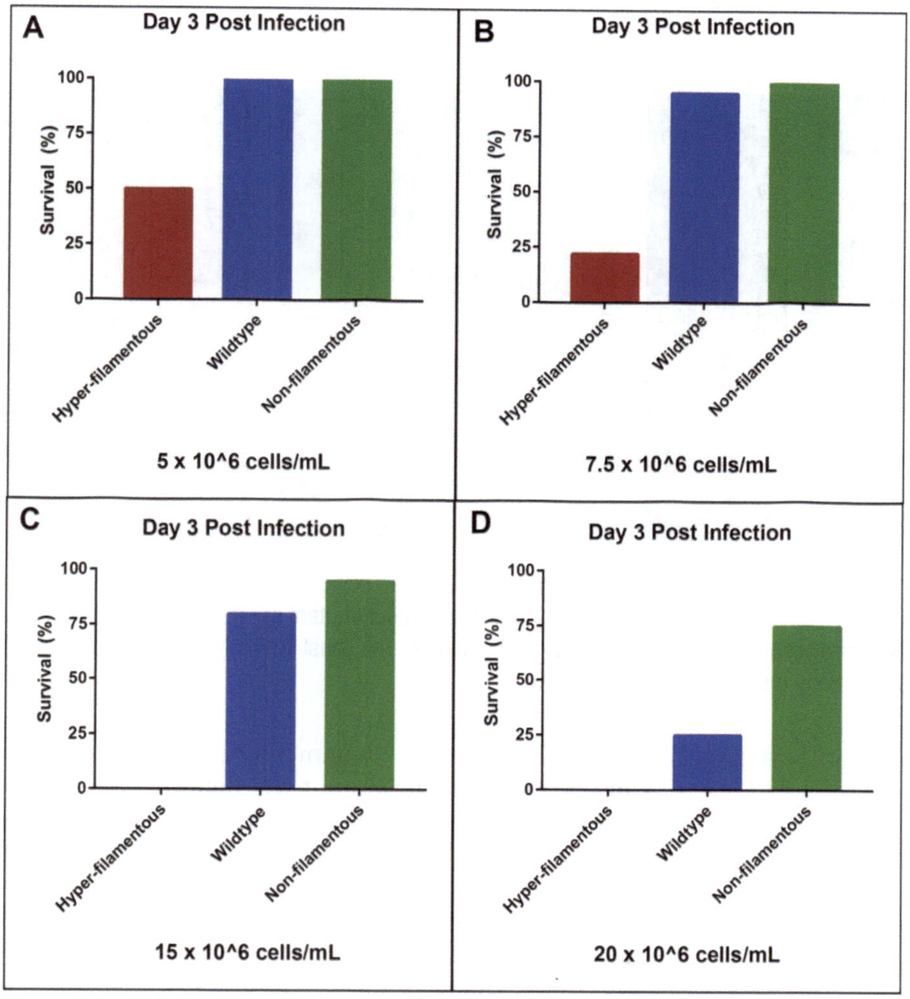

Fig. 3 Planarian survival upon infection with different *C. albicans* strains (wild-type, hyper-filamentous, and nonfilamentous) 3 days post infection. (**a**) Planarian survival after infection with 5 × 10^6 cells/mL of *C. albicans* cells of the three different strains. (**b**) Planarian survival after infection with 7.5 × 10^6 cells/mL of *C. albicans* cells of the three different strains. (**c**) Planarian survival after infection with 15 × 10^6 cells/mL of *C. albicans* cells of the three different strains. (**d**) Planarian survival after infection with 20 × 10^6 cells/mL of *C. albicans* cells of the three different strains

6. Recommendations on Immunohistochemistry. Use a black background underneath the plate to help in visualizing the bleached planarians. When removing liquid from the plate, tilt the plate at a 45° angle and retrieve the solution slowly from the top of the well, being careful not to touch the planarians.
7. Recommendations on Visualization of *C. albicans*. Depending on the infection method chosen, visualization of the infection will vary (*see* Fig. 2). Using the injection method, very few wild-type *C. albicans* cells are observed to adhere to the planarians

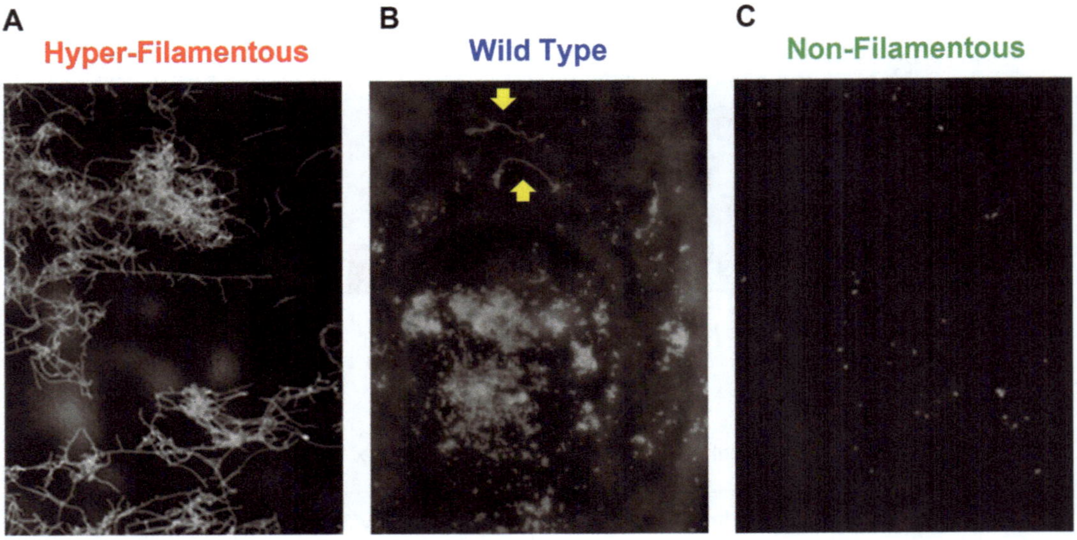

Fig. 4 Different *C. albicans* cellular morphologies are observed using the soaking infection method 3 days post infection. (**a**) Hyphal cells are observed in the hyper-filamentous mutant strain. (**b**) Both yeast-form and hyphal cells (yellow arrows) are observed in the wild-type strain. (**c**) The yeast-form growth morphology is observed in the nonfilamentous mutant strain (*see* **Note 8**)

(*see* Fig. 2b). In the feeding method, wild-type *C. albicans* cells can be observed throughout the digestive tract in the yeast-form cell morphology that was introduced (*see* Fig. 2c). In the soaking method, wild-type *C. albicans* can be observed adhered to the epithelial layer of planarians and can be observed to transition from the yeast form to the hyphal form (*see* Fig. 2d).

8. Recommendation on Observing *C. albicans* Yeast and Hyphal Morphologies Using the Planarian Soaking Infection Method. An isogenic *C. albicans* wild-type strain should be used as a reference strain to compare to the mutant strains of interest. In this example, we use the isogenic wild-type strain SN250, the hyper-filamentous strain TF125, and the nonfilamentous strain TF156 to infect the planarians using the soaking method. This method is optimal for visualizing the infection process in the planarians as well as changes in *C. albicans* cellular morphologies during the infection (*see* Fig. 4).

Acknowledgments

We thank Edelweiss Pfister for lab management and planarian maintenance in the Oviedo lab and all members of the Oviedo and Nobile labs for insightful discussions on protocols and optimization procedures. This work was supported by the National Institutes of Health (NIH), National Institute of Allergy and Infectious

Diseases (NIAID) award R21AI125801, and the National Institute of General Medical Sciences (NIGMS) award R35GM124594 and by the Kamangar family in the form of an endowed chair to CJN. This work was also supported by the NIGMS award R01GM132753 to NJO. AVA was supported by diversity supplement fellowship R21AI125801-02S1 to parent grant R21AI125801. EIM was supported by the National Science Foundation (NSF) graduate fellowship award 1744620. The funders had no role in the study design, data collection and interpretation, or the decision to submit the work for publication.

References

1. Oviedo NJ, Nicolas CL, Adams DS, Levin M (2008) Planarians: a versatile and powerful model system for molecular studies of regeneration, adult stem cell regulation, aging, and behavior. CSH Protoc (2008):pdb.emo101
2. Abnave P, Aboukhatwa E, Kosaka N, Thompson J, Hill MA, Aboobaker AA (2017) Epithelial-mesenchymal transition transcription factors control pluripotent adult stem cell migration in vivo in planarians. Dev Camb Engl 144(19):3440–3453
3. Forsthoefel DJ, James NP, Escobar DJ, Stary JM, Vieira AP, Waters FA et al (2012) An RNAi screen reveals intestinal regulators of branching morphogenesis, differentiation, and stem cell proliferation in planarians. Dev Cell 23(4): 691–704
4. Rink JC, Vu HT-K, Sánchez AA (2011) The maintenance and regeneration of the planarian excretory system are regulated by EGFR signaling. Dev Camb Engl. 138(17):3769–3780
5. Gentile L, Cebrià F, Bartscherer K (2011) The planarian flatworm: an in vivo model for stem cell biology and nervous system regeneration. Dis Model Mech 4(1):12–19
6. Rink JC (2013) Stem cell systems and regeneration in planaria. Dev Genes Evol 223(1–2): 67–84
7. Pellettieri J, Fitzgerald P, Watanabe S, Mancuso J, Green DR, Sánchez AA (2010) Cell death and tissue remodeling in planarian regeneration. Dev Biol 338(1):76–85
8. Aboobaker AA (2011) Planarian stem cells: a simple paradigm for regeneration. Trends Cell Biol 21(5):304–311
9. Scimone ML, Kravarik KM, Lapan SW, Reddien PW (2014) Neoblast specialization in regeneration of the planarian Schmidtea mediterranea. Stem Cell Rep 3(2):339–352
10. Arnold CP, Merryman MS, Harris-Arnold A, McKinney SA, Seidel CW, Loethen S et al (2016) Pathogenic shifts in endogenous microbiota impede tissue regeneration via distinct activation of TAK1/MKK/p38. elife 21:5
11. Maciel EI, Oviedo NJ (2018) Platyhelminthes: molecular dissection of the planarian innate immune system. In: Cooper EL (ed) Advances in comparative immunology [internet]. Springer International Publishing, Cham, pp 95–115. Available from: http://link.springer.com/10.1007/978-3-319-76768-0_4
12. Abnave P, Mottola G, Gimenez G, Boucherit N, Trouplin V, Torre C et al (2014) Screening in planarians identifies MORN2 as a key component in LC3-associated phagocytosis and resistance to bacterial infection. Cell Host Microbe 16(3): 338–350
13. Peiris TH, Hoyer KK, Oviedo NJ (2014) Innate immune system and tissue regeneration in planarians: an area ripe for exploration. Semin Immunol 26(4):295–302
14. Maciel EI, Jiang C, Barghouth PG, Nobile CJ, Oviedo NJ (2019) The planarian Schmidtea mediterranea is a new model to study host-pathogen interactions during fungal infections. Dev Comp Immunol 93:18–27
15. Maciel EI, Valle Arevalo A, Ziman B, Nobile CJ, Oviedo NJ (2020) Epithelial infection with Candida albicans elicits a multi-system response in planarians. Front Microbiol 11: 629526
16. Forsthoefel DJ, Cejda NI, Khan UW, Newmark PA (2020) Cell-type diversity and regionalized gene expression in the planarian intestine. elife 2:9
17. Jabra-Rizk MA, Kong EF, Tsui C, Nguyen MH, Clancy CJ, Fidel PL et al (2016) Candida albicans pathogenesis: fitting within the host-microbe damage response framework. Maurelli AT, editor. Infect Immun 84(10):2724–2739
18. Mayer FL, Wilson D, Hube B (2013) Candida albicans pathogenicity mechanisms. Virulence 4(2):119–128

19. Mukaremera L, Lee KK, Mora-Montes HM, Gow NAR (2017) Candida albicans yeast, Pseudohyphal, and hyphal morphogenesis differentially affects immune recognition. Front Immunol 7(8):629
20. Homann OR, Dea J, Noble SM, Johnson AD (2009) A phenotypic profile of the Candida albicans regulatory network. PLoS Genet 5: e1000783
21. Kadosh D, Lopez-Ribot JL (2013) Candida albicans: adapting to succeed. Cell Host Microbe 14(5):483–485
22. Gulati M, Lohse MB, Ennis CL, Gonzalez RE, Perry AM, Bapat P et al (2018) *In vitro* culturing and screening of *Candida albicans* biofilms. Curr Protoc Microbiol 50(1):e60

Chapter 15

TUNEL Staining in Sections of Paraffin-Enabled Planarians

Maria Rossello and Teresa Adell

Abstract

Planarians are a model animal for the study of regeneration and homeostasis. Understanding how planarians control their cellular balance is key to the knowledge of their plasticity. Both apoptotic and mitotic rates can be quantified in "whole mount" planarians. Apoptosis is usually analyzed through terminal deoxynucleotidyl transferase dUTP nick end labeling (TUNEL), a technique that detects cell death by identifying DNA breaks. In this chapter we detail a protocol to analyze apoptotic cells in paraffin sections of planarians, which enables a more accurate cellular visualization and quantification than in "whole mount."

Key words TUNEL, Paraffin sections, Planarian, Apoptosis, Cell death, DNA fragmentation

1 Introduction

Planarians are flatworms, which are able to regenerate any body part and to change their size upon nutrient availability. These amazing features are controlled by changes in cell proliferation and cell death. During planarian regeneration, two peaks of apoptosis occur: one initial response near the wound at 1 to 4 h after injury and a second systemic increase of apoptosis after three days of injury [1]. Changes in size related to the nutrient status is also related to cell death, since there is a proportional decrease in cell death to the nutritional abundance [1, 2]. All these features make cell death an essential process for understanding planarian biology.

Terminal deoxynucleotidyl transferase dUTP nick end labeling (TUNEL) is a technique that permits to observe apoptotic cells directly in the tissue. This technique is based in the DNA breaks produced when cells undergo apoptosis [3]. This oligos leave 3′--hydroxyl termini (3'-OH) that are targeted by the enzyme terminal deoxynucleotidyl transferase (TdT). The TdT catalyzes the addition of labeled dUTP to the free 3'-OH. These newly modified ends are finally targeted by specific antibodies allowing the immunostaining of cells that undergo cell death.

Luca Gentile (ed.), *Schmidtea Mediterranea: Methods and Protocols*, Methods in Molecular Biology, vol. 2680,
https://doi.org/10.1007/978-1-0716-3275-8_15,
© The Author(s), under exclusive license to Springer Science+Business Media, LLC, part of Springer Nature 2023

TUNEL has been used in whole-mount planarians [4]. However, this protocol presents technical difficulties, since in uninjured animals the poor permeability of the tissue impedes the detection of all apoptotic cells. Here we provide a new TUNEL protocol in planarian paraffin sections, which allows the detection of apoptotic cells inside the planarian tissues, providing a detailed view of all apoptotic cells and their localization.

The TUNEL staining in paraffin sections has three main steps: (1) fixation of the animals; (2) paraffin inclusion of planarians, which consists in their paraffin embedding, sectioning, and deparaffinization [5]; and (3) the tunnel staining, using the ApopTag® Red In Situ Apoptosis Detection Kit.

2 Materials

2.1 Planarian Fixation

- 2% HCl solution: HCl diluted in ultrapure cold H_2O (*see* **Note 1**).
- 4% PFA solution: 4% paraformaldehyde powder diluted in 1× PBS. To dilute warm the solution at no more than 50 °C and agitate manually. Let the solution to set at room temperature before using (*see* **Note 2**).
- 1× PBS: 10× PBS stock diluted to 1× with ultrapure H_2O.
- 10× PBS: 1.37 M NaCl, 27 mM KCl, 100 mM Na2HPO4, and 20 mM KH2PO4 in ultrapure H_2O. Adjust pH to 7.4 with 37% HCl solution.

2.2 Paraffin Embedding, Sectioning, and Deparaffinization

- 100% ethanol.
- 96% ethanol (diluted in ultrapure H_2O).
- 70% ethanol (diluted in ultrapure H_2O).
- 100% xylene (*see* **Note 3**).
- 1:1 xylene/EtOH (1:1 mixture of xylene and EtOH).
- 100% paraffin for histopathology (melting T 56 °C).
- 50% paraffin (1:1 mixture of paraffin and xylene, preheated to 56 °C).

2.3 Tunnel Staining Using ApopTag® Red in Situ Apoptosis Detection Kit

- Proteinase K solution: 20 μg/mL of Proteinase K diluted in PBS.
- ApopTag® Red In Situ Apoptosis Detection Kit (Sigma-Aldrich S7165).
- TdT enzyme solution: 70% reaction buffer from ApopTag® kit and 30% TdT Enzyme from ApopTag® kit.

- Stop/wash solution: 2 mL of stop/wash buffer from ApopTag® kit in 68 ml of ultrapure H_2O.
- Anti-DIG solution: 53% blocking solution from ApopTag® kit and 47% anti-digoxigenin-rhodamine from ApopTag® kit.
- Nuclear staining: 0.5 μg/mL DAPI in PBS. Conveniently prepared as a 1:10.000 dilution of a 5 mg/mL stock in ultrapure H_2O (can be frozen at −20 °C).
- Mounting solution: glycerol-based commercial reagent or 4% n-propyl gallate in 70% glycerol.

3 Methods

3.1 Planarian Fixation

1. Transfer planarians to a 15-mL tube. Replace planarian salt solution with pre-chilled 2% HCl solution to kill the animals and to remove the mucus. Incubate for 4 min alternating one minute in ice and one minute in gentle manual agitation.
2. Replace the HCl with the fixative 4% PFA and incubate planarian during 4 h at 4 °C in agitation.
3. Wash fixed planarians once with 1× PBS and continue washing with 1× PBS overnight at 4 °C in agitation.
4. Discard the 1× PBS and replace it with 70% EtOH. Incubate during 10 min at room temperature (*see* **Note 4**).

3.2 Paraffin Embedding, Sectioning, and Deparaffinization

1. Dehydrate planarians performing the following washes at room temperature under agitation:
   - 10 min 70% EtOH
   - 10 min 96% EtOH
   - 2 × 5 min 100% EtOH
   - 10 min 1:1 xylene/EtOH
   - 2 × 5 min 100% xylene.
2. Bring planarians to 56 °C replacing xylene with 1:1 xylene/paraffin solution. Incubate 5 min at 56 °C.
3. Transfer planarians from tubes to small glass containers and fill them with paraffin. This process must be done in the 56 °C oven; otherwise, the paraffin will solidify.
4. Incubate for 20 min with paraffin.
5. Replace the paraffin in the glass containers with new paraffin. Incubate for 20 min.
6. Repeat the previous step.
7. After the last paraffin incubation, put a Petri dish on a warm plate a 56 °C. Fill it with paraffin.

8. Transfer planarians from the oven to the Petri dish using a preheated 56 °C Pasteur pipette.
9. Set up the planarians among the Petri dish separated from each other (*see* **Note 5**).
10. Remove the Petri dish from the warm plate and let it sit for at least 30 min.
11. Store at 4 °C overnight (*see* **Note 6**).
12. Use a scalpel blade to cut each planarian in cubes (*see* **Notes 7** and **8**). Attach the cubes to the microtome support (*see* **Note 9**).
13. Section the planarian with a microtome (*see* **Note 10**).
14. Place the ribbon of histological sections on a poly-l-lysine-coated slide. The slide must be covered with ultrapure H_2O and on the hot plate set to 37 °C.
15. Dry the slices overnight at 37 °C.
16. Deparaffinize and rehydrate planarian sections performing the following washes at room temperature in glass jars:
    - 2 × 5 min 100% xylene
    - 2 × 5 min 100% EtOH
    - 2 × 5 min 96% EtOH
    - 2 × 5 min 70% EtOH
    - 2 × 5 min 1× PBS.

3.3 Tunnel Staining Using ApopTag® Red in Situ Apoptosis Detection Kit

1. If a positive control of the experiment is required, slides can be treated with agents that produce DNA breaks. This steep is optional and must only be done if we want to induce cell death. We have two options:
    - Incubate slides with 2 N HCl during 40 min at room temperature.
    - Treat slides with 5 U/mL of DNAse diluted in 1× PBS for 1 h at room temperature.
2. For antigen unmasking incubate slides with 20 μg/mL Proteinase K during 10 min at room temperature. This steep can be done in glass jars.
3. Wash the slides twice with 1× PBS during 2 min in glass jars.
4. Take out the slides from the jar and dry the excess liquid by capillarity. Quickly pipette 150 μL of equilibration buffer from the ApopTag® Kit and incubate for at least 10 seconds (*see* **Note 11**).
5. Remove the equilibration buffer and rapidly add 50 μL of TdT enzyme solution (*see* **Note 12**) on top of the slide. Cover the slide with a plastic cover provided in the kit (*see* **Note 13**). Incubate in a humidified chamber (*see* **Note 14**) at 37 °C for 2 h.

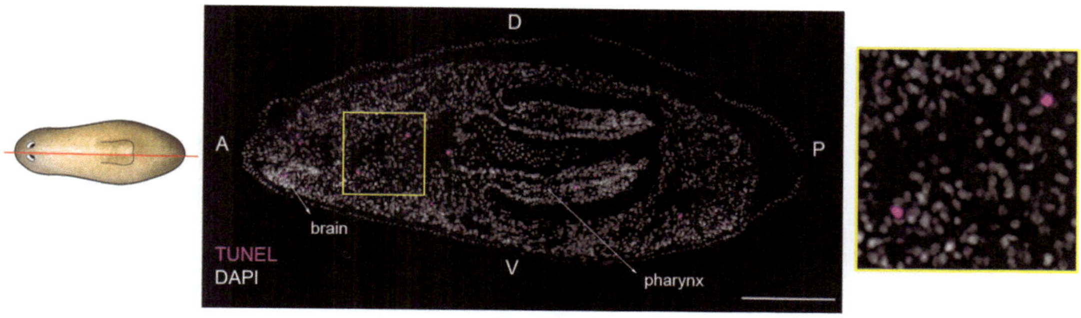

Fig. 1 TUNEL staining in a section of paraffin-embedded planarian. Magenta dots are cells undergoing apoptosis, since they are positive for TUNEL staining. Nucleus are stained with DAPI (white) and show the different planarian tissues, as the pharynx and the brain. A magnification of the yellow squared area is shown (x 3). (A, anterior; P, posterior; V, ventral; D, dorsal). The drawing on the left indicates the level of the sagittal section performed. Scale is 0,5 mm

6. After TdT incubation submerge the slides in a glass jar with stop/wash buffer. Incubate at room temperature for 10 min.
7. Wash the slides for 2 min in 1× PBS three times.
8. Remove the slides from the jar and rapidly add 80 μL of anti-DIG solution on top of the slide. Cover the slide with a plastic cover provided in the kit. Slides must never be completely dry. Process one slide at a time, as the slices must not dry out (*see* **Note 11**). Put slides in a humidified chamber and incubate at 4 °C overnight (*see* **Note 15**).
9. Submerge slides in a glass jar with 1× PBS for 2 min.
10. Wash 3× with 1× PBS, 2 min each time.
11. Incubate with a nuclear staining for at least 30 min (*see* **Note 16**).
12. Mount using a mounting solution and image (*see* Fig. 1).

4 Notes

1. The solution must be made just before use.
2. The 4% PFA solution can be prepped the day before and kept in the fridge until use. Avoid using solutions that had been stored for a long time because it can impact the fixation process.
3. Xylene is toxic, and the commercial substitute Neo-Clear can be used.
4. Other protocols suggest that animals can be stored at −20 °C in 70% EtOH. For the TUNEL this steep is not recommended because it can affect the final staining.

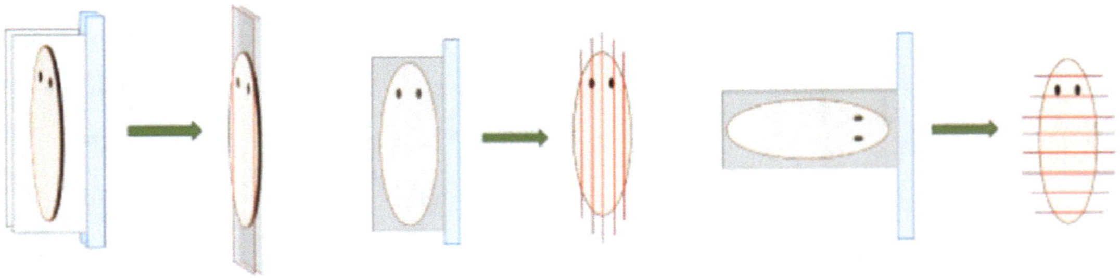

Fig. 2 Schematics showing the orientation of the cube on the microtome support according to the type of sections that will be made. Sagittal, longitudinal, and traverse sections are shown

5. In order to spread the planarians in the plate, it is useful to use the punch at 54 °C. The punch, with wooden or plastic handle, can be kept in the 56 °C chamber in order to keep it warm. If the punch is not warm, the planarians will stick to it.
6. Paraffin blocks can be stored at 4 °C for short-term storage or at 20 °C for long-term storage.
7. To cut the planarians individually, use a warm blade to avoid cracking. The blade can be from a scalpel or an old microtome blade. To warm the blade, a Bunsen flame can be used. When cutting the paraffin, avoid using excessive force. The warm metal of the blade will melt the paraffin making the process easier. Do not heat very near or directly on the animal, only in the surrounding paraffin.
8. The paraffin cubes must be a few centimeters wider than the planarian to avoid damaging the animal. In terms of height, the zone that will be facing the microtome support should be thicker. See the scheme in **Note 9** on how to orient planarians.
9. The orientation of the cube on the microtome support is determining the type of cut that will be made. See the scheme in Fig. 2.
10. The thickness of the microtome slides is usually set at 10 μm but this parameter can vary depending on the experimental needs.
11. Slides must never be completely dry. In some steeps, low quantities of reagents are used. To avoid drying the tissue, process one slide at a time; finish the process in one slide before continuing with the next one.
12. The TdT enzyme solution must be freshly made and kept in ice until use.
13. Plastic covers can be washed and reused in further experiments. Wash the covers with abundant distillated H_2O and let them dry covered with absorbent paper.

14. A humidified chamber can be obtained with wet absorbent paper in a big Petri dish, to create a humid ambient to avoid solution evaporation.
15. Antibody incubation can also be performed at room temperature during 3 h. Room temperature incubation sometimes can lead to the appearance of unspecific staining.
16. Several nuclear staining can be used. We use DAPI because it has no interference with the red fluorescence of the TUNEL antibody.

References

1. Pellettieri J, Fitzgerald P, Watanabe S, Mancuso J, Green DR, Sanchez Alvarado A (2010) Cell death and tissue remodeling in planarian regeneration. Dev Biol 338(1):76–85. https://doi.org/10.1016/j.ydbio.2009.09.015
2. Baguña J (1976) Mitosis in the intact and regenerating planarian Dugesia mediterranea sp. Mitotic studies during growth, feeding and starvation. J Exp Zool 195(1):53–64. https://doi.org/10.1002/jez.1401950106
3. Bortner CD, Oldenburg NB, Cidlowski JA (1995) The role of DNA fragmentation in apoptosis. Trends Cell Biol 5(1):21–26. https://doi.org/10.1016/s0962-8924(00)88932-1
4. Stubenhaus B, Pellettieri J (2018) Detection of apoptotic cells in planarians by whole-mount TUNEL. Methods Mol Biol 1774:435–444. https://doi.org/10.1007/978-1-4939-7802-1_16
5. Adell T, Barberan S, Sureda-Gomez M, Almuedo-Castillo M, de Sousa N, Cebria F (2018) Immunohistochemistry on paraffin-embedded planarian tissue sections. Methods Mol Biol 1774:367–378. https://doi.org/10.1007/978-1-4939-7802-1_11

References

Chapter 16

Quantitative Analysis of Planarian Pigmentation

Matthew Pittendreigh, Kaleigh Powers, Meenalosini Vimal Cruz, and Jason Pellettieri

Abstract

The ommochrome and porphyrin body pigments that give freshwater planarians their brown color are produced by specialized dendritic cells located just beneath the epidermis. During embryonic development and regeneration, differentiation of new pigment cells gradually darkens newly formed tissue. Conversely, prolonged light exposure ablates pigment cells through a porphyrin-based mechanism similar to the one that causes light sensitivity in rare human disorders called porphyrias. Here, we describe a novel program using image-processing algorithms to quantify relative pigment levels in live animals and apply this program to analyze changes in bodily pigmentation induced by light exposure. This tool will facilitate further characterization of genetic pathways that affect pigment cell differentiation, ommochrome and porphyrin biosynthesis, and porphyrin-based photosensitivity.

Key words Platyhelminthes, Pigment, Ommochrome, Porphyrin

1 Introduction

The Platyhelminthes are best known for their capacity to regenerate lost body parts [1], but their wide variety of pigment hues and patterns is equally striking [2]. Marine polyclad and terrestrial flatworms have colors ranging from bright pinks and yellows to deep blacks, whereas the more commonly studied freshwater species, such as *Schmidtea mediterranea* and *Dugesia japonica,* typically exhibit a light or dark brown color. As in other animals, body pigments in the flatworms may afford at least some degree of photoprotection and/or camouflage [2]. However, there is reason for caution in making functional inferences, as pigments can also act as photosensitizers and have been ascribed remarkably diverse roles [3].

A combination of ultrastructural, biochemical, and molecular studies have provided significant insight into pigment cell biology in freshwater planarians [2]. The body pigment cells of

Luca Gentile (ed.), *Schmidtea Mediterranea: Methods and Protocols*, Methods in Molecular Biology, vol. 2680, https://doi.org/10.1007/978-1-0716-3275-8_16,
© The Author(s), under exclusive license to Springer Science+Business Media, LLC, part of Springer Nature 2023

S. mediterranea, for example, differentiate from progenitors through a mechanism requiring a putative FGF receptor, as well as Forkhead, FOX, and ETS family transcription factors [4, 5]. Mature pigment cells reside immediately below the body wall muscle and project dendrites upward through the muscle fibers to carry pigment granules toward the epidermis [2, 6]. The pigment molecules themselves consist of tryptophan metabolites called ommochromes and tetrapyrroles called porphyrins [7]. While the precise chemical structure of the body pigment is unknown, disruption of either underlying biosynthetic pathway leads to a color change or complete color loss [5, 7]. Available evidence supports comparable pigment biochemistry and cell biology in other freshwater species [8–11], though it is clear at least some differences (e.g., in porphyrin levels) exist [7].

Changes in planarian pigmentation can occur in a variety of contexts. Most notably, pigmentation increases as new pigment cells begin to differentiate during late stages of embryonic development [12, 13] and also within the stem cell-derived mass of new tissue, or blastema, that forms at sites of amputation in regenerating animals [4]. Conversely, pigment cells are rapidly eliminated when *S. mediterranea* is exposed to bright light [7]. This porphyrin-dependent response is due to a physiological bottleneck in the heme biosynthesis pathway, mirroring the pathological one that leads to severe light sensitivity in human disorders called porphyrias [7, 14].

Analyses of pigmentation changes like the ones noted above, or those caused by RNAi knockdown of pigmentation genes, have been almost exclusively qualitative to date. Quantitative approaches are possible [7], but currently require biochemical pigment extraction, precluding analyses of live animals. The uneven appearance of bodily pigmentation, both at the organismal level (e.g., lighter above the pharynx) and within local regions of the epidermis, limits the accuracy of simple image analysis approaches. Here, we present a new program capable of quantifying both pigment intensity and area in live animals that can be used to track changes in relative pigmentation over time. RGB images obtained with a dissecting microscope are (1) uploaded and masked to remove the background and (2) subjected to automated pixel intensity and Otsu thresholding analyses based on user-defined parameters. Results are returned in a downloadable format. We apply this program to quantify depigmentation induced by light exposure, demonstrating its utility for analyses of pigment cell development, pigment biosynthesis, and photosensitivity in the planarian porphyria model.

2 Materials

1. Stereomicroscope and digital camera (*see* **Note 1**).
2. Python: https://www.python.org/downloads/.
3. PlanaraChrome: https://github.com/matthewpittendreigh/planarachrome.

3 Methods

Follow the steps below to quantify relative pigment levels in images of live planarians using the PlanaraChrome app and any web browser as a user interface. The program code is open source and available at the GitHub download page linked above.

1. Launch PlanaraChrome (*see* **Note 2**) and then upload the image(s) to be analyzed (*see* **Note 3**), using the *Browse* and *Upload* buttons at the top left of the *Upload* page. Thumbnails of all uploaded images will be displayed. This step can be repeated to upload images from multiple folders, as necessary. Uploaded images can be removed using the *Reset Files* button.
2. Go to the *Mask* page and adjust the threshold setting for each image to remove the background. The slider changes the threshold value (*see* **Note 4**), and the *Submit* and *Apply All* buttons apply this value to the current or all image(s), respectively. If the setting is too high, an error message will prompt selection of a lower value (*see* **Note 5**). Use the arrow buttons at the bottom of the page to navigate from one image to the next and the *Raw* or *Segmented* view options to toggle between the original color image and a grayscale version with the current mask applied.
3. Go to the *Settings* page to adjust the parameters for analyzing relative pigment intensity (*Intensity Quantification—see* **step 4** below) and defining pigmented and unpigmented regions of the epidermis (*Area Quantification—see* **step 5** below).
4. Enter a number between 0 and 100 in the *Percentile* box under *Intensity Quantification* to determine whether the pigment intensity score is based upon the darkest pixel(s) in the masked image (percentile = 0), the brightest pixel(s) in the masked image (percentile = 100), or pixels with intermediate intensities (e.g., setting the percentile to 50 selects the median pixel intensity) (*see* **Note 6**). Click on the *Apply Changes* button at the top left of the page if the default setting is changed.
5. Quantification of the pigmented area of the epidermis requires accurate recognition of pigmented and unpigmented regions in the masked image. This distinction, made on the basis of the

minimum and maximum threshold values set for each color channel (*see* **Note 7**), may require empirical optimization. First, evaluate the default settings by entering an image number in the *Image Index* box and clicking on the *Test Threshold Values* button. This will generate a side-by-side view of the original image and a processed image with any areas that have been scored as pigmented colored green (*see* Fig. 1h). If desired, make threshold adjustments by returning to the *Settings* page, entering new values, and clicking on the *Apply Changes* button. Repeat this process with representative images from all experimental conditions. Unlike the mask settings, which should be optimized for individual images (*see* **step 2** above), color threshold values will be universally applied. Thus, the objective is to arrive at settings that will generate the most accurate possible assessment of pigment area across all analyzed images (*see* **Note 8**).

6. Histograms depicting the spread of the pigment intensity and area scores are automatically generated in each analysis. Enter the desired number of bins (i.e., intervals) under *Histogram* to specify how the data are displayed.
7. Click on the *Apply Changes* button before leaving the *Settings* page to save any adjustments made to the pigment quantification settings.
8. Go to the *Results* page to obtain relative pigment intensity and area measurements for each uploaded image. Both results are expressed on a scale of 0.00 to 1.00, with higher numbers denoting greater pigmentation (*see* **Note 9**). The histograms provide a graphical representation of the degree to which scores cluster or diverge within the analyzed dataset. Click on the download links to save the results and histograms as .csv and . png files, respectively.
9. Results can be plotted and/or subjected to statistical analyses according to individual user goals. As an example, we quantified the light-induced depigmentation response that we previously reported in *S. mediterranea* [7] to demonstrate the utility of the program. Animals were continuously exposed to red light for a total of seven days and photographed at 24-h intervals (*see* **Note 10**). Images were then processed according to the steps above. While visual inspection of the photographs from each timepoint was sufficient to establish an obvious decline in overall pigmentation (*see* Fig. 2a), the quantitative data revealed statistical significance for incremental changes and also provided insight into trends that might have otherwise been missed. Most notably, while the decline in pigment intensity was nearly constant over the analyzed time course (*see* Fig. 2b), there was an abrupt drop in the pigmented area of

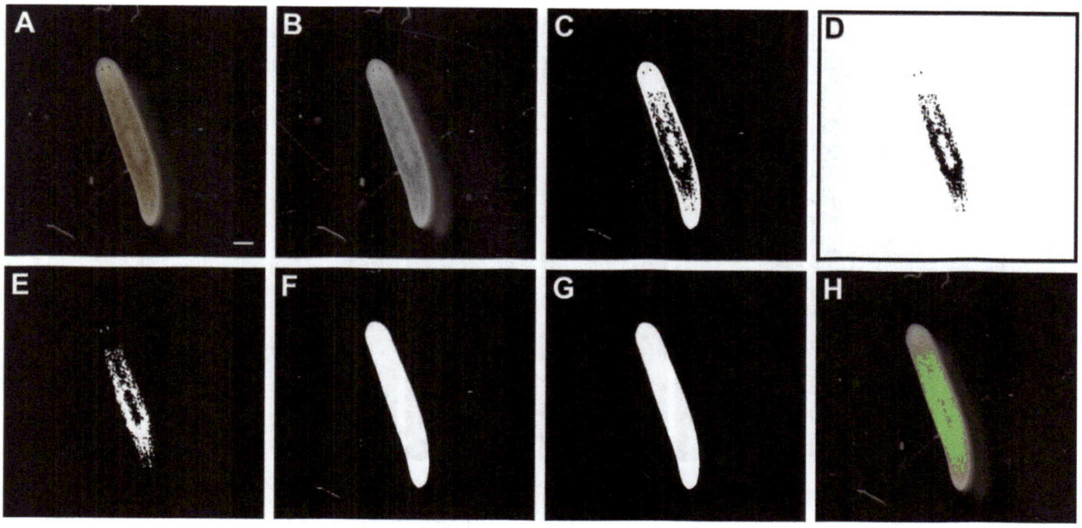

Fig. 1 Image processing for pigment quantification. (**a**) Raw image of a live planarian subjected to 72 h of red light exposure (this image, part of the dataset analyzed in Fig. 2, was selected to illustrate the performance of the program with a partially pigmented animal). Scale bar: 500 μm. (**b–h**) Processed images generated during masking and pigment area quantification: (**b**) grayscale, (**c**) binary threshold, (**d**) flood fill, (**e**) inverted flood fill, (**f**) rough mask, (**g**) refined mask, and (**h**) pigment area. (Note that the area quantification settings used to generate panel H correspond to those applied across all timepoints in Fig. 2. *See* **Notes 4** and **7** for further details)

the epidermis between 96 h and 120 h of light exposure (*see* Fig. 2c). This may reflect a corresponding increase in the rate of pigment cell death during this interval [7], a hypothesis that could easily be addressed in follow-up experiments. In summary, these results demonstrate the utility of the Planara-Chrome app for resolving small, but significant changes in planarian pigmentation and for informing our understanding of pigment cell biology in the Platyhelminthes.

4 Notes

1. All analyzed images were obtained with an Olympus SZX16 stereomicroscope equipped with a DP72 digital camera. Comparable equipment from other manufacturers should produce similar results. We used a fiber-optic ring light (Schott, A08625) to minimize shadows and to standardize illumination across conditions and experiments. Gooseneck lamps are not ideal because they introduce unavoidable variability, though we did not find this to be a major issue in preliminary analyses. It is imperative that microscope and camera settings (e.g., magnification and exposure time), as well as light intensity, be constant within any given experiment.

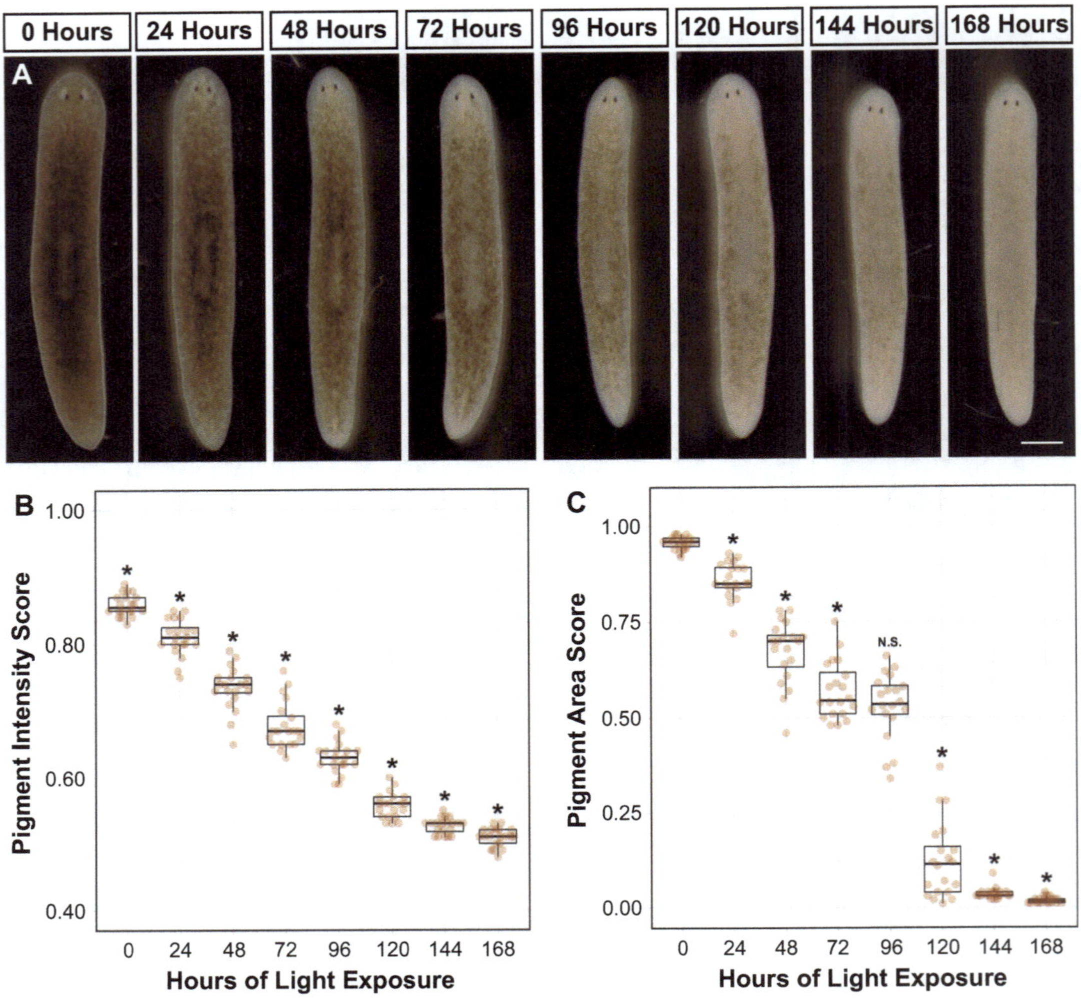

Fig. 2 Quantitative analysis of light-induced depigmentation in *S. mediterranea*. (**a**) Representative images of live animals photographed after the indicated number of hours of red light exposure (*see* **Note 10** for details). No brightness, contrast, or gamma adjustments were made for any of the images in this experiment. Scale bar: 500 μm. (**b**, **c**) Pigment intensity (**b**) and area (**c**) scores for 20 light-exposed animals. Boxes, whiskers, and horizontal lines denote the interquartile range (IQR), values within 1.5X the IQR, and medians, respectively. Asterisks denote p-values <5E-3 for T-tests comparing each timepoint with the previous timepoint; (*NS* not significant)

2. After downloading PlanaraChrome from GitHub, follow the README instructions for your operating system to launch the program using Terminal, Python, and any web browser as a user interface. When your browser window opens, you will automatically be directed to the *Upload* page referenced in **step 1**.
3. The program accepts RGB images in .tif or .png formats. Do not include a scale bar or text (or black them out prior to processing using an application such as Fiji/ImageJ [15]). Our standard image size is 1360×1024 pixels, but other

resolutions are supported. Images should have a background that is darker than the animal being analyzed. Nonlinear (i.e., gamma) adjustments must be avoided, and any linear (i.e., brightness and/or contrast) adjustments must be applied evenly across all conditions.

4. Thresholding begins with conversion of the original RGB image to grayscale (*see* Fig. 1a and b), followed by application of a binary threshold to generate a solid outline corresponding to the perimeter of the animal (*see* Fig. 1c). After a subsequent flood-fill operation originating from the upper right and lower left corners (*see* Fig. 1d), the resulting image is inverted (*see* Fig. 1e) and combined with the binary threshold image to produce a rough mask that distinguishes the animal from its background (*see* Fig. 1f). Finally, the mask is refined by morphological opening (the largest connected component is isolated after two iterations with a 9 by 9 kernel). The refined mask (*see* Fig. 1g) is applied when the *Submit* or *Apply All* buttons are clicked on the *Mask* page with the *Segmented* view option selected.

5. We are often able to identify a single threshold setting that will clearly distinguish animals from their backgrounds for most of the images in a given experimental condition. However, refinements for selected images are often necessary to achieve the best possible results (e.g., to completely remove animal reflections). In rare cases, it may be necessary to black out background features prior to processing if they cannot be removed in the initial mask attempt.

6. The pigment intensity measurement is generated from the masked grayscale image, based upon the percentile specified in **step 4**. We typically choose the 15th percentile, as this targets part of the pixel intensity distribution corresponding to dark (i.e., pigmented) regions of the epidermis, while avoiding exceptionally dark outliers (e.g., pixels within the eyespots). It is important to reiterate that this approach quantifies relative, rather than absolute pigment levels.

7. The pigment area measurement is generated from the red, green, and blue channels of the masked color image, based upon the minimum and maximum thresholds entered in **step 5**. These variables set the boundaries for an Otsu thresholding function, which is used to differentiate pigmented and unpigmented regions of the epidermis, while accounting for the possibility of differences in both pigment intensity and area between experimental conditions (e.g., *see* Fig. 2). In order for a pixel to be scored as pigmented, there must be agreement between the results for at least two of the three color channels. Outputs, which can be visualized using the *Test Threshold Values* function on the *Settings* page (Fig. 1h), are expressed

as the ratio of pigmented to unpigmented pixels within the unmasked region of each image.

8. Default values for the color thresholding step are based upon our experience of processing images from preliminary datasets and should be reevaluated for each new experiment. When pigmentation varies widely (*see*, for example, Fig. 2), the optimal settings may vary to some degree from one condition or timepoint to the next. However, it should be possible to achieve a reasonable compromise by testing representative images with the *Test Threshold Values* function. In practical terms, increasing the minimum threshold for any color channel limits the darkest pixel intensity that can be scored as the boundary between pigmented and nonpigmented regions. Decreasing the minimum threshold allows darker pixel intensities to be selected. Similarly, decreasing the maximum threshold for any color channel limits the lightest pixel intensity that can be scored as the boundary between pigmented and unpigmented regions, while increasing this variable allows lighter pixel intensities to be selected.
9. In order to express relative pigment intensity scores on a 0.00 to 1.00 scale, with higher numbers denoting higher pigment levels, the initial values (*see* **Note 6**) are divided by 255 and the quotient is subtracted from 1.00.
10. For the experiment in Fig. 2, size-matched, seven-day-starved animals were placed under a 625-nm LED (BML Horticulture) in a 48-well plate (one or two animals per well in 0.6 mL of water) and continuously exposed for a total of one week. Every animal was photographed at 24-h intervals, using identical microscope, light, and camera settings at each timepoint.

Acknowledgements

This work was supported by New Hampshire-INBRE through an Institutional Development Award (P20GM103506) from the National Institute of General Medical Sciences of the National Institutes of Health and by NIH 1R15DK127370-01. We thank Sarah McGregor, Mahad Ahmad, and Spencer Lynch for assistance with preliminary experiments.

References

1. Reddien PW (2018) The cellular and molecular basis for planarian regeneration. Cell 175:327–345
2. Lindsay-Mosher N, Pearson BJ (2019) The true colours of the flatworm: mechanisms of pigment biosynthesis and pigment cell lineage development in planarians. Semin Cell Dev Biol 87:37–44
3. Hill HZ (1992) The function of melanin or six blind people examine an elephant. BioEssays 14:49–56

4. He X, Lindsay-Mosher N, Li Y, Molinaro AM, Pellettieri J, Pearson BJ (2017) FOX and ETS family transcription factors regulate the pigment cell lineage in planarians. Development 144:4540–4551
5. Wang C, Han X-S, Li F-F, Huang S, Qin Y-W, Zhao X-X, Jing Q (2016) Forkhead containing transcription factor albino controls tetrapyrrole-based body pigmentation in planarian. Cell Discov 2:16029
6. Brubacher JL, Vieira AP, Newmark PA (2014) Preparation of the planarian Schmidtea mediterranea for high-resolution histology and transmission electron microscopy. Nat Protoc 9:661–673
7. Stubenhaus BM, Dustin JP, Neverett ER, Beaudry MS, Nadeau LE, Burk-McCoy E, He X, Pearson BJ, Pellettieri J (2016) Light-induced depigmentation in planarians models the pathophysiology of acute porphyrias. Elife 5:e14175
8. Krugelis-MacRae E (1956) The occurrence of porphyrin in the planarian. Biol Bull 110:69–76
9. Krugelis-MacRae E (1961) Localization of porphyrin fluorescence in planarians. Science 134:331–332
10. Palladini G, Medolago-albani L, Margotta V, Conforti A, Carolei A (1979) The pigmentary system of planaria. I. Morphology. Cell Tissue Res 199:197–202
11. Hase S, Wakamatsu K, Fujimoto K, Inaba A, Kobayashi K, Matsumoto M, Hoshi M, Negishi S (2006) Characterization of the pigment produced by the planarian, Dugesia ryukyuensis. Pigment Cell Res 19:248–249
12. Martín-Durán JM, Monjo F, Romero R (2012) Planarian embryology in the era of comparative developmental biology. Int J Dev Biol 56:39–48
13. Davies EL, Lei K, Seidel CW, Kroesen AE, McKinney SA, Guo L, Robb SMC, Ross EJ, Gotting K, Sánchez Alvarado A (2017) Embryonic origin of adult stem cells required for tissue homeostasis and regeneration. Elife 6: e21052
14. Phillips JD (2019) Heme biosynthesis and the porphyrias. Mol Genet Metab 128:164–177
15. Schindelin J, Arganda-Carreras I, Frise E, Kaynig V, Longair M, Pietzsch T, Preibisch S, Rueden C, Saalfeld S, Schmid B, Tinevez JY, White DJ, Hartenstein V, Eliceiri K, Tomancak P, Cardona A (2012) Fiji: an open-source platform for biological-image analysis. Nat Methods 9:676–682

Chapter 17

mRNA Transfection of *S. mediterranea* for Luminescence Analysis

Uri Weill, Richard Nelson Hall, Leonard Drees, Bo Wang, and Jochen C. Rink

Abstract

Planarians have become a powerful model system for stem cell research and regeneration. While the tool kit for mechanistic investigations has been steadily expanding over the last decade, robust genetic tools for transgene expression are still lacking. We describe here methods for in vivo and in vitro mRNA transfection of the planarian species *Schmidtea mediterranea*. These methods utilize the commercially available *Trans*IT-mRNA transfection reagent to efficiently deliver mRNA encoding a synthetic nanoluciferase reporter. Using a luminescent reporter overcomes the bright autofluorescent background of planarian tissues and allows quantitative measurements of protein expression levels. Collectively, our methods provide the means for heterologous reporter expression in planarian cells and the basis for future development of transgenic techniques.

Key words mRNA transfection, Transgenics, Luminescence, Live imaging, Reporter expression, Planaria

1 Introduction

Planarians provide unique experimental access to questions of stem cell homeostasis, tissue dynamics or regeneration, and scaling of a metazoan body plan. The community has a range of molecular genetic tools to probe the underlying molecular mechanisms, including RNA interference (RNAi), fluorescence in situ hybridization (FISH), and various next-generation sequencing techniques [1–4]. However, transgenic tools commonly used in many model organisms are still lacking, at least in parts due to the lack of single-celled zygotes in the commonly studied asexual strains that reproduce by fission.

In general, the de novo establishment of transgene expression in a model system requires navigating a substantial parameter space of transfection and expression conditions. While microinjection

Luca Gentile (ed.), *Schmidtea Mediterranea: Methods and Protocols*, Methods in Molecular Biology, vol. 2680, https://doi.org/10.1007/978-1-0716-3275-8_17,
© The Author(s), under exclusive license to Springer Science+Business Media, LLC, part of Springer Nature 2023

provides a species-independent delivery method, it can be slow and cumbersome to apply. Lipofection and electroporation allow efficient construct delivery into cells or tissues, though the precise protocols and reagents tend to be optimized for specific species. Delivery of mRNA, instead of DNA constructs, can be beneficial to establish successful transfection in a novel species, as mRNA is rapidly expressed and only requires delivery into the cytoplasm of cells. Another important variable is the choice of the reporter for monitoring transgene expression. Although fluorescent proteins can achieve high signal-to-noise (SNR) ratios, strong autofluorescence of some tissues or animals (e.g., planarians) may obscure weak transgene expression or present as a false-positive. Luciferases produce light as the result of an enzymatic reaction, which eliminates autofluorescence elicited by the excitation light and can thus achieve even higher SNR and the robust detection of extremely low-level reporter expression. Recent advances here include the development of nanoluciferase (Nluc), which is a highly compact (19 kDa), stable, and orders of magnitude brighter engineered derivative of wild-type luciferase [5].

This protocol describes transgenic reporter expression in *Schmidtea mediterranea* via Nluc mRNA reporter transfection with *Trans*IT-mRNA reagent [6]. Step-by-step instructions for both, the transfection of planarian cells in vitro and of living planarians via microinjection, are provided. The protocol can be used to investigate posttranscriptional regulation by comparing different regulatory elements such as untranslated region (UTR) sequences, codon bias, mRNA modifications, and others. Overall, the protocol not only provides a positive control to catalyze the future development of planarian transgenesis but also provides a new route toward the manipulation of planarian gene expression in vitro and in vivo.

2 Materials

2.1 *In Vitro* Transcription

1. T7 mScript™ Standard mRNA Production System (CELLSCRIPT) (*see* **Note 1**).
2. Ethanol, Absolute, Molecular Biology Grade (e.g., Fisher BioReagents).
3. Cleaning reagent for RNase (e.g., RNaseZap).
4. Microcentrifuge tubes RNase-free, 1.5 (e.g., Eppendorf).
5. Qubit fluorometer system (Thermo Fisher Scientific) or bioanalyzer system (Agilent).

2.2 Dissociation of Planaria into Individualized Cells

1. Transfer pipettes, polyethylene.
2. 6-cm-diameter cell culture dishes, sterile.

3. CMF (Ca/Mg-free medium): NaH2PO4 480 mg/L, NaCl 960 mg/L, KCl 1.44 g/L, NaHCO3 960 mg/L, HEPES (4-(2-hydroxyethyl)-1-piperazineethanesulfonic acid) 3.57 g/L, D-glucose 0.24 g/L, BSA (bovine serum albumin) 1 g/L, 10 mg/L ciprofloxacin, pH 7.4, in ultrapure H2O. Filter and store at room temperature, protected from light. A 10× solution can be made and stored at 4 °C.
4. Razor blades.
5. Microcentrifuge tubes, 1.5 (e.g., Eppendorf).
6. Tube rotator (e.g., Roth).
7. 100-, 70-, 40-, and 30-μm cell strainers (e.g., Corning).
8. Iso-L15 (isotonic Leibovitz's L-15 medium): 1:1 Leibovitz's L-15 to ultrapure H2O, 1× MEM nonessential amino acids (glycine 0.75 mg/L, L-alanine 0.89 mg/L, L-asparagine 1.32 mg/L, L-aspartic acid 1.33 mg/L, L-glutamic acid 1.47 mg/L, L-proline 1.15 mg/L, L-serine 1.05 mg/L), 1× antibiotic-antimycotic, 1× MEM vitamin solution (choline chloride 0.1 mg/L, D-calcium pantothenate 0.1 mg/L, folic acid 0.1 mg/L, nicotinamides 0.1 mg/L, pyridoxal hydrochloride 0.1 mg/L, riboflavin 0.01 mg/L, thiamine hydrochloride 0.1 mg/L, i-inositol 0.2 mg/L, sodium chloride [NaCl] 8.5 mg/L), 1 mM sodium pyruvate, 2.5 g/L HEPES (4-(2-hydroxyethyl)-1-piperazineethanesulfonic acid), 5% FBS (fetal bovine serum), 10 mg/L ciprofloxacin, buffer to pH 7.8 with 0.2 M NaOH, then filter, and store at 4 °C.

2.3 In Vitro Cell mRNA Transfection

1. Disposable cytometer (e.g., C-chip).
2. Microcentrifuge tubes RNase-free, 1.5 (e.g., Eppendorf).
3. Inverted light microscope (e.g., Olympus IXplore).
4. Iso-L15 (isotonic Leibovitz's L-15 medium): 1:1 Leibovitz's L-15 to ultrapure H2O, 1× MEM nonessential amino acids (glycine 0.75 mg/L, L-alanine 0.89 mg/L, L-asparagine 1.32 mg/L, L-aspartic acid 1.33 mg/L, L-glutamic acid 1.47 mg/L, L-proline 1.15 mg/L, L-serine 1.05 mg/L), 1× antibiotic-antimycotic, 1× MEM vitamin solution (choline chloride 0.1 mg/L, D-calcium pantothenate 0.1 mg/L, folic acid 0.1 mg/L, nicotinamides 0.1 mg/L, pyridoxal hydrochloride 0.1 mg/L, riboflavin 0.01 mg/L, thiamine hydrochloride 0.1 mg/L, i-inositol 0.2 mg/L, sodium chloride [NaCl] 8.5 mg/L), 1 mM sodium pyruvate, 2.5 g/L HEPES (4-(2-hydroxyethyl)-1-piperazineethanesulfonic acid), 5% FBS (fetal bovine serum), 10 mg/L ciprofloxacin, buffer to pH 7.8 with 0.2 M NaOH, then filter, and store at 4 °C.
5. Clear or white flat-bottom 96-well microplates (e.g., Greiner).

6. *Trans*IT-mRNA transfection mix: 2 μL *Trans*IT-mRNA transfection reagent (Mirus), 1 μL *Trans*IT Boost reagent, 1.5 μg Nluc mRNA; add L-15 (no serum) up to 25 μL in total (*see* **Note** 7).

2.4 Dissociated Cell Luminescence Analysis

1. Nano-Glo Live mix: 1.25 μL Nano-Glo Live Cell Substrate (Promega) with 23.75 μL Nano-Glo Live Cell Buffer.
2. Plate reader (e.g., BioTek Synergy™ Neo2).
3. A bioluminescence imaging system (e.g., Olympus LV200).

2.5 Live Worm mRNA Transfection by Injection

1. 12-cm-diameter cell culture dishes, sterile.
2. Transfer pipettes, polyethylene.
3. Planarian water (1× "Montjuic salts"): 1.6 mM NaCl, 1.0 mM CaCl2, 1.0 mM MgSO4, 0.1 mM MgCl2, 0.1 mM KCl, 1.2 mM NaHCO3 in deionized water, autoclaved.
4. Gentamicin antibiotics.
5. Borosilicate glass capillaries (e.g., Drummond).
6. Micropipette puller (e.g., Sutter Instruments P-1000).
7. Microinjector (e.g., Eppendorf FemtoJet).
8. Micromanipulator (e.g., ASI MM33).
9. Fine forceps.
10. Peltier cooling block (or another cooling device).
11. Stereomicroscope (e.g., Olympus SZX10).
12. Microcentrifuge tubes RNase-free, 1.5 (e.g., Eppendorf).
13. *Trans*IT-mRNA transfection mix: 2 μL *Trans*IT-mRNA transfection reagent (Mirus), 1 μL *Trans*IT Boost reagent, 1.5 μg Nluc mRNA; add L-15 (no serum) up to 25 μL in total (*see* **Note** 7).
14. Delicate task wipers (e.g., Kimtech).
15. Grade 551 Filter Paper, black (e.g., VWR).

2.6 Luminescence Analysis of Lysed Transfected Worms by a Plate Reader

1. Planarian water (1× "Montjuic salts"): 1.6 mM NaCl, 1.0 mM CaCl2, 1.0 mM MgSO4, 0.1 mM MgCl2, 0.1 mM KCl, 1.2 mM NaHCO3 in deionized water, autoclaved.
2. Transfer pipettes, polyethylene.
3. 6-cm-diameter cell culture dishes, sterile.
4. Razor blades.
5. Iso-L15 (isotonic Leibovitz's L-15 medium): 1:1 Leibovitz's L-15 to ultrapure H2O, 1× MEM nonessential amino acids (glycine 0.75 mg/L, L-alanine 0.89 mg/L, L-asparagine 1.32 mg/L, L-aspartic acid 1.33 mg/L, L-glutamic acid

1.47 mg/L, L-proline 1.15 mg/L, L-serine 1.05 mg/L), 1× antibiotic-antimycotic, 1× MEM vitamin solution (choline chloride 0.1 mg/L, D-calcium pantothenate 0.1 mg/L, folic acid 0.1 mg/L, nicotinamides 0.1 mg/L, pyridoxal hydrochloride 0.1 mg/L, riboflavin 0.01 mg/L, thiamine hydrochloride 0.1 mg/L, i-inositol 0.2 mg/L, sodium chloride [NaCl] 8.5 mg/L), 1 mM sodium pyruvate, 2.5 g/L HEPES (4-(2-hydroxyethyl)-1-piperazineethanesulfonic acid), 5% FBS (fetal bovine serum), 10 mg/L ciprofloxacin, buffer to pH 7.8 with 0.2 M NaOH, then filter, and store at 4 °C.

6. White flat-bottom 96-well microplate (e.g., Greiner).
7. Nano-Glo Lysis mix for worms: 2 μL Nano-Glo Luciferase Assay Substrate (Promega), 98 μL Nano-Glo Luciferase Assay Buffer.
8. Multi-pipettor.
9. Plate reader (e.g., BioTek Synergy™ Neo2).

2.7 Luminescence In Vivo Imaging of mRNA Transfected Worms

1. Planarian water (1× "Montjuic salts"): 1.6 mM NaCl, 1.0 mM CaCl2, 1.0 mM MgSO4, 0.1 mM MgCl2, 0.1 mM KCl, 1.2 mM NaHCO3 in deionized water, autoclaved.
2. Linalool.
3. Nano-Glo Live mix for worms: 5 μL Nano-Glo Live Cell Substrate (Promega), 95 μL planarian water (1% DMSO).
4. 3.5-cm-diameter glass-bottom cell culture dishes, sterile.
5. Fine brush.
6. Transfer pipettes, polyethylene.
7. Microcentrifuge tubes, 1.5 (e.g., Eppendorf).
8. Agarose mix for worms: 1 mL planarian water with 10 mg SeaPlaque™ Agarose (Lonza) and 1:5000 linalool. Heat to 85 °C until agarose is dissolved. Cool the agarose mix at room temperature until it is warm but does not harden. Add 50 μL of Nano-Glo Live Cell Substrate (Promega).
9. Coverslips, 22 × 55 mm.
10. A bioluminescence imaging system (e.g., Olympus LV200).

3 Methods

All the steps are performed at room temperature unless otherwise stated.

3.1 In Vitro Transcription

1. Generate capped and polyadenylated Nluc single-strand mRNA via the T7 mScript™ Standard mRNA Production System and according to the manufacturer's protocol (*see* **Note 1**).
2. Incubate 1 h for the T7 transcription, 2 h for 5′ capping, and 1 h for poly-A tailing.
3. In the protocol's purification steps, perform the mRNA precipitation step by adding 600 μL of ethanol to 50 μL of 10 M ammonium acetate.
4. Measure mRNA concentration using a Qubit fluorometer system or a bioanalyzer system.
5. Aliquot the mRNA into separate 1.5-mL tubes.
6. Store the mRNA at −80 °C until use (*see* **Note 2**).

3.2 Dissociation of Planaria into Individualized Cells

1. Select planarian worms of the ClW4 asexual strain of *Schmidtea mediterranea*. Animals should be approximately 6 mm in length.
2. Transfer the worms into a 6-cm cell culture dish.
3. Wash worms twice in CMF and then remove as much liquid as possible.
4. Prepare dissociated cells by finely mincing the worms with a razor blade (*see* **Note 3**).
5. Suspend the cells with 1 mL CMF using a P1000 pipette with a cut tip, and move the cells into a 1.5-mL tube.
6. Gently pipette the cells ten times up and down with the cut tip to further dissociate them.
7. Rotate the cells on a tube rotator for 5 min, followed by gentle pipetting ten times up and down with a non-cut 1-ml tip.
8. Repeat **step** 7 once, or until most tissue pieces have dissociated.
9. Centrifuge the cells at 250 × g for 4 min, then remove the supernatant, and replace with 1.5 mL of fresh CMF.
10. Serially filter the cells through 100-, 70-, 40-, and then 30-μm mesh strainers to get a clean cell suspension.
11. Centrifuge the cell suspension at 250 × g for 4 min, then remove the supernatant, and replace with 1.5 mL of Iso-L15 medium (*see* **Note 4**).

3.3 In Vitro Cell mRNA Transfection

1. Count the dissociated cells under a microscope using a disposable cytometer (*see* **Note 5**).
2. Dilute the cells to a concentration of 0.88×10^6 cells/mL in the Iso-L15 medium.

3. Place 225 μL of cell suspension into each well of a flat-bottom 96-well microplate for a total of ~200,000 cells per well (*see* **Note 6**).
4. Combine the reagents for the transfection mix in a 1.5-ml tube—per reaction combine: 2 μL *Trans*IT-mRNA transfection reagent (Mirus), 1 μL *Trans*IT Boost reagent, 1.5 μg Nluc mRNA, and then add L-15 (no serum) up to 25 μL in total.
5. Incubate at room temperature for 5 min (*see* **Note 7**).
6. Add 25 μL of the transfection mix into each well, and gently pipette up and down once to mix.
7. Incubate the cells at 20 °C in the dark before assaying luminescence (*see* **Note 8**).

3.4 Dissociated Cell Luminescence Analysis

1. Preparation of the Nano-Glo Live mix for cells: Per reaction, combine 1.25 μL Nano-Glo Live Cell Substrate with 23.75 μL Nano-Glo Live Cell Buffer. Immediately mix 25 μL into each well with cells by pipetting up and down once (use of a multipipettor is recommended) (*see* **Note 9**).
2. Promptly measure luminescence using a plate reader with an integration time of 1 s (*see* **Note 10**). Or proceed to microscopy imaging using a bioluminescence imaging system (*see* **Note 11**) (*see* Fig. 1).

3.5 Live Worm mRNA Transfection by Injection

1. Select planarian worms of the ClW4 asexual strain of *Schmidtea mediterranea*. Animals should be approximately 6 mm in length (*see* **Note 12**).
2. Transfer the worms into a 12-cm cell culture dish.
3. Wash worms twice in planarian water (+50 μg/ml gentamicin).

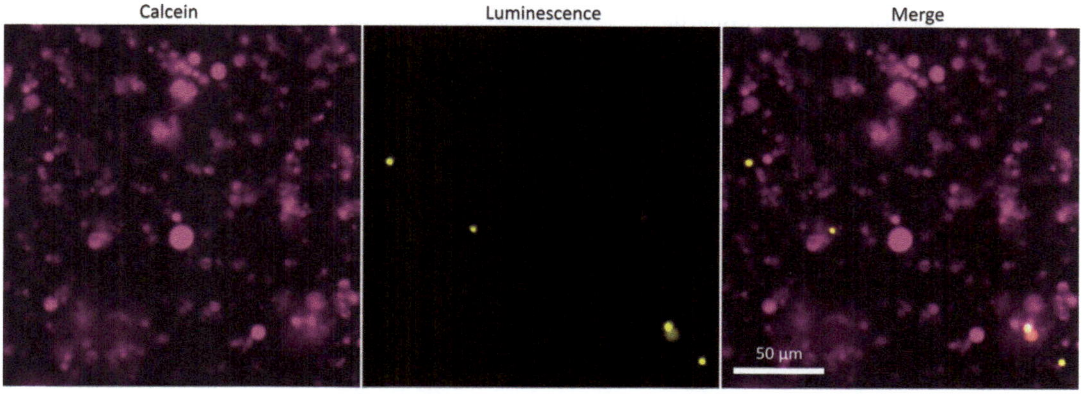

Fig. 1 Imaged planarian cells transfected with Nluc mRNA using *Trans*IT-mRNA. Cells were imaged on an Olympus LV200 luminescence microscope with a 20× air objective (NA 0.8, WD 0.6 mm, Olympus). Images were acquired at 12 h post-transfection with Nluc mRNA and *Trans*IT-mRNA mix. Cells were stained with 10 ng/μL calcein (magenta) with luminescence shown in yellow. The luminescence image was acquired using an exposure of 1 min. Calcein images were acquired using an exposure of 100 ms.

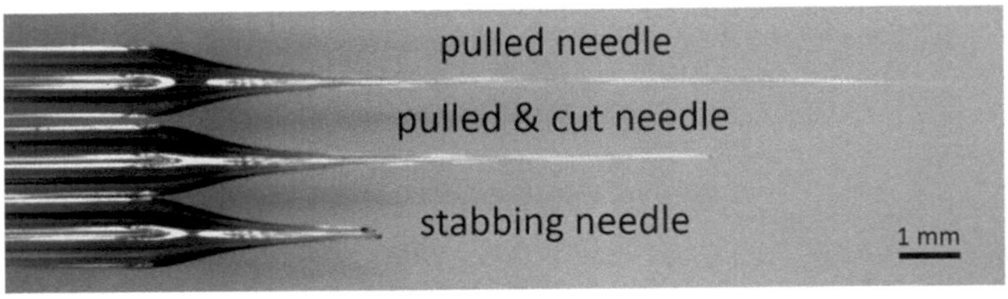

Fig. 2 Pulling and cutting of injection needles. Needles were pulled on a Sutter Instruments P-1000 micropipette puller with the following settings: Heat (530), Pull (150), Vel (150), Delay (0), Pressure (100), Ramp (484), Delay Mode (On), and Safe Heat (On). For injections, the tips were cut to a length of about 1 cm. The tips of stabbing needles were cut to 3–4 mm

4. Pull borosilicate glass needles using the following settings on a Sutter Instruments P-1000 micropipette puller: Heat (530), Pull (150), Vel (150), Delay (0), Pressure (100), Ramp (484), Delay Mode (On), and Safe Heat (On). These settings may need to be optimized for different microcapillary pullers, heating filaments, and glass capillaries (*see* Fig. 2).
5. Break the tips of the needles 1 cm from the base of the needle using fine forceps.
6. Connect a needle to the micromanipulator and microinjector and adjust the microinjector settings so that 0.9 μl is injected per worm (*see* **Note 13**).
7. Combine the reagents for the transfection mix in a 1.5-mL-tube: 2 μL *Trans*IT-mRNA transfection reagent, 1 μL *Trans*IT Boost reagent, 1.5 μg Nluc mRNA, and then add L-15 (no serum) up to 25 μL in total. Incubate at room temperature for 5 min (*see* **Note 7**).
8. Load 8 μL of the transfection mixture onto the mounted needle.
9. Place a Peltier cooling block (or another cooling device) under the stereoscope and set it to 4 °C (*see* **Note 14**).
10. Place a delicate task wiper on the cooling block (*see* **Note 15**) and wet it with cooled planarian water (*see* **Note 16**).
11. Cut a black filter paper into a rectangle of 2×2 cm, soak it in cooled planarian water, and place it on the wet delicate task wiper (*see* Fig. 3).
12. Move a worm to chilled planarian water (+50 μg/mL gentamicin) on ice for 2 min.
13. Mount the worm on the black chilled filter paper ventral side up. Make sure the worm is stretched and not moving.
14. Using a broken stabbing needle, pre-wound the worm at the posterior end about 1 mm from the tail tip at the center of the

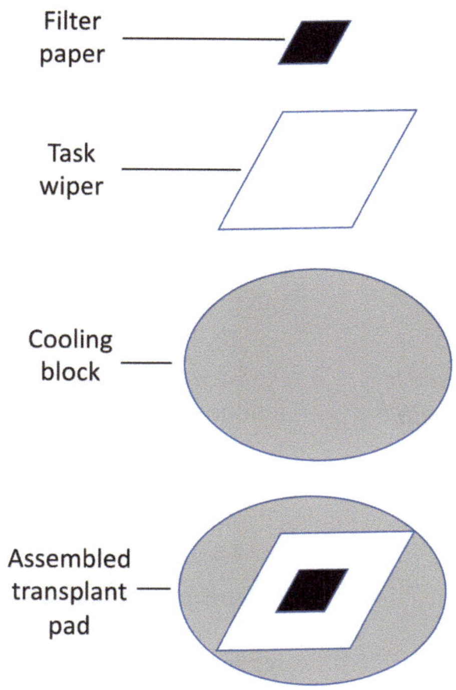

Fig. 3 Assembling of an injection pad for planarian worms. 2×2 cm black filter paper and a task wiper are soaked in cold planarian water and placed on a cooling block at 4 °C. A planarian worm can then be placed on the filer paper and injected

body. A shallow wound is recommended, and it is important to make sure not to pierce the worm through the dorsal side (*see* **Note 17**).

15. Using the micromanipulator, place the tip of the needle into the wound site until only the tip of the needle is not visible, and inject 0.9 μL of the transfection mix.
16. Gently take out the needle from the wound and move the worm into a 12-cm cell culture dish with planarian water (+50 μg/mL gentamicin).
17. Repeat **steps 12** to **16** for additional worms.
18. Incubate the worms at 20 °C in the dark before assaying luminescence by plate reader or microscopy (*see* **Note 8**).

3.6 Luminescence Analysis of Lysed Transfected Worms by a Plate Reader

1. Wash worms twice in planarian water.
2. Transfer an injected worm into a 6-cm cell culture dish and remove as much liquid as possible.
3. Finely mince the worm with a razor blade (*see* **Note 3**).
4. Suspend the dissociated tissue with 250 μL Iso-L15 medium using a P1000 pipette with a cut tip, and move the dissociated tissue into a well of a white flat-bottom 96-well microplate.

5. Repeat **steps 2–4** for additional worms.
6. Prepare the Nano-Glo Lysis mix for worms: for each reaction combine 2 μL Nano-Glo Luciferase Assay Substrate (Promega) with 98 μL Nano-Glo Luciferase Assay Buffer. Immediately mix 100 μL into each well with dissociated tissue by pipetting up and down ten times (use of a multi-pipettor is recommended).
7. Promptly measure luminescence using a plate reader with an integration time of 1 second (*see* **Note 10**).

3.7 Luminescence In Vivo Imaging of mRNA Transfected Worms

1. Anesthetize injected worms in planarian water with 1:5000 linalool for 2 h at 20 °C in the dark.
2. Incubate up to four injected worms in 100 μL Nano-Glo Live mix for worms: Combine 5 μL Nano-Glo Live Cell Substrate (Promega) with 95 μL planarian water (1% DMSO). Incubate for 15 min at 20 °C in the dark.
3. Mount injected worms on 3.5-cm glass-bottom dishes ventral side down, and position them with a fine brush so they are separate and straight.
4. Using a transfer pipette, cover the worms with a warm (but not hot) agarose mix for worms and then gently cover them with a cover slip.
5. Cool dishes in 4 °C for 5 min until the agarose hardens.
6. Image the worms for luminescence using a bioluminescence imaging system (*see* **Note 11**) (*see* Fig. 4).

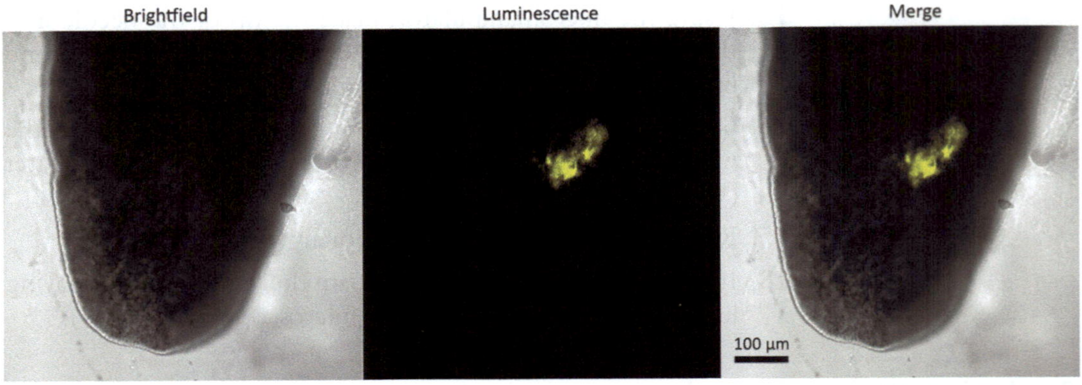

Fig. 4 Live planarian luminescence imaging. A live planarian imaged on the Olympus LV200 luminescence microscope through a 20× air objective (NA 0.8, WD 0.6 mm, Olympus). Images were acquired at 12 h post-transfection injection with Nluc mRNA and *TransI*T-mRNA mix. The luminescence image was acquired using an exposure of 1 min. Brightfield images were acquired using an exposure of 10 ms.

4 Notes

1. Single-stranded mRNA is highly susceptible to degradation. To reduce the degradation of the mRNA, it is recommended to plate the cells and mix the transfection reagents in a clean environment such as a laminar flow hood and to use RNase-free reagents, devices, filter tips, and cleaning reagents for RNase (e.g., RNaseZap) on gloves and pipettes. Further details on the T7 mScript™ Standard mRNA Production System can be found here:
   https://www.cellscript.com/products/007pl0114CS.pdf.
2. Do not thaw and refreeze the mRNA as it will lead to rapid degradation.
3. When dissociating worms, it is important to minimize the amount of mucus by using a small blade and regularly cleaning the blade with delicate task wipers. It is recommended, for the first dissociation step, to use a pipette tip that is cut at the end to give a wider bore. This helps in picking up the dissociated cells from the plate. Later dissociation steps should be done with non-cut tips, to promote the further mechanical dissociation of the remaining cell clumps.
4. It is recommended to move from the cell dissociation step to the transfection step as fast as possible, as planarian cells tend to rapidly die in culture.
5. For cell counting using a Neubauer improved C-chip, transfer 10 μL of cells to the C-chip, and count cells with bright cytoplasm from larger corner squares. Cells per mL = cells in 1 large square x 10^4.
6. For plate reader assays, use an opaque white-bottom 96-well plate. For imaging, use a clear-bottom 96-well plate.
7. The *Trans*IT reagents should be warmed to room temperature and vortexed before use. When preparing the *Trans*IT mix, use RNase-free reagents, devices, filter tips, and cleaning reagents for RNase (e.g., RNaseZap) on gloves and pipettes to prevent mRNA degradation. Per reaction it is important to first add 2 μL of *Trans*IT-mRNA transfection reagent and 1 μL *Trans*IT Boost reagent, then add the L-15 without serum, and mix well by pipetting. Finally, add 1.5 μg of Nluc mRNA and then mix well again by pipetting. This order insures the correct complex assembly between the transfection reagents and the mRNA. When tested, this ratio between the reagents showed the highest luminescence signal from cells after 12 h of incubation, yet mRNA amounts as low as 1 μg or as high as 3 μg with different rations to *Trans*IT reagents give a clear yet lower signal.

8. Luminescence in mRNA transfected cells has been detected in vitro from 4 h after transfection up to 72 h after transfection while rising in signal over time. For in vivo transfection the signal is strongest at 12 h but is still detectable at 72 h post-transfection. In addition, planarian cell viability under our culture conditions decrease over time with 72 h after dissociation as the furthest time point tested with viable cells [7].
9. If additional live cell dyes such as Hoechst or calcein are included, they should be added and incubated with the cells before the Nano-Glo mix is added. Once Nano-Glo reagents have been moved to room temperature, mix them as fast as possible with the cells and proceed to the plate reader or microscopy. This is because the substrate degrades rapidly at room temperature leading to a 50% decrease of signal within 50 min, which may negatively affect the luminescence output.
10. The specific gain setting on the plate reader might have to be adjusted depending on the specific device used. Choose an ultrasensitive gain setting if available, as the luminescence output can be weak.
11. We have successfully imaged luminescence with 4×–100× objectives. Generally choose the highest numerical aperture (NA) objectives available, as the higher the NA, the higher the photon collection efficiency. While high magnification objectives often have higher NA, they also have a much more limited field of view. This may be a problem when transfection efficiencies are low, as most fields of view will not contain transfected cells.
12. In vivo transfection in planarians was found to vary dramatically with some injected worms showing a high luminescent signal above background and others only background signals. For this reason, it is recommended to inject at least ten worms for each condition and to assay them separately by plate reader or microscopy.
13. Calibrating the injection amount is required for each microinjection system. Testing of different pressures and injection durations is needed to achieve a volume of 0.9 μL per injection. The volume can be assessed by injecting water into mineral oil and comparing the size of the resulting droplet to droplets of a known volume generated with a pipettor. In some cases, several pulses of the microinjector may be needed to reach the target volume.
14. It is recommended to use a cooling block with temperature control at 4 °C.
15. Suitably dimensioned blocks of ice can provide a simple substitute, provided that they fit under the stereoscope and provide a flat surface for arranging the animals to be injected.

16. The task paper should be soaked with planarian water to keep the worm moist but no excess water should form above the paper; otherwise, the worm may move.
17. It is recommended to pre-wound to ease the insertion of the injecting needle; otherwise, the injecting needle can be broken and so change the amount injected. Still, injection without pre-wounding is also possible.

Acknowledgments

We thank Dr. Tobias Boothe for his support in imaging.

References

1. Reddien PW, Bermange AL, Murfitt KJ, Jennings JR, Sánchez Alvarado A (2005) Identification of genes needed for regeneration, stem cell function, and tissue homeostasis by systematic gene perturbation in planaria. Dev Cell 8:635–649. https://doi.org/10.1016/J.DEVCEL.2005.02.014
2. Wagner DE, Wang IE, Reddien PW (2011) Clonogenic neoblasts are pluripotent adult stem cells that underlie planarian regeneration. Science 332:811–816. https://doi.org/10.1126/science.1203983
3. Zeng A, Li H, Guo L, Gao X, McKinney S, Wang Y, Yu Z, Park J, Semerad C, Ross E, Cheng LC, Davies E, Lei K, Wang W, Perera A, Hall K, Peak A, Box A, Sánchez Alvarado A (2018) Prospectively isolated Tetraspanin + Neoblasts are adult pluripotent stem cells underlying Planaria regeneration. Cell 173:1593–1608.e20. https://doi.org/10.1016/j.cell.2018.05.006
4. Grohme MA, Schloissnig S, Rozanski A, Pippel M, Young GR, Winkler S, Brandl H, Henry I, Dahl A, Powell S, Hiller M, Myers E, Rink JC (2018) The genome of Schmidtea mediterranea and the evolution of core cellular mechanisms. Nature 554:56–61. https://doi.org/10.1038/nature25473
5. England CG, Ehlerding EB, Cai W (2016) NanoLuc: a small luciferase is brightening up the field of bioluminescence. https://doi.org/10.1021/acs.bioconjchem.6b00112
6. Hall RN, Weill U, Drees L, Leal-Ortiz S, Li H, Khariton M, Chai C, Xue Y, Rosental B, Quake SR, Sánchez Alvarado A, Melosh NA, Fire AZ, Rink JC, Wang B (2022) Heterologous reporter expression in the planarian Schmidtea mediterranea through somatic mRNA transfection. Cell Reports Methods 2:100298. https://doi.org/10.1016/J.CRMETH.2022.100298
7. Lei K, McKinney SA, Ross EJ, Lee H-C, Alvarado AS (2019) Cultured pluripotent planarian stem cells retain potency and express proteins from exogenously introduced mRNAs bioRxiv:573725. https://doi.org/10.1101/573725

Index

A

B

C

D

E

Luca Gentile (ed.), *Schmidtea Mediterranea: Methods and Protocols*, Methods in Molecular Biology, vol. 2680,
https://doi.org/10.1007/978-1-0716-3275-8,
© The Editor(s) (if applicable) and The Author(s), under exclusive license to Springer Science+Business Media, LLC, part of Springer Nature 2023

F

G

H

I

L

M

N

O

P

Q

R

S

T

MIX
Papier aus verantwortungsvollen Quellen
Paper from responsible sources
FSC® C105338
FSC
www.fsc.org

If you have any concerns about our products,
you can contact us on
ProductSafety@springernature.com

In case Publisher is established outside the EU,
the EU authorized representative is:
Springer Nature Customer Service Center GmbH
Europaplatz 3, 69115 Heidelberg, Germany

Printed by Libri Plureos GmbH
in Hamburg, Germany